Integrated Circuits and Systems

Series Editor
Anantha Chandrakasan, Massachusetts Institute of Technology
Cambridge, Massachusetts

For other titles published in this series, go to
http://www.springer.com/series/7236

Yuan Xie · Jason Cong · Sachin Sapatnekar
Editors

Three-Dimensional Integrated Circuit Design

EDA, Design and Microarchitectures

Springer

Editors
Yuan Xie
Department of Computer Science and Engineering
Pennsylvania State University
yuanxie@cse.psu.edu

Jason Cong
Department of Computer Science
University of California, Los Angeles
cong@cs.ucla.edu

Sachin Sapatnekar
Department of Electrical and Computer Engineering
University of Minnesota
sachin@umn.edu

ISBN 978-1-4419-0783-7 e-ISBN 978-1-4419-0784-4
DOI 10.1007/978-1-4419-0784-4
Springer New York Dordrecht Heidelberg London

Library of Congress Control Number: 2009939282

Printed on acid-free paper

Springer is part of Springer Science+Business Media (www.springer.com)

Foreword

We live in a time of great change. In the electronics world, the last several decades have seen unprecedented growth and advancement, described by Moore's law. This observation stated that transistor density in integrated circuits doubles every 1.5–2 years. This came with the simultaneous improvement of individual device performance as well as the reduction of device power such that the total power of the resulting ICs remained under control. No trend remains constant forever, and this is unfortunately the case with Moore's law.

The trouble began a number of years ago when CMOS devices were no longer able to proceed along the classical scaling trends. Key device parameters such as gate oxide thickness were simply no longer able to scale. As a result, device off-state currents began to creep up at an alarming rate. These continuing problems with classical scaling have led to a leveling off of IC clock speeds to the range of several GHz. Of course, chips can be clocked higher but the thermal issues become unmanageable. This has led to the recent trend toward microprocessors with multiple cores, each running at a few GHz at the most. The goal is to continue improving performance via parallelism by adding more and more cores instead of increasing speed. The challenge here is to ensure that general purpose codes can be efficiently parallelized.

There is another potential solution to the problem of how to improve CMOS technology performance: three-dimensional integrated circuits (3D ICs). By moving to a technology with multiple active "tiers" in the vertical direction, a number of significant benefits can be realized. Global wires become much shorter, interconnect bandwidth can be greatly increased, and latencies can be significantly decreased. Large amounts of low-latency cache memory can be utilized and intelligent physical design can help mitigate thermal and power delivery hotspots. Three-dimensional IC technology offers a realistic path for maintaining the progress defined by Moore's Law *without requiring classical scaling.* This is a critical opportunity for the future.

The Defense Advanced Research Project Agency (DARPA) has recognized the significance of 3D IC technology a number of years ago and began carefully targeted investments in this area based on the potential for military relevance and applications. There are also many potential commercial benefits from such a technology. The Microsystems Technology Office at DARPA has launched a number of 3D IC-based Programs in recent years targeting areas such as intelligent imagers, heterogeneously integrated 3D stacks, and digital performance enhancement. The research results in a number of the chapters in this book were made possible by DARPA-sponsored programs in the field of 3D IC.

Three-dimensional IC technology is currently at an early stage, with several processes just becoming available and more in the early development stage. Still, its potential is so great that a dedicated community has already begun to seriously study the EDA, design, and architecture issues associated with 3D IC, which are well summarized in this book. Chapter 1 provides a good introduction to this field by an expert from IBM well versed in both design and technology aspects. Chapter 2 provides an excellent overview of key 3D IC technology issues by process technology researchersfrom IBM and can be beneficial to any designer or architect. Chapters 3–6 cover important 3D IC electronic design automation (EDA) issues by researchers from the University of California, Los Angeles and the University of Minnesota. Key issues covered in these chapters include methods for managing the thermal, electrical, and layout challenges of a multi-tier electronic stack during the modeling and physical design processes. Chapters 7–9 deal with 3D design issues, including the 3D processor design by authors from the Georgia Institute of Technology, a 3D network-on-chip (NoC) architecture by authors from Pennsylvania State University, and a 3D architectural approach to energy efficient server designs by authors from the University of Michigan and Intel. The book concludes with a system–level analysis of the potential cost advantages of 3D IC technology by researchers at Pennsylvania State University.

As I mentioned in the beginning we live at a time of great change. Such change can be viewed as frightening, as long-held assumptions and paradigms, such as Moore's Law, lose relevance. Challenging times are also important opportunities to *try new ideas.*

Three-dimensional IC technology is such a new idea and this book will play an important and pioneering role in ushering in this new technology to the research community and the IC industry.

DARPA Microsystems Technology Office
Arlington, Virginia
March 2009

Michael Fritze, Ph.D.

Preface

To the observer, it would appear that New York city has a special place in the hearts of integrated circuit (IC) designers. Manhattan geometries, which mimic the blocks and streets of the eponymous borough, are routinely used in physical design: under this paradigm, all shapes can be decomposed into rectangles, and each wire is either parallel and perpendicular to any other. The advent of 3D circuits extends the analogy to another prominent feature of Manhattan – namely, its skyscrapers – as ICs are being built upward, with stacks of active devices placed on top of each other. More precisely, unlike conventional 2D IC technologies that employ a single tier with one layer of active devices and several layers of interconnect above this layer, 3D ICs stack multiple tiers above each other. This enables the enhanced use of silicon real estate and the use of efficient communication structures (analogous to elevators in a skyscraper) within a stack.

Going from the prevalent 2D paradigm to 3D is certainly not a small step: in more ways than one, this change adds a new dimension to IC design. Three-dimensional design requires novel process and manufacturing technologies to reliably, scalably, and economically stack multiple tiers of circuitry, design methods from the circuit level to the architectural level to exploit the promise of 3D, and computer-aided design (CAD) techniques that facilitate circuit analysis and optimization at all stages of design. In the past few years, as process technologies for 3D have neared maturity and 3D circuits have become a reality, this field has seen a flurry of research effort. The objective of this book is to capture the current state of the art and to provide the readers with a comprehensive introduction to the underlying manufacturing technology, design methods, and computer-aided design (CAD) techniques. This collection consists of contributions from some of the most prominent research groups in this area, providing detailed insights into the challenges and opportunities of designing 3D circuits.

The history of 3D circuits goes back many years, and some of its roots can be traced to a major government-funded program in Japan from a couple of decades ago. It is only in the past few years that the idea of 3D has gained major traction, so that it is considered a realistic option today. Today, most major players in the semiconductor industry have dedicated significant resources and effort to this area. As a result, 3D technology is at a stage where it is poised to make a major leap. The context and motivation for this technology are provided in Chapter 1.

The domain of 3D circuits is diverse, and various 3D technologies available today provide a wide range of tradeoffs between cost and performance. These include silicon-carrier-like technologies with multiple dies mounted on a substrate, wafer stacking with intertier spacings of the order of hundreds of microns, and thinned die/wafer stacks with intertier distances of the order of ten microns. The former two have the advantage of providing compact packaging and higher levels of integration but often involve significant performance overheads in communications from one tier to another. The last, with small intertier distances, not only provides increased levels of integration but also facilitates new architectures that can actually improve significantly upon an equivalent 2D implementation. Such advanced technologies are the primary focus of this book, and a cutting-edge example within this class is described in detail in Chapter 2.

In building 3D structures, there are significant issues that must be addressed by CAD tools and design techniques. The change from 2D to 3D is fundamentally topological, and therefore it is important to build floorplanning, placement, and routing tools for 3D chips. Moreover, 3D chips see a higher amount of current per unit footprint than their 2D counterparts, resulting in severe thermal and power delivery bottlenecks. Any physical design system for 3D must incorporate thermal considerations and must pay careful attention to constructing power delivery networks. All of these issues are addressed in Chapters 3–6.

At the system level, 3D architectures can be used to build new architectures. For sensor chips, sensors may be placed in the top tier, with analog amplifier circuits in the next layer, and digital signal processing circuitry in the layers below: such ideas have been demonstrated at the concept or implementation level for image sensors and antenna arrays. For processor design, 3D architectures allow memories to be stacked above processors, allowing for fast communication between the two and thereby removing one of the most significant performance bottlenecks in such systems. Several system design examples are discussed in Chapters 7–9. Finally, Chapter 10 presents a methodology for cost analysis of 3D circuits.

It is our hope that the book will provide the readers with a comprehensive view of the current state of 3D IC design and insights into the future of this technology.

Sachin Sapatnekar

Contents

1 **Introduction** 1
Kerry Bernstein

2 **3D Process Technology Considerations** 15
Albert M. Young and Steven J. Koester

3 **Thermal and Power Delivery Challenges in 3D ICs** 33
Pulkit Jain, Pingqiang Zhou, Chris H. Kim, and Sachin S. Sapatnekar

4 **Thermal-Aware 3D Floorplan** 63
Jason Cong and Yuchun Ma

5 **Thermal-Aware 3D Placement** 103
Jason Cong and Guojie Luo

6 **Thermal Via Insertion and Thermally Aware Routing in 3D ICs** 145
Sachin S. Sapatnekar

7 **Three-Dimensional Microprocessor Design** 161
Gabriel H. Loh

8 **Three-Dimensional Network-on-Chip Architecture** 189
Yuan Xie, Narayanan Vijaykrishnan, and Chita Das

9 **PicoServer: Using 3D Stacking Technology to Build Energy Efficient Servers** 219
Taeho Kgil, David Roberts, and Trevor Mudge

10 **System-Level 3D IC Cost Analysis and Design Exploration** 261
Xiangyu Dong and Yuan Xie

Index 281

Contributors

Kerry Bernstein Applied Research Associates, Inc., Burlington, VT, kbernstein@ara.com

Jason Cong Department of Computer Science, University of California; California NanoSystems Institute, Los Angeles, CA 90095, USA, cong@cs.ucla.edu

Chita Das Pennsylvania State University, University Park, PA 16801, USA, das@cse.psu.edu

Xiangyu Dong Pennsylvania State University, University Park, PA 16801, USA

Pulkit Jain Department of Electrical and Computer Engineering, University of Minnesota, Minneapolis, MN 55455, USA, jainx104@umn.edu

Taeho Kgil Intel, Hillsboro, OR 97124, USA, taeho.kgil@intel.com

Chris H. Kim Department of Electrical and Computer Engineering, University of Minnesota, Minneapolis, MN 55455, USA, chriskim@umn.edu

Steven J. Koester IBM Thomas J. Watson Research Center, Yorktown Heights, NY 10598, USA, skoester@us.ibm.com

Gabriel H. Loh College of Computing, Georgia Institute of Technology, Atlanta, GA, USA, loh@cc.gatech.edu

Guojie Luo Department of Computer Science, University of California, Los Angeles, CA 90095, USA, gluo@cs.ucla.edu

Yuchun Ma Department of Computer Science and Technology, Tsinghua University, Beijing, P.R.China 100084, myc@mail.tsinghua.edu.cn

Trevor Mudge University of Michigan, Ann Arbor, MI 48109, USA, tnm@eecs.umich.edu

David Roberts University of Michigan, Ann Arbor, MI 48109, USA, daverobe@eecs.umich.edu

Sachin S. Sapatnekar Department of Electrical and Computer Engineering, University of Minnesota, Minneapolis, MN 55455, USA, sachin@umn.edu

Narayanan Vijaykrishnan Pennsylvania State University, University Park, PA 16801, USA, vijay@cse.psu.edu

Yuan Xie Pennsylvania State University, University Park, PA 16801, USA, yuanxie@cse.psu.edu

Albert M. Young IBM Thomas J. Watson Research Center, Yorktown Heights, NY 10598, USA, yalbert@us.ibm.com

Pingqiang Zhou Department of Electrical and Computer Engineering, University of Minnesota, Minneapolis, MN 55455, USA, pingqiang@umn.edu

Chapter 1
Introduction

Kerry Bernstein

Much as the development of steel girders suddenly freed skyscrapers to reach beyond the 12-story limit of masonry buildings [6], achievements in four key processes have allowed the concept of 3D integrated circuits [2], proposed more than 20 years ago by visionaries (such as Jim Meindl in the United States and Mitsumasa Koyanagi in Japan), to actually begin to become realized. These factors are (1) low-temperature bonding, (2) layer-to-layer transfer and alignment, (3) electrical connectivity between layers, and (4) an effective release process. These are the cranes which will assemble our new electronic skyscrapers. As these emerged, the contemporary motivation to create such an unusual electronic structure remained unresolved. That argument finally appeared in a casual magazine article that certainly was not immediately recognized for the prescience it offered [5]. Doug Matzke from TI recognized in 1997 that, even traveling at the speed of light in the medium, signal locality would ultimately limit performance and throughput gains in processors. It was clear at that time that wire delay improvements were not tracking device improvements, and that to keep up, interconnects would need a constant infusion of new materials and structures. Indeed, history has proven this argument correct. Figure 1.1 illustrates the pressure placed on interconnects since 1995. In the figure, the circle represents the area accessible within one cycle, and it is clear that its radius has shrunk with time, implying that fewer on-chip resources can be reached within one cycle. Three trends have conspired to monotonically reduce this radius [1]:

1. *Wire Nonscaling.* Despite the heroic efforts of metallurgists and back-end-of-line engineers, at best, chip interconnect delay has remained constant from one generation to the next. This is remarkable, as each generation has asserted new materials such as reduced permittivity dielectrics, copper, and extra levels of metal. Given that in the same period device performance has improved by the scaling factor, the accessible radius was bound to be reduced.

K. Bernstein (✉)
Applied Research Associates, Inc., Burlington, VT
e-mail: kbernstein@ara.com

Y. Xie et al. (eds.), *Three-Dimensional Integrated Circuit Design*,
Integrated Circuits and Systems, DOI 10.1007/978-1-4419-0784-4_1,

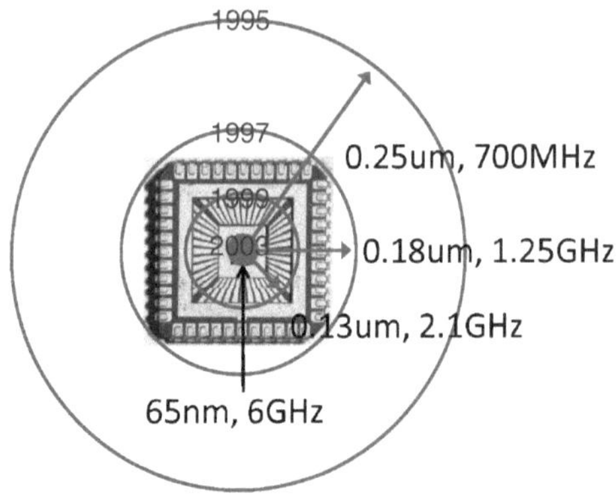

Fig. 1.1 The perfect storm for uniprocessor cross-chip latency: (1) wire nonscaling; (2) die size growth; (3) shorter FO4 stages. The power cost of cross-chip latency increases [1]

2. *Die Size Growth.* The imbalance between device and wire delays would be hard enough if die sizes were to track scaling and correspondingly reduce in each generation. Instead, the trend has actually been the opposite, to grow relative die size in order to accommodate architectural approaches to computer throughput improvement. It takes longer time for signals, even those traveling at the speed of light in the medium, to get across chip. Even without dies size growth, the design would be strained to meet the same cycle time constraints as in the previous process generation.
3. *Shorter Cycles.* The argument above is complicated by the fact that cycle time constraints have not remained fixed, but continually reduced. In order to more fully utilize on-chip resource, over each successive generation, architects have reduced the number of equivalent "inverter-with fan-out-of 4" delays per cycle, so that pipeline bubbles would not idle on-chip functional units as seriously as they would under longer cycle times. Under this scenario, not only does a signal have farther to go now on wires that have not improved, it also has less time to get there than before.

The result as shown in Fig. 1.1 indicates that uniprocessors have lost the ability to access resources across a chip within one cycle. One fix, which adds multiple identical resources to the uniprocessor so that at least one is likely to be accessible within one cycle, only makes the problem worse.

The trend suggested above is in fact borne out by industry data. Figure 1.2 shows the ratio of area to the SpecInt2000 (a measure of microprocessor performance) performance for processors described in conferences over the past 10 years and projected ahead. An extrapolation of the trend indicates that there is a limit to this practice.

While architectures can adapt to enable performance improvements, such an approach is expensive in its area overhead and comes at the cost of across-chip signal latency. An example of the overhead is illustrated in Fig. 1.3. As the number of stages per cycle drops (as described in item 3 above), the processor must store a larger number of intermediate results, requiring more latches and registers.

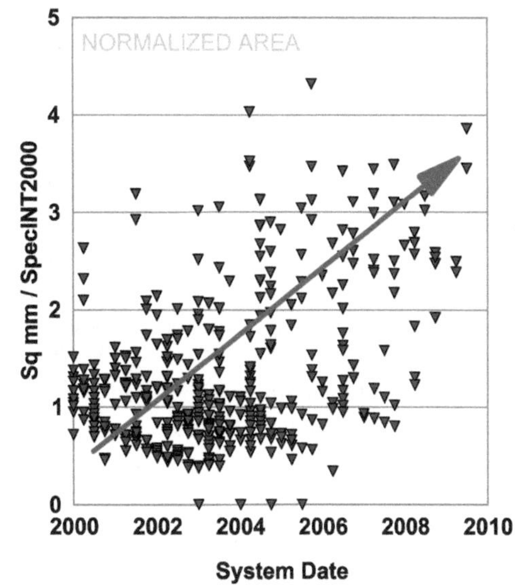

Fig. 1.2 The ratio of area to the SpecInt2000 performance

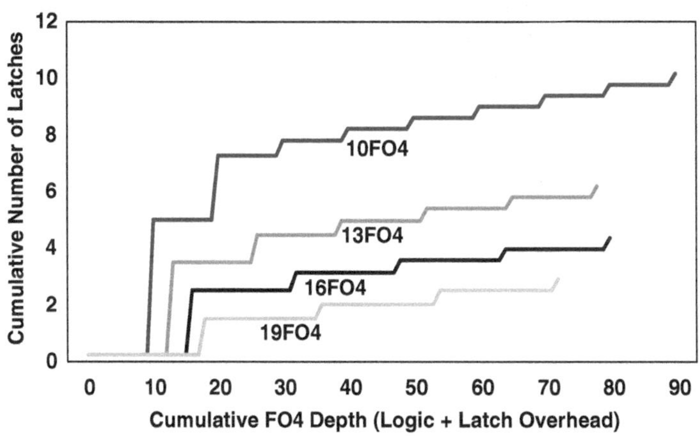

Fig. 1.3 Trends showing a superlinear rise in the latch count with pipeline depth. © 2002 IEEE. Reprinted, with permission, from [7]

Srinivasan shows that for a fixed cumulative logic depth, as the number of FO4-equivalent delay stages drops per cycle, the number of required registers increases [7]. Not only do the added registers consume area, but they also require a greater percentage of the cycle to be allocated for timing margin.

It is appropriate to qualitatively explain some of the reasons that industry has been successful with technology scaling, in spite of the limitation cited above. As resources grew farther apart electrically, microprocessor architectures began favoring multicore, SMP machines, in which each core was a relatively simple processor which executed instructions in order. In these multicore systems, individual cores have dispensed with much of the complexity of their more convolved uniprocessor predecessor. As we shall discuss in a moment, increasing the number of on-chip processor cores sustains microprocessor performance improvement as long as the bandwidth in and out of the die can feed each of the cores with the data they need. In fact, this has been true in the early life of multicore systems, and it is no coincidence that multicore processors dominated high-performance processing precisely when interconnect performance has become a significant limitation for designers, even as device delays have continued to improve.

As shown qualitatively in Fig. 1.4, the multicore approach will continue to supply performance improvements as shown by the black curve, until interconnect bandwidth becomes the performance bottleneck again. Overcoming the bandwidth limitation at this point will require a substantial paradigm change beyond the material changes that succeeded in improving interconnect latency in 2D designs. Three-dimensional integration provides precisely this ability: if adopted, this technology will continue to extend microprocessor throughput until its advantages saturate, as shown in the upper dashed curve. Without 3D, one may anticipate limitations to multicore processing occurring much earlier, as shown in the lower dashed line. This limitation was recognized as far back as 2001 [2]. Their often-quoted work showed that one could "scale-up" or "scale-out" future designs and that interconnect asserts an unacceptable limitation upon design. Figure 1.5 from their work showed that either an improbable 90 wiring levels would eventually be required or that designs would need to be kept to under 10 M devices per macro in order to remain fully wirable. Neither solution is palatable.

Let us return to examine the specific architectural issues that make 3D integration so timely and useful. We begin by examining what we use processors for and how we organize them to be most efficient.

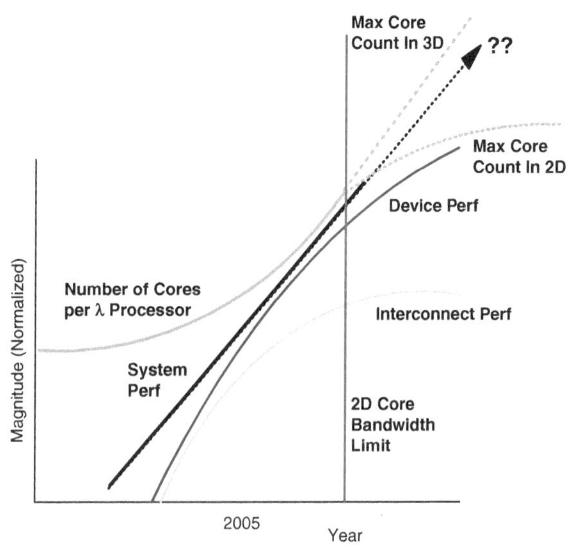

Fig. 1.4 Microprocessor performance extraction

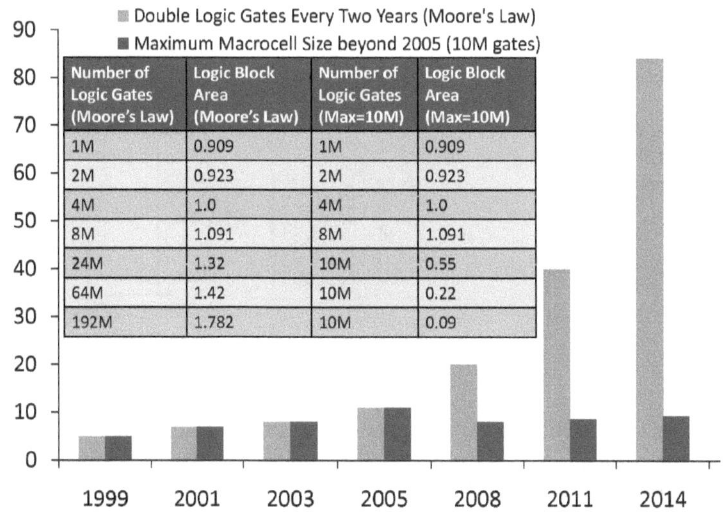

Number of Logic Gates (Moore's Law)	Logic Block Area (Moore's Law)	Number of Logic Gates (Max=10M)	Logic Block Area (Max=10M)
1M	0.909	1M	0.909
2M	0.923	2M	0.923
4M	1.0	4M	1.0
8M	1.091	8M	1.091
24M	1.32	10M	0.55
64M	1.42	10M	0.22
192M	1.782	10M	0.09

Fig. 1.5 Early 3D projections: Scale-up ("Beyond 2005, unreasonable number of wiring levels."), Scale-out ("Interconnect managed if gates/macro < 10 M."), Net ("Interconnect has ability to assert fundamental limitation.") [2] © 2001 IEEE. Reprinted, with permission, from Interconnect Limits on Gigascale Integration (GSI) in the 21st Century J. Davis, et al., Proceedings of the IEEE, Vol 89 No 3, March 2001

Generally, processors are state machines constructed to most efficiently move a microprocessor system from one machine state to the next, as efficiently as possible. The state of the machine is defined by the contents of its registers; the machine state it moves to is designated by the instructions executed between registers. Transactions are characterized by sets of instructions performed upon sets of data brought in close proximity to the processor. If needed data is not stored locally, the call for it is said to be a "miss." A sequence of instructions, known as the processor workload, are generally either scientific or commercial in nature. These two divergent workloads utilize the resources of the processor very differently. Enablements such as 3D are useful in general purpose processors only if they allow a given design to retire both types of operations gracefully.

To appreciate the composition of performance, let us examine the contributors to microprocessor throughput delay. Figure 1.6a shows the fundamental components of a microprocessor. Pictured is an instruction unit ("I-Unit") which interprets and dispatches instructions to the processor, an execution unit ("E-Unit") which executes these instructions, and the L1 cache array which stores the operands [3]. If the execution unit had no latency-securing operands, the number of cycles required to retire an instruction would be captured in the lowest line on the plot in Fig. 1.6b, labeled E-busy. Data for the execution unit however must be retrieved from the L1 cache which hopefully has been predictively filled with the data needed. In the so-called infinite cache scenario, the cache is infinitely large and contains every possible word

Fig. 1.6 Components of processor performance. Delay is sequentially determined by (**a**) ideal processor, (**b**) access to local cache, and (**c**) refill of cache. © 2006 IEEE. Reprinted, with permission, from [3]

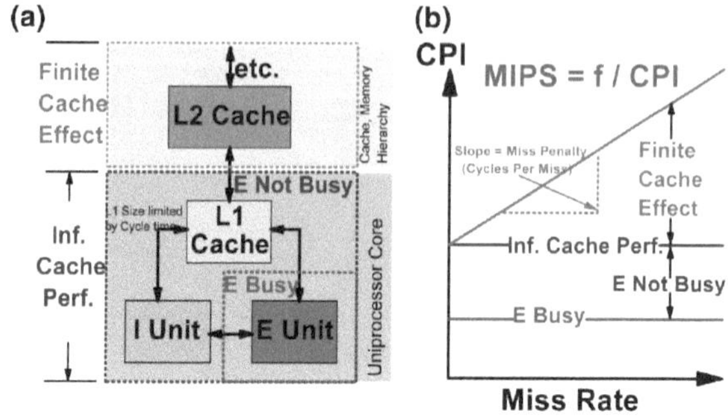

which may be requested. The performance in this case would be defined by the second, blue line, which includes the delay of the microprocessor as well as the access latency to the L1 array. L1 caches are not infinite however, and too often, data is requested which has not been predictively preloaded into the L1 array. The number of cycles required per instruction in the "finite cache" reality includes the miss time penalty required to get data.

Whether scientific or commercial, microprocessor performance is as dependent on effective data delivery as it is on high-performance logic. Providing data to the memory as quickly as possible is accomplished with latency, bandwidth, and cache array. The delay between when data is requested and when it becomes available is known as latency. The amount of temporary cache memory on chip alleviates some of the demand for off-chip delivery of data from main store. Bandwidth allows more data to be brought on chip in parallel at any given time. Most importantly, these three attributes of a processor memory subsystem are interchangeable.

When the on-chip cache can no longer be increased on chip, then technologies that improve the bandwidth and the latency to main store become very important. Figure 1.7 shows the hypothetical case where the number of threads or separate computing processes running in parallel on a given microprocessor chip has been doubled. To hold the miss rate constant, it has been observed that the amount of data made available to the chip must be increased [3]. Otherwise, it makes no sense to increase the number of threads because they will encounter more misses and any potential advantage will be lost. As shown in the bottom left of Fig. 1.7, a second cache may be added for the new thread, requiring a doubling of the bandwidth. Alternatively, if the bandwidth is not doubled, then each cache must be quadrupled to make up for the net bandwidth loss per thread, as seen in the bottom right of Fig. 1.7. Given that the miss rate goes as the square root of the on-chip cache size, the interchangeability of bandwidth and memory size may be generalized as

$$2^{a+b}T = (2^a B) \times (8^b C) \qquad (1)$$

where T is the number of threads, B is the bandwidth, and C is the amount of on-chip cache available. The exponents a and b may assume any combination of values

Fig. 1.7 When the number of threads doubles, if the total bandwidth is kept the same, the capacity of caches should be quadrupled to achieve similar performance for each tread. © 2006 IEEE. Reprinted, with permission, from [3]

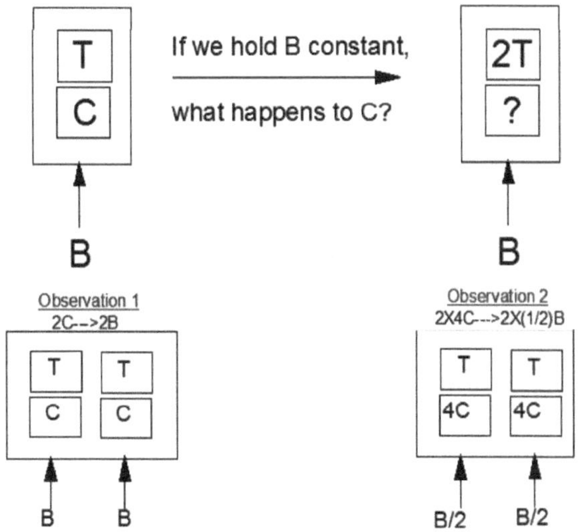

which maintain the equality (and hence a fixed miss rate), demonstrating the fungibility of bandwidth and memory size. Given the expensive miss rate dependence on memory in the second term on the right, it follows that bandwidth, as provided by 3D integration, is potentially lucrative in multithreaded future processors.

We now consider the impact of misses on various types of workloads. *Scientific workloads*, such as those at large research installations, are highly regular patterns of operations performed upon massive data sets. The data needed in the processor's local cache memory array from main store is very predictable and sequential and allows the memory subsystem to stream data from the memory cache to the processor with a minimum of specification, interruption, or error. Misses occur very infrequently. In such systems, performance is directly related to the bandwidth of the bus piping the data to the processor. This bus has intrinsically high utilization and is full at all times. System throughput, in fact, is degraded if the bus is not full. *Commercial workloads*, on the other hand, have unpredictable, irregular data patterns, as the system is called upon to perform a variety of transactions. Data "misses" occur frequently, with their rate of occurrence following a Poisson distribution. Figure 1.8 shows a histogram of the percent of total misses as a function of the number of instructions retired between each miss. Although it is desirable for the peak to be to the right end of the *X*-axis in this plot, in reality misses are frequent and a fact of life. High throughput requires low bus utilization to avoid bus "clog-ups" in the event of a burst of misses. Figure 1.9 shows this dependence and how critical it is for commercial processors to have access to low-utilization busses. When bus utilization exceeds approximately 30%, the relative performance plummets.

In short, both application spaces need bandwidth, but for very different reasons. Given that processors often are used in general purpose machines that are used in both scientific and commercial settings, it is important to do both jobs well. While a

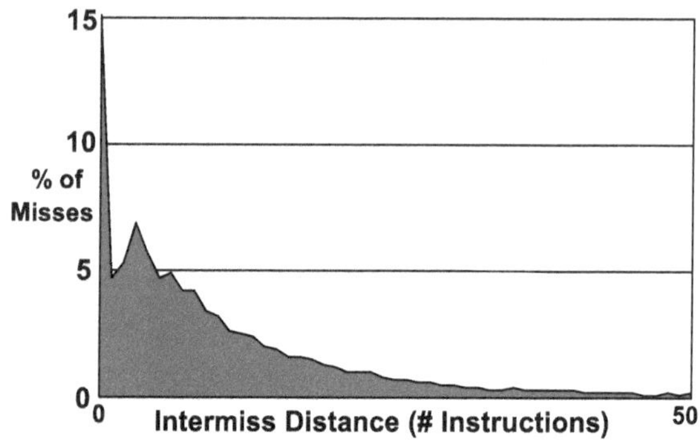

Fig. 1.8 A histogram of the percent of total misses as a function of the number of instructions retired between each miss. © 2006 IEEE. Reprinted, with permission, from [3]

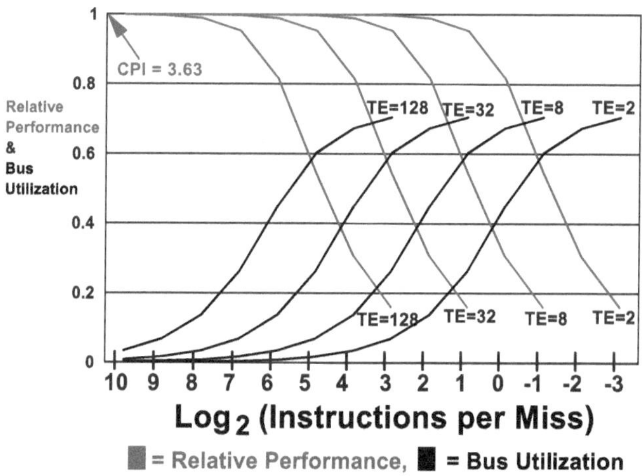

Fig. 1.9 Relative performance and bus utilization. © 2006 IEEE. Reprinted, with permission, from [3]

number of technical solutions address one or the other, one common solution is 3D integration for its bandwidth benefits.

As shown in Fig. 1.6, there are a number of contributors to delay. All of them are affected by interconnect delay, but to varying degrees. The performance in the "infinite cache scenario" is defined by the execution delay of the processor itself. The processor delay itself is, of course, improved with less interconnect latency, but in the prevailing use of 3D as a conduit for memory to the processor, the processor execution delay realizes little actual improvement. The finite cache scenario, where we take into account the data latency associated with finite cache and bandwidth, is what demonstrates when 3D integration is really useful. In Fig. 1.10, we see the improvement when a 1-core, a 2-core, and a 4-core processor are realized in 3D integration technology. The relative architecture performance of the system increases until the point where the system is adequately supplied with data such that the miss rate is under control. Beyond this point there is no advantage in providing

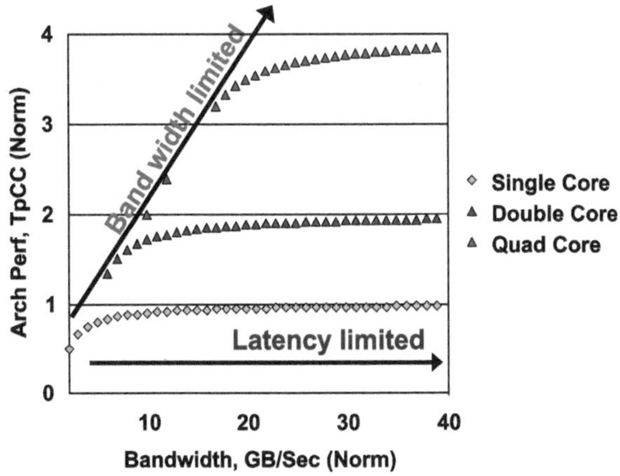

Fig. 1.10 Bandwidth and latency boundaries for different cores

additional bandwidth. The reader should note that this saturation point slides farther over as the number of cores is increased. The lesson is that while data delivery is an important attribute in future multicore processors, designers will still need to ensure that core performance continues to be improved.

One final bus concept needs to be treated, which is central to 3D integration. There is the issue of "bus occupancy." The crux of the problem is as follows: the time it takes to get a piece of needed data to the processor defines the system data latency. Latency, as we learned above, directly impacts the performance of the processor waiting for this data. The other, insidious impact to performance associated with the bus, however, is in the time for which the bus is tied up delivering this data. If one is lucky, the whole line of data coming in will be needed and is useful. Too often, however, it is only the first portion of the line that is useful. Since data must come into the processor in "single line address increments," this means that at least one entire line must be brought in. Microprocessor architects obsess about how long a line of data being fetched from main store should be. If it is too short, then the desired bus traffic can be more precisely specified, but the on-chip address directory of data contained in microprocessor's local cache explodes in size. Too long a line reduces the book-keeping and memory management overhead in the system, but ties up the bus downloading longer lines when subsequent misses require new addresses to be fetched. Herein lies the rub. The line length is therefore determined probabilistically based on the distribution of workloads that the microprocessor may be called upon to execute in the specific application. This dynamic is illustrated in Fig. 1.11. When an event causes a cache miss, there is a delay as the memory subsystem rushes to deliver data back to the cache [3]. Finally, data begins to arrive with a latency characterized by the time to the arrival of the leading edge of the line. However, the bus is no longer free until the entire line makes it in, i.e., until the trailing edge arrives. The longer the bus is tied up, the longer it will be until the next miss may be serviced. The value of 3D is realized in that greater bandwidth (through wider busses) relieves the latency dependence on this trailing edge effect.

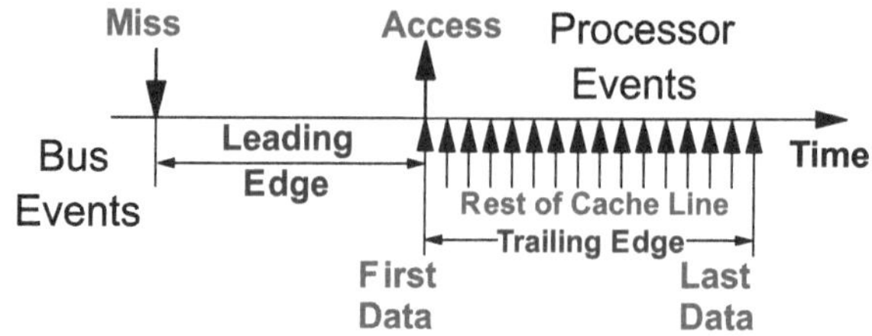

Fig. 1.11 A cache miss occupies the bus till the whole cache line required is transferred and blocks the following data requests. © 2006 IEEE. Reprinted, with permission, from [3]

Going forward, it should be pointed out that the industry trend appears to quadratically increase the number of cores per generation. Figure 1.12 shows the difference in growth rates between microprocessor clock rates (which drive data appetite), memory clock rate, and memory bus width. Data bus frequency has traditionally followed MPU frequency at a ratio of approximately 1:2, doubling every 18–24 months. Data bus bandwidth, however, has increased at a much slower rate (note the *Y*-axis scale is logarithmic!), implying that the majority of data bus transfer rate improvement is due to frequency. If and when the clock rate slows, data bus traffic is in trouble unless we supplant its bandwidth technology! It can be said that the leveraging of this bandwidth is the latest in a sequence of architecture paradigms asserted by designers to improve transaction rate. Earlier tricks, shown in Fig. 1.13, tended to stress other resources on the chip, i.e., increase the number of registers required, as described in Fig. 1.3. This time, it is the bandwidth. Integration into the *Z*-plane again postpones interconnect-related limitations to extending classic scaling, but along the way, something is really changed. The solution the rest of this book explores, that of expansion into another dimension, is one that nature has already shown us is essential for cognitive biological function.

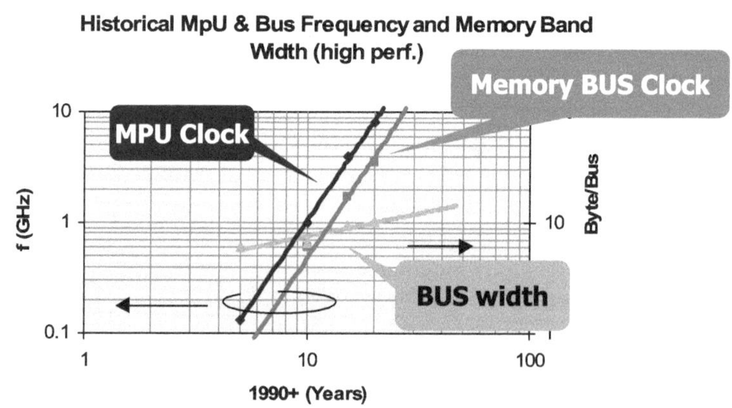

Fig. 1.12 Frequency drives data rate: data bus frequency follows MPU frequency at a ratio 1:2, roughly doubling every 18–24 months, while data bus bandwidth shows only a moderate increase. The data bus transfer rate is basically scaled by bus frequency: when the clock growth slows, the bus data rate growth will slow too

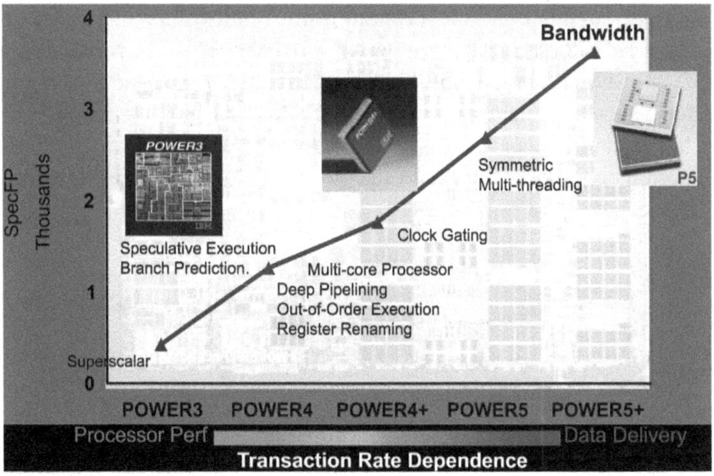

Fig. 1.13 POWER series: architectural performance contributions

To summarize the architecture issue, the following points should be kept in mind:

- The frequency is no longer increasing.
  - Logic speeds scale faster than the memory bus.
  - Processor clocks and bus clocks consume bandwidth.
- More speculation multiplies the number of prefetch attempts.
  - Wrong guesses increase miss traffic.
- Reducing line length is limited by directory as cache grows.
  - However, doubling the line size doubles the bus occupancy.
- The number of cores per die, N, is increasing in each generation.
  - This multiplies off-chip bus transactions by $N/2 \times \text{Sqrt}(2)$.
- This results in more threads per core and an increase in virtualization.
  - This multiplies off-chip bus transactions by N.
- The total number of processors/SMP is increasing.
  - This aggravates queuing throughout the system.
- Growing the number of cores/chip increases the demand for bandwidth.
  - Transaction retirement rate dependence on data delivery is increasing.
  - Transaction retirement rate dependence on uniprocessor performance is decreasing.

The above discussion treats the technology of 3D integration as a single entity. In practice, the architectural advantages we explored above will be realized incrementally as 3D improves. "Three Dimension" actually refers to a spectrum of processes and capabilities, which in time evolves to smaller via pitches, higher via densities, and lower via impedances. Figure 1.14 gives a feel for the distribution of 3D implementations and the applications that become eligible to use 3D once certain thresholds in capability are crossed.

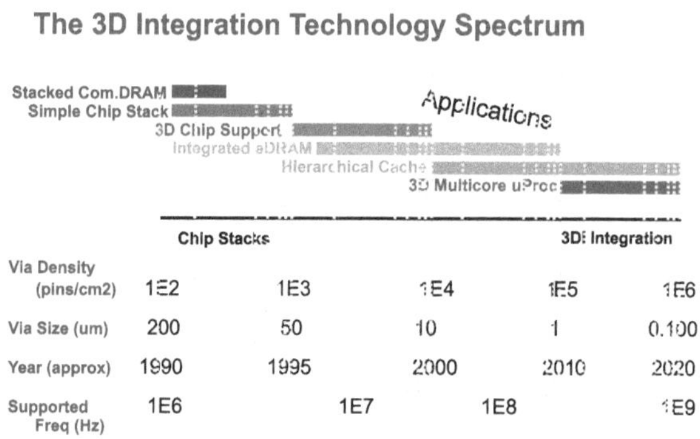

Fig. 1.14 The 3D integration technology spectrum

In the remainder of this book, several 3D integration experts show us how to harness the capabilities of this technology, not just in memory subsystems as discussed above, but in any number of new, innovative applications. The challenges of the job ahead are formidable; the process and technology recipes need to be established of course, but also the hidden required infrastructure: EDA, test, reliability, packaging, and the rest of the accoutrements we have taken for granted in 2D VLSI must be reworked. But executed well, the resulting compute densities and the new capabilities they support will be staggering. Even just in the memory management example explored earlier in this chapter, real-time access to the massive amounts of storage enabled by 3D will be seen later as a watershed event for our industry.

The remainder of the book starts with a brief introduction on the 3D process in Chapter 2, which serves as a short reference for designers to understand the 3D fabrication approaches (for comprehensive details on the 3D process, one can refer to [4]). The next part of the book focuses on design automation tools for 3D IC designs, including thermal analysis and power delivery (Chapter 3), thermal-aware 3D floorplanning (Chapter 4), thermal-aware 3D placement (Chapter 5), and thermal-aware 3D routing (Chapter 6). Following the discussion of the 3D EDA tools, the next three chapters present the 3D microprocessor design (Chapter 7), 3D network-on-chip architecture (Chapter 8), and the application of 3D stacking for energy efficient server design (Chapter 9). Finally the book is concluded with a chapter on the cost implication for 3D IC technology.

References

1. S. Amarasinghe, Challenges for Computer Architects: Breaking the Abstraction Barrier, *NSF Future of Computer Architecture Research Panel*, San Diego, CA, June 2003.
2. J. Davis, R. Venkatesan, A. Kaloyeros, M. Beylansky, S. J. Souri, K. Banerjee, K. C. Saraswat, A. Rahman, R. Reif, and J. D. Meindl, Interconnect Limits on Gigascale Integration (GSI) in the 21st Century, *Proceedings of the IEEE*, 89(3): 305–324, March 2001.
3. P. Emma, The End of Scaling? Revolutions in Technology and Microarchitecture as We Pass the 90 Nanometer Node, *Proceedings of the 33rd International Symposium on Computer Architecture*, IBM T. J. Watson Research Center, pp. 128–128, June 2006.
4. P. Garrou, C. Bower, and P. Ramm, Handbook of 3D Integration, Wiley-VCH, 2008.
5. D. Matzke, Will Physical Scalability Sabotage Performance Gains? *IEEE Computer*, 30(9): 37–39, September 1977.
6. F. Mujica, *History of the Skyscraper*, Da Capo Press, New York, NY, 1977.
7. V. Srinivasan, D. Brooks, M. Gschwind, P. Bose, V. Zyuban, P. Strenski, and P. Emma, "Optimizing Pipelines for Performance and Power," *Proceedings of the 35th Annual IEEE/ACM International Symposium on Microarchitecture*, pp. 333–344, November 2002.

Chapter 2
3D Process Technology Considerations

Albert M. Young and Steven J. Koester

Abstract Both form-factor and performance-scaling trends are driving the need for 3D integration, which is now seeing rapid commercialization. While overall process integration schemes are not yet standardized across the industry, it is now important for 3D circuit designers to understand the process trends and tradeoffs that underlie 3D technology. In this chapter, we outline the basic process considerations that designers need to be aware of: strata orientation, inter-strata alignment, bonding-interface design, TSV dimensions, and integration with CMOS processing. These considerations all have direct implications on design and will be important in both the selection of 3D processes and the optimization of circuits within a given 3D process.

2.1 Introduction

Both form-factor and performance-scaling trends are driving the need for 3D integration, which is now seeing rapid commercialization. While overall process integration schemes are not yet standardized across the industry, nearly all processes feature key elements such as vertical through-silicon interconnect, aligned bonding, and wafer thinning with backside processing. In this chapter we hope to give designers a better feel of the process trends that are affecting the evolution of 3D integration and their impact on design.

The last few decades have seen an astonishing increase in the functionality of computational systems. This capability has been driven by the scaling of semiconductor devices, from fractions of millimeters in the 1960s to the tens of nanometers in present-day technologies. This scaling has enabled the number of transistors on a single chip to correspondingly grow at a geometric rate, doubling roughly every

A.M. Young (✉)
IBM Thomas J. Watson Research Center, Yorktown Heights, NY 10598, USA
e-mail: yalbert@us.ibm.com

Y. Xie et al. (eds.), *Three-Dimensional Integrated Circuit Design*,
Integrated Circuits and Systems, DOI 10.1007/978-1-4419-0784-4_2,

18 months, a trend originally predicted by Gordon Moore and now referred to as Moore's law [1]. The impact of this trend cannot be underestimated, and the resulting increase in computational capacity has greatly influenced nearly every facet of society.

The tremendous success of Moore's law, and in particular, the scaling of Si MOSFETs [2], drives the ongoing efforts to continue this trend into the future. However, several serious roadblocks exist. The first is the difficulty and expense of continued lithographic scaling, which could make it economically impractical to scale devices beyond a certain pitch. The second roadblock is that, even if lithographic scaling can continue, the power dissipated by the transistors will bring clock frequency scaling to a halt. In fact, it could be argued that clock frequency scaling has already stopped as microprocessor designs have increasingly relied upon new architectures to improve performance. These arguments suggest that, in the near future, it will no longer be possible to improve system performance through scaling alone, and that additional methods to achieve the desired enhancement will be needed. Three-dimensional (3D) integration technology offers the promise of being a new way of increasing system performance even in the absence of scaling. This promise is due to a number of characteristic features of 3D integration, including (a) decreased total wiring length, and thus reduced interconnect delay times; (b) dramatically increased number of interconnects between chips; and (c) the ability to allow dissimilar materials, process technologies, and functions to be integrated.

Overall, 3D technology can be broadly defined as any technology that stacks semiconductor elements on top of each other and utilizes vertical, as opposed to peripheral, interconnects between the wafers. Under this definition, 3D technology casts a wide net and could include simple chip stacks, silicon chip carriers and interposers, chip-to-wafer stacks, and full wafer-level integration. Each of these individual technologies has benefits for specific applications, and the technology appropriate for a particular application is driven in large part by the required interconnect density. For instance, for wireless communications, only a few through-silicon vias (TSVs) may be needed per chip in order to make low-inductance contacts to a backside ground plane. On the other hand, high-performance servers and stacked memories could require extremely high densities (10^5–10^6 pins/cm^2) of vertical interconnects. Applications such as 3D chips for supply voltage stabilization and regulation reside somewhere in the middle, and a myriad of applications exist which require the full range of interconnect densities possible.

A wide range of 3D integration approaches are possible and have been reviewed extensively elsewhere [3]. These schemes have various advantages and trade-offs and ultimately a variety of optimized process flows may be needed to meet the needs of the various applications targeted. However, nearly all 3D ICs have three main process components: (a) a vertical interconnect, (b) aligned bonding, and (c) wafer thinning with backside processing. The order of these steps depends on the integration approach chosen, which can depend strongly on the end application. The process choices which impact overall design points will be discussed later on in this chapter. However, to more fully appreciate where 3D IC technology is heading, it is helpful to first have an understanding of work that has driven early commercial adoption of 3D integration today.

2.2 Background: Early Steps in the Emergence of 3D Integration

Early commercialization efforts leading to 3D integration have been fueled by mobile-device applications, which tended to be primarily driven by form-factor considerations. One key product area has been the CMOS image-sensor market (camera modules used in cellular handsets), which has driven the development of wafer-level chip-scale packaging (WL-CSP) solutions. Shellcase (later bought by Tessera) was one company that had strong efforts in this area. Many of these solutions can be contrasted with 3D integration in that they (1) do not actually feature circuit stacking and (2) often use wiring routed around the edge of the die to make electrical connections from the front to the back side of the wafer. However, these WL-CSP products did help drive significant advances in technologies, such as silicon-to-glass wafer bonding and subsequent wafer thinning, that are used in many 3D integration process flows today. In a separate market segment, multichip packages (MCPs) that integrated large amounts of memory in multiple silicon layers have also been heavily developed. This product area has also contributed to the development of reliable wafer-thinning technology. In addition, it has helped to drive die-stacking technology as well, both with and without spacer layers between thinned die. Many of these packages have made heavy use of wirebonding between layers, which is a cost-effective solution for form-factor-driven components. However, as mobile devices continue to evolve, new solutions will be needed.

Portable devices are taking on additional functionality, and product requirements are extending beyond simple form-factor reductions to deliver increased performance per volume. More aggressive applications are demanding faster speeds than wire bonding can support, and the need for more overall bandwidth is forcing a transition from peripheral interconnect (such as die-edge or wirebond connection) to distributed area-array interconnect. The two technology elements that are required to deliver this performance across stacked circuits are TSVs and area-array chip-to-chip connections. TSV adoption in these product areas has been limited to date, as cost sensitivities for these types of parts have slowed TSV introductions. So far, the use of TSVs has been dominated by fairly low I/O-count applications, where large TSVs are placed at the periphery of the die, which tends to limit their inherent advantages. However, as higher levels of performance are being demanded, a transition from die-periphery TSVs to TSVs which are more tightly integrated in the product area is likely to take hold. The advent of deep reactive-ion etching for the micro-electro-mechanical systems (MEMS) market will help enable improved TSVs with reduced footprints. Chip-on-chip area-array interconnect such as that used by Sony in their PlayStation Portable (PSP) can improve bandwidth between memory and processor at reasonable cost. One version of their microbump-based technology offers high-bandwidth (30-μm solder bumps on 60-μm pitch) connections between two dies assembled face-to-face in a flip-chip configuration. This type of solution can deliver high-bandwidth between two dies; however, without TSVs it is not immediately extendable to support communication in stacks with more than two dies. Combining fine-pitch TSVs with area-array inter-tier connectivity is clearly the direction for the next generation of 3D IC technologies.

Wafer bonding to glass, wafer thinning, and TSVs-at-die-periphery are all technologies used in the manufacturing of large volumes of product today. However, the next generation of 3D IC technologies will continue to build on these developments in many different ways, and the impact of process choices on product applications needs to be looked at in detail. In the next section, we identify important process factors that need consideration.

2.3 Process Factors That Impact State-of-the-Art 3D Design

The interrelation of 3D design and process technology is important to understand, since the many process integration schemes available today each have their own factors which impact 3D design in different ways. We try to provide here a general guide to some of the critical process factors which impact 3D design. These include strata orientation, alignment specifications, and bonding-interface design, as well as TSV design point and process integration.

2.3.1 Strata Orientation: Face-to-Back vs. Face-to-Face

The orientation of the die in the 3D stack has important implications for design. The choice impacts the distances between the transistors in different strata and has electronic design automation (EDA) impact related to design mirroring. While multiple die stacks can have different combinations of strata orientations within the stack, the face-to-back vs. face-to-face implications found in a two-die stack can serve to illustrate the important issues. These options are illustrated schematically in Fig. 2.1.

2.3.1.1 Face-to-Back

The "face-to-back" method is based on bonding the front side of the bottom die with the back side (usually thinned) of the top die. Similar approaches were originally developed at IBM for multi-chip modules (MCMs) used in IBM G5 systems [4], and later this same approach was demonstrated on wafer level for both CMOS and MEMS applications. Figure 2.1a schematically depicts a two-layer stack assembled in the face-to-back configuration. The height of the structure and therefore the height of the interconnecting via depend on the thickness of the thinned top wafer. If the aspect ratio of the via is limited by processing considerations, the substrate

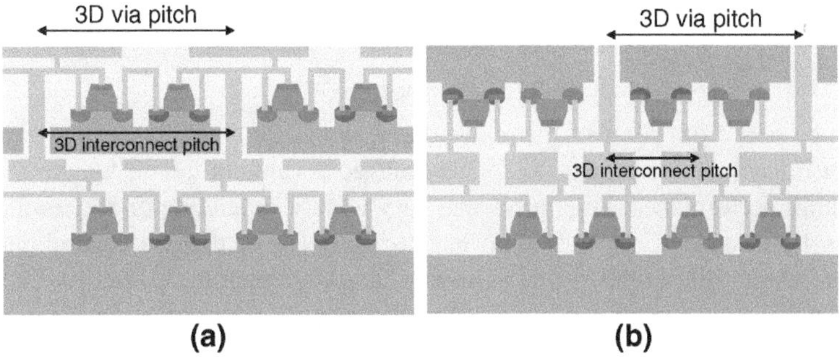

Fig. 2.1 Schematic illustrating face-to-back (**a**) and face-to-face (**b**) orientations for a two-strata 3D stack. Three-dimensional via pitch is contrasted with 3D interconnect pitch between strata

thickness can then directly impact the number of possible interconnects between the two wafers. It is also important to note that in this scheme, the total number of interconnects between the wafers cannot be larger than the number of TSVs. In order to construct such a stack, a handle wafer must be utilized, and it is well known that handle wafers can induce distortions in the top wafer that can make it difficult to achieve tight alignment tolerances. In previous work we have found that distortions as large as 50 ppm can be induced by handle wafers, though strategies such as temperature-compensated bonding can be used to reduce them. For advanced face-to-back schemes, typical thicknesses of the top wafer are on the order of 25–50 μm, which limit the via and interconnect pitch to values on the order of 10–20 μm. Advances in both wafer thinning and via-fill technologies could reduce these numbers in the future.

2.3.1.2 Face-to-Face

Figure 2.1b shows the "face-to-face" approach which focuses on joining the front sides of two wafers. This method was originally utilized [5] at IBM to create MCMs with sub-20-μm interconnect pitch with reduced process complexity compared to the face-to-back scheme. A key potential advantage of face-to-face assembly is the ability to decouple the number of TSVs from the total number of interconnections between the layers. Therefore, it could be possible to achieve much higher interconnect densities than allowed by face-to-back assembly. In this case, the interconnect pitch is only limited by the alignment tolerance of the bonding process (plus the normal overlay error induced by the standard CMOS lithography steps). For typical tolerances of 1–2 μm for state-of-the-art aligned-bonding systems, it is therefore conceivable that interconnect pitches of 10 μm or even smaller can be achieved with face-to-face assembly. However, the improved inter-level connectivity achievable can only be exploited for two-layer stacks; for multi-layer stacks, the TSVs will still limit the total 3D interconnect density.

2.3.2 Inter-strata Alignment: Tolerances for Inter-layer Connections

As can be seen from some of the above discussions, alignment tolerances can have a direct impact on the density of connections achievable in the 3D stack, and thus the overall performance. Tolerances can vary widely depending on the tooling and process flow selected, so it is important to be aware of process capabilities. For example, die-to-die assembly processes can have alignment tolerances varying from the 1-μm range to the 20-μm range, depending on the speed of assembly required. Fine-pitch capability in these systems is possible, but transition to manufacturing can be challenging because of the time required to align each individual die. It is certainly plausible that these alignment-throughput challenges will be solved in coming years; when married with further scaling of chip-on-chip area connections to fine pitch, this could enable high-performance die-stack solutions. Today, wafer-to-wafer alignment offers an alternative, where a wafer fully populated with die (well registered to one another) can be aligned in one step to another wafer. This allows more time and care to be used in achieving precise alignment at all chip sites. Advanced wafer-to-wafer align-and-bond systems today can typically achieve tolerances in the 1- to 2-μm range. Although the set of issues for different processes is complex, a deeper discussion of aligned wafer bonding below can help highlight some of the issues found in both wafer- and die-stacking approaches.

Aligned wafer bonding for 3D integration is fundamentally different than blanket wafer bonding processes that are used. For instance, in silicon on insulator (SOI) substrate manufacturing, the differences are several-fold. First of all, alignment is required since patterns on each wafer need to be in registration, in order to allow the interconnect densities required to take advantage of true 3D system capabilities. Second, wafers for 3D integration typically have significant topography on them, and these surface irregularities can make high-quality bonding significantly more difficult than for blanket wafers, particularly for oxide-fusion bonding. Finally, due to the fact that CMOS circuits (usually with BEOL metallization) already exist on the wafers, the thermal budget restriction on the bonding process can be quite severe, and typically the bonding process needs to be performed at temperatures less than 400°C.

Wafer-level alignment is fundamentally different from stepper-based lithographic alignment typically used today in CMOS fabrication. This is because the alignment must be performed over the entire wafer, as opposed to on a die-by-die basis. This requirement makes overlay control much more difficult than in die-level schemes. Non-idealities, such as wafer bow, lithographic skew or run-out, and thermal expansion can all lead to overlay tolerance errors. In addition, the transparency or opacity of the substrate can also affect the wafer alignment. Tool manufacturers have developed alignment tools for both full 200-mm and 300-mm wafers, with alignment accuracy in the ~1–2 μm range.

Due to the temperature excursions and potential distortions associated with the bonding process itself, it is standard procedure in the industry to first use aligner tools (which have high throughput) and then move wafers to specialized bonding tools for good control of temperature and pressure across the wafers and through the 3D stack. The key to good process control is the ability to separate the alignment and pre-bonding steps from the actual bonding process. Such integration design allows for a better understanding of the final alignment error contributions. That said, the actual bonding process and the technique used can affect overall alignment overlay, and understanding of this issue is critical to the proper choice of bonding process.

For example, an alignment issue arises for Cu–Cu bonding when the surrounding dielectric materials from both wafers are recessed. In this scenario, if there is a large misalignment prior to bonding, the wafers can still be clamped for bonding. In addition, this structure cannot inhibit the thermal misalignment created during thermo-compression and is not resistant to additional alignment slip due to shear forces induced during the thermo-compression process. One way to prevent such slip is to use lock-and-key structures across the interface to limit the amount of misalignment by keeping the aligned wafers registered to one another during the steps following initial align and placement. This could be a significant factor in maintaining the ability to extend the 3D process to tighter interconnect pitches, since the allowable pitch is often ultimately limited by alignment and bonding tolerances.

Wafer-thinning processes which use handle wafers and lamination can also add distortion to the thinned silicon layer. This distortion can be caused by both differences in coefficients of thermal expansion between materials and by the use of polymer lamination materials with low elastic modulus. As one example, if left uncontrolled, the use of glass-handle wafers can introduce alignment errors in the range of 5 μm at the edge of a 200-mm wafer, a value which is significantly larger than the errors achievable in more direct silicon-to-silicon alignments. So in any process using handle wafers, control and correction of these errors is an important consideration. In practice, these distortions can often be modeled well as global magnification errors. This allows the potential to correct most of the wafer-level distortion using methods based on control of temperature, handle-wafer materials, and lamination polymers.

Alignment considerations are somewhat unique in SOI-based oxide-fusion bonding. This process is often practiced where the SOI wafer is laminated to a glass-handle wafer and the underlying silicon substrate is removed, leaving a thin SOI layer attached to the glass prior to alignment. Unlike other cases, where either separate optical paths are used to image the surfaces to be aligned or where IR illumination is required to image through the wafer stack, one can see through this type of sample at visible wavelengths. This allows very accurate direct optical alignment to an underlying silicon wafer, in a manner similar to wafer-scale contact aligners. Wafer contact and a preliminary oxide-fusion bond must be initiated in the alignment tool itself, but once this is achieved, there is minimal alignment distortion introduced by downstream processing [6, 7].

2.3.3 *Bonding-Interface Design*

Good design of the bonding interface between the stacked strata involves careful analysis of mechanical, electrical, and thermal considerations. In the next section, we briefly describe three particular technologies for aligned 3D wafer bonding that have been investigated at IBM: (i) Cu–Cu compression bonding, (ii) transfer-join bonding (hybrid Cu and adhesive bonding), and (iii) oxide-fusion bonding. Similar attention to inter-strata interface design is required for die-scale stacking technologies, which often feature solder and underfill materials between silicon die.

2.3.3.1 Copper-to-Copper Compression Bonding

Attachment of two wafers is possible using a thermo-compression bond created by applying pressure to two wafers with Cu metallized surfaces at elevated temperatures. For 3D integration, the Cu–Cu join can serve the additional function of providing electrical connection between the two layers. Optimization of the quality of this bonding process is a key issue being addressed and includes provision of various surface preparation techniques, post-bonding straightening, thermal annealing cycles, as well as use of optimized pattern geometry [8–10].

Copper thermo-compression bonding occurs when, under elevated temperatures and pressures, the microscopic contacts between two Cu regions start to deform, further increase their contact area, and finally diffuse into each other to complete the bonding process. Key parameters of Cu bonding include bonding temperature, pressure, duration, and Cu surface cleanliness. Optimization of all of these parameters is needed to achieve a high-quality bond. The degree of surface cleanliness is related to not only the pre-bonding surface clean but also the vacuum condition during bonding [8]. In addition, despite the fact that bonding temperature is the most significant parameter in determining bond quality, the temperature has to be compatible with BEOL process temperatures in order to not affect device performance.

The quality of patterned Cu bonding at wafer-level scales has been investigated for real device applications [9, 10]. The design of the Cu-bonding pattern influences not only circuit placement but also bond quality since it is related to the available area that can be bonded in a local region or across the entire wafer. Cu bond pad size (interconnect size), pad pattern density (total bond area), and seal design have also been studied. Studies based on varying Cu-bonding pattern densities have shown that higher bond densities result in better bond quality and can reach a level where they rarely fail during dicing tests. In addition, a seal design which has extra Cu bond area around electrical interconnects, chip edge, and wafer edge could prevent corrosion and provide extra mechanical support [9].

2.3.3.2 Hybrid Cu/Adhesive Bonding (Transfer-Join)

A variation on the Cu–Cu compression-bonding process can be accomplished by utilizing a lock-and-key structure along with an intermediate adhesive layer to improve bond strength. This technology was originally developed for MCM thin film modules and underwent extensive reliability testing during the build and qualification [11, 12]. However, as noted previously, this scheme is equally suitable for wafer-level 3D integration and could have significant advantages over direct Cu–Cu-based schemes.

In the transfer-join assembly scheme, the mating surfaces of the two device wafers that are to be joined together are provided with a set of protrusions (keys) on one side that are matched to receptacles (locks) on the other, as shown in Fig. 2.2a. A protrusion, also referred to as a stud, can be the extension of a TSV or a specially fabricated BEOL Cu stud. The receptacle is provided at the bottom with a Cu pad which will be bonded to the Cu stud later. At least one of the mating surfaces (in Fig. 2.2a, the lower one) is provided with an adhesive atop the last passivation dielectric layer. Both substrates can be silicon substrates or optionally one of them could be a thinned wafer attached to a handle substrate. The studs and pads are optionally connected to circuits within each wafer by means of 2D wiring and/or TSVs as appropriate.

These substrates can be aligned in the same way as in the direct Cu–Cu technique and then bonded together by the application of a uniform and moderate pressure at a temperature in the 350–400°C range. The height of the stud and thickness of the adhesive/insulator layer are typically adjusted such that the Cu stud to Cu pad contact is established first during the bonding process. Under continued bonding pressure, the stud height is compressed and the adhesive is brought into contact and bonded with the opposing insulator surface. The adhesive material is chosen

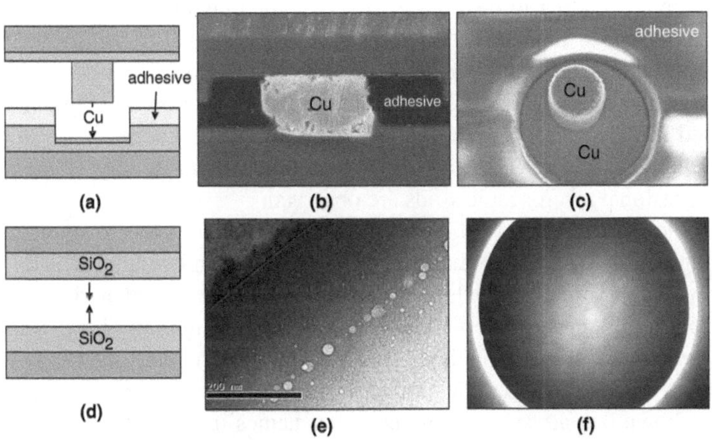

Fig. 2.2 Bonding schemes: (**a**) cross section of the transfer-join bonding scheme; (**b**) polished cross section of a completed transfer-join bond; (**c**) a top–down scanning-electron micrograph (SEM) view of a transfer-join bond after delayering, showing lock-and-key structure and surrounding adhesive layer; (**d**) oxide-fusion bonding scheme; (**e**) cross-sectional transmission-electron micrograph (TEM) of an oxide-fusion bond; (**f**) whole-wafer infrared image of two wafers joined using oxide-fusion bonding

to have the appropriate rheology required to enable flow and bonds the two wafers together by filling any gaps between features. Additionally, the adhesive is tailored to be thermally stable at the bonding temperature and during any subsequent process excursions required (additional layer attachment, final BEOL wiring on 3D stack, etc.). Depending upon the wafers bonded together, either a handle wafer removal or back side wafer thinning is performed next. The process can be repeated as needed if additional wafer layer attachment is contemplated.

A completed Cu–Cu transfer-join bond with polymer adhesive interlayer is shown in Fig. 2.2b. Figure 2.2c additionally shows the alignment of a stud to a pad in a bonded structure after the upper substrate has been delayered for the purpose of constructional analysis. This lock-and-key transfer join approach can be combined with any of the 3D integration schemes described earlier. The fact that the adhesive increases the mechanical integrity of the joined features in the 3D stack means that the copper-density requirements that were needed to ensure integrity in direct Cu–Cu bonding can be relaxed or eliminated.

2.3.3.3 Oxide-Fusion Bonding

Oxide-fusion bonding can be used to attach two fully processed wafers together. At IBM we have published extensively on the use of this basic process capability to join SOI wafers in a face-to-back orientation [13], and other schemes for using oxide-fusion bonding in 3D integration have been implemented by others [14]. General requirements include low-temperature bonding-oxide deposition and anneal for compatibility with integrated circuits, extreme planarization of the two surfaces to be joined, and surface activation of these surfaces to provide the proper chemistry to allow robust bonding to take place. A schematic diagram of the oxide-bonding process is shown in Fig. 2.2d, along with a cross-sectional transmission-electron micrograph (TEM) of the bonding interface (Fig. 2.2e) and a whole-wafer IR image of a typical bonded pair (Fig. 2.2f). The transmission-electron micrograph shows a distributed microvoiding pattern, while the plan-view IR image shows that, after post-bonding anneals of 150 and 280°C, excellent bond quality is maintained, though occasional macroscopic voids are observed.

The use of multiple levels of back-end wiring typically leads to significant surface topography. This creates challenges for oxide-fusion bonding, which requires extremely planar surfaces. While it is possible to reduce non-planarity by aggressively controlling metal-pattern densities in mask design, we have found that process-based planarization methods are also required. As described in [15], typical wafers with back-end metallization have significant pattern-induced topography. We have shown that advanced planarization schemes incorporating the deposition of thick SiO_2 layers followed by a highly optimized chemical-mechanical polishing (CMP) protocol can dramatically reduce pattern-dependent variations, which is needed to achieve good bonding results. The development of this type of advanced planarization technology will be critical to the commercialization of oxide-bonding schemes, where pattern-dependent topographic variations will be encountered on a routine basis. While bringing this type of technology into manufacturing poses many challenges, joining SOI wafers using oxide-fusion bonding can help enable very small distances between the device strata and likewise can lead to very high-density interconnect.

2.3.4 TSV Dimensions: Design Point Selection

Perhaps the most important technology element for 3D integration is the vertical interconnect, i.e., the TSV. Early TSVs have been introduced into the production environment by companies such as IBM, Toshiba, and ST Microelectronics, using a variety of materials for metallization including tungsten and copper. A high-performance vertical interconnect is necessary for 3D integration to truly take advantage of 3D for system-level performance, since interconnects limited to the periphery of the chip do not provide densities significantly greater than in conventional planar technology. Methods to achieve through-silicon interconnections within the product area of the chip often have similarities to back-end-of-the-line (BEOL) semiconductor processes, with one difference being that a much deeper hole typically has to be created vertically through the silicon material using a special etch process.

The dimensions of the TSV are key to 3D circuit designers since they directly impact exclusion zones where designers cannot place transistors, and in some cases, back-end-of-the-line wiring as well. However, the dimensions of the TSV are very dependent on the 3D process technology used to fabricate them and more specifically are a function of silicon thickness, aspect ratio and sidewall taper, and other process considerations. These dimensions are also heavily dependent on the metallization used to fill the vias. Here we will take a look at the impact of these process choices when dealing with two of the most important metallization alternatives, tungsten (W) and copper (Cu).

2.3.4.1 Design Considerations for Tungsten and Copper TSVs

Impact of Wafer Thinning

Wafer thinning is a necessary component of 3D integration as it allows the interlayer distance to be reduced, therefore allowing a high density of vertical interconnects. The greatest challenge in wafer thinning is that the wafer must be thinned to ~5–10% of its original thickness, with a typical required uniformity of <1–2 μm. In bulk Si, this thinning is especially challenging since there is no natural etch stop. The final thickness depends on thinning process control capabilities and is limited by the thickness uniformity specifications of the silicon removal process (that being mechanical grinding and polishing, wet or dry etching). Successful thinning to a uniform Si thickness of a few microns has been demonstrated, but typically thicknesses of greater than 20 μm are necessary for a robust process. Standard process steps for Si thinning are as follows: First a coarse grinding step is performed in order to thin the wafer from its original thickness (~700–800 μm) to a thickness of

125–150 μm. This type of process is usually performed using a 400 mesh grinding surface. This step is followed by a fine grinding step to thin to ~100 μm using an 1800–2000 mesh surface. Next a mechanical polishing step can be performed down to the desired thickness of 30–60 μm. For most processes it is desirable for these grinding steps to be performed only on uniform regions of the silicon, due to the fact that stresses associated with mechanical grind/polish steps can damage fine features in the silicon. If it is necessary to expose the TSV from the backside, it is typically desirable to complete the thinning process using a plasma-based etching process (such as reactive-ion etching) to expose the TSVs. Limitations on the uniformity of the backside thinning sets the practical limit on the TSV depth, and therefore, the via pitch that can be achieved using this type of blind-thinning process. After vias are exposed from the back side of the wafer, a redistribution layer is often fabricated using process technology similar to that used in wafer-level packaging. In some cases more than one level of wiring can be fabricated.

Different wafer thicknesses can lead to the use of different via geometries, as seen in Fig. 2.3. For the aggressively thinned case where the silicon thickness is <30 μm, fully filled vias made of a single-conductor metallized with either tungsten or copper can be realized in a fairly straightforward manner. However, in the case where wafer thickness is larger, say >100 μm, the via possibilities take on different forms. For copper vias, the silicon thickness has a fairly direct impact on the size of the TSV footprint at the wafer surface; thicker silicon substrates force larger copper-TSV footprints, which will, at some point, become economically infeasible. Also, copper-based vias need to deal with the low aspect ratios that are reliably possible with copper plating. As wafer thicknesses get larger, there tends to be a transition from full-plating to partial-plating methods in order to maintain manufacturable copper-plating thicknesses. The footprint of tungsten vias tends to be less sensitive to silicon thickness. Although the dimensions of tungsten-filled vias can be limited by deposition-film thickness, the aspect ratios achievable using tungsten deposition can be quite impressive. A wide range of silicon thicknesses are generally possible within a small TSV footprint; at IBM we routinely fabricate tungsten TSVs spanning silicon thicknesses in excess of 100 μm.

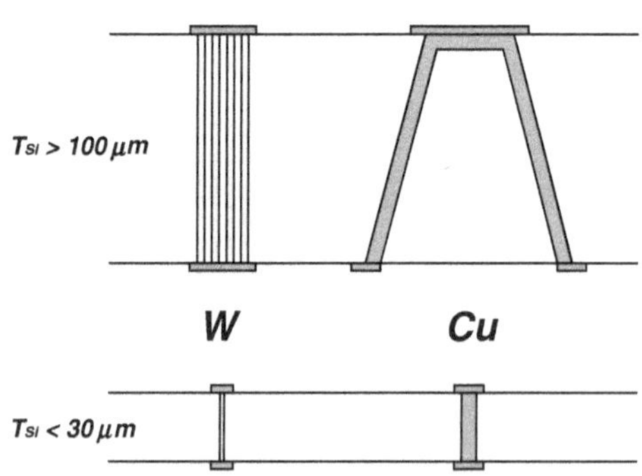

Fig. 2.3 Schematic illustration of the effect of different silicon thicknesses on via geometry for tungsten (W) and copper (Cu) via cases

Impact on Via Resistance and Capacitance

The choice of conductor metallization used in the TSV has a direct impact on important via parameters such as resistance and capacitance. Not only do different metals have different intrinsic values of resistivity, but their respective processing limitations also dictate the range of via geometries that are possible. Since the choice of metallization directly impacts via aspect ratio (i.e., the ratio of via depth to via width), it also has a direct impact on via resistance. Tungsten vias are typically deposited as thin films with very high aspect ratio (>>20:1) and hence are narrow and tend to have relatively high resistance. To mitigate this effect, multiple tungsten conductors can be strapped together in parallel to provide an inter-strata connection of suitably low overall resistance, at the cost of increased area, as shown schematically at the top left of Fig. 2.3. Copper has a better intrinsic resistivity than tungsten. The allowable geometries of plated-copper vias also have low aspect ratios (typically limited from 6:1 to 10:1) and therefore lend themselves well to low-resistance connections.

Via capacitance can be impacted strongly by the degree of sidewall taper introduced into the via design. Tungsten vias typically have nearly vertical sidewalls and no significant taper; copper vias may more readily benefit from sloped sidewalls. Although sidewall taper enlarges the via footprint at the wafer surface which is undesirable, the introduction of sidewall taper can help to improve copper plating quality and increase the deposition rate of via-isolation dielectrics. These deposition rates are strongly impacted by via geometry, and methods which help to increase final via-isolation thickness will enable lower via capacitance levels and thus improve inter-strata communication performance.

2.3.4.2 Ultra-High Density Vias Using SOI-Based 3D Integration

It is also possible to utilize the buried oxide of an SOI wafer as a way of enhancing the 3D integration process, and this scheme is shown in Fig. 2.4. This so-called SOI scheme has been described extensively in previous publications [13, 16] and is only briefly summarized here. Unlike other more conventional 3D processes, in our SOI-based 3D integration scheme, the buried oxide can act as an etch stop for the final wafer thinning process. This allows the substrate to be completely removed before the two wafers are combined. A purely wet chemical etching process can be used; for instance, TMAH (Tetramethylammonium hydroxide) can remove 0.5 μm/min of silicon with excellent selectivity to SiO_2. In our process, we typically remove ~600 μm of the silicon wafer by mechanical techniques and then employ 25% TMAH at 80°C to etch (40 μm/h etch rate) the last 100 μm of silicon down to the buried oxide layer. The buried oxide has a better than 300:1 etch selectivity relative to silicon and therefore acts as a very efficient etch stop layer. Overwhelming advantages of such an approach are that all of the Si can be uniformly removed, leaving a very smooth (<10 nm) surface for the application of bonding oxide films, and the fact that the layer of transferred circuits is automatically a minimal thickness across the wafer, facilitating the fabrication of high-density inter-strata connections later in

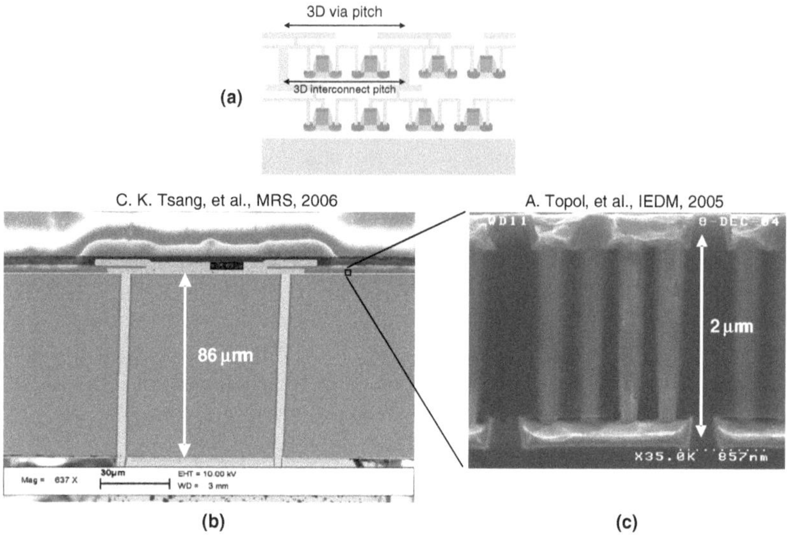

Fig. 2.4 (**a**) Schematic illustrating face-to-back SOI-3D assembly; (**b**) annular TSV with a fairly conventional silicon thickness, shown for size comparison with (**c**) interlayer 3D-via in SOI-3D technology

the process flow. Because the distance between the layers is much smaller than in conventional TSV schemes, the via pitch and size can be dramatically reduced. The minimum height of an interconnecting via could be as low as 1–2 μm, allowing via dimensions as small as 0.2–0.25 μm [13]. If extremely tight wafer-level alignment can also be achieved, then via pitches on the order of 1–2 μm are conceivable, opening the door to numerous new system-level opportunities not achievable with looser interconnect pitches. Figure 2.4b, c shows a comparison of more conventional TSVs (fabricated with tungsten deposited in an annular-shaped via region) with Cu-filled vias fabricated using our SOI-based integration scheme. The size comparison shows the potential advantages of SOI-based 3D integration.

2.3.5 Via-Process Integration and a Reclassification of Via Types

The specific process flow used to fabricate the TSV is of importance to the circuit designer since the method of via-process integration drives specific design rules that must be adhered to. As one example, certain via processes inherently require exclusion rules in BEOL wiring to allow the TSV to pass through, whereas others do not. Another example follows from the fact that alignment tolerances differ for front-side and back-side via processing, which drives via-process-specific design rules. In much of the early literature, the use of the terms "vias-first" and "vias-last" has been common, but such terms have led to significant confusion in the industry. At times these designations have been used to denote whether vias are processed before or after the base integrated-circuit wafers have been completed (e.g., for vias processed from the front side of the wafer, these terms have been used to distinguish between vias processed before and after back-end-of-line wiring), and in other cases these terms have been used to refer to whether via formation on completed base wafers occurs before or after wafer thinning (e.g., from the front side or the back side of the thinned wafer).

An alternate classification scheme that is based on the timing of the TSV via etch in the process flow can be used to improve clarity. This redefined classification is based on the two most important practical considerations to the designer: (1) Does the process use frontside or backside via etch? (2) If it uses frontside via etch, is this etch done before or after back-end-of-line (BEOL) wiring? This classification leads us to three primary cases of interest:

(i) Pre-backend frontside via (type-F1)
(ii) Post-backend frontside via (type-F2), and
(iii) Backside via etch (after wafer thinning, type-B).

Since backside via etches generally land on the lowest available BEOL metal layer, we ignore here the case where the backside via is etched deeper into the BEOL. Properties of the three primary classes of TSVs are outlined in Table 2.1 and are discussed in further detail below.

Table 2.1 Characteristics of various TSV types

Via type	F1	F2	B
Via etch	Frontside	Frontside	Backside
Via formation	Via before BEOL	Via after BEOL	Via after thinning
High-temperature materials compatibility	+	−	−
Reduced via dimensions	+	−	+
Low circuit blockage	+	−	+
Ease of process-integration	−	+	+

2.3.5.1 Pre-backend Frontside Via (Type F1)

One approach to TSVs that has been demonstrated at IBM and elsewhere is a type of "vias-first" process flow. In this approach, F1 vias are formed before BEOL processing, and thus the approach has the advantage of allowing higher-temperature dielectric and metal fill processes and allows high-aspect-ratio vias to be formed. The F1 via tends to be more challenging to integrate with conventional CMOS, but has the advantage of low-wiring blockage in the BEOL. One particular approach that has been described is an annular-via geometry for large-area contacts [17]. This structure utilizes an etched ring-shaped via geometry, where the ring width is narrow enough to allow complete fill of the structure utilizing a variety of materials, including doped polysilicon, electroplated copper, or CVD tungsten. For higher-density 3D integration applications, annuli with small central cores (the region defined by the inner diameter of the annulus) can be used, where the annular region is filled with isolation dielectrics and the central core subsequently etched and metallized. The term "pre-backend frontside via" can be extended to all cases where via integration occurs immediately before any particular level of metal (e.g., pre-M1 frontside via, pre-M2 frontside via). It is distinguished from the F2 via in that F2 vias are processed after CMOS fabrication is complete.

2.3.5.2 Post-backend Frontside Via (Type F2)

TSVs can be etched from the frontside of the wafer, taking advantage of frontside alignment capabilities, and yet still be fabricated after completion of BEOL wiring. This F2 via will be preferred in many cases since it is easier to integrate with CMOS process technology and will have the yield advantages that come from not interfering with the standard CMOS process flow. It does have the downside of having slightly larger vias due to the requirement of extending the vias through the BEOL and, more importantly, has the significant disadvantage of blocking wiring channels through the entire BEOL stack. For this reason, unless via dimensions or the total number of TSVs required are small, this type of via can become difficult to implement, especially for designs of high complexity.

2.3.5.3 Backside Via (After Wafer Thinning, Type B)

A variety of vertical through-silicon-interconnect technologies have been developed by IBM [17–19]. Often the TSV is formed only after wafer thinning by utilizing a backside deep reactive ion etch. Like the F2 via, this type B via also has the significant advantage that the CMOS technologies used in the base wafers do not need to be modified. After backside deep reactive ion etching, insulation can subsequently be applied to the interior of the via and selectively removed from the bottom to allow electrical contact to the front side of the wafer. Metallization of large vias prepared in this manner has been demonstrated by various means. For example, the use of an initial partial fill with plated copper followed by evaporation of Cr/Cu BLM and Pb/Sn solder has been demonstrated. Backside vias typically have aspect ratios limited by process capabilities for dielectric and metal fill. They will also be significantly hampered by the alignment capabilities of backside lithography on thinned wafers, and thus may not be well suited for high-density 3D integration.

2.4 Conclusion

In order to be an effective 3D circuit designer, it is important to understand the process considerations that underlie 3D technology. In this chapter, we have tried to outline the basic process considerations that 3D circuit designers need to be aware of: strata orientation, inter-strata alignment, bonding-interface design, TSV dimensions, and integration with CMOS processing. These considerations all have direct implications on design and will be important in both the selection of 3D processes and the optimization of circuits within a given 3D process.

Technology developments in the areas of CMOS image-sensors, wafer-level chip-scale packages, multichip packages for memory applications, TSVs, and chip-on-chip area–array interconnect are all well on their way and are pushing us rapidly toward the development of full 3D circuit integration. We hope that circuit designers reading this volume will be inspired to learn how to best take advantage of the unique aspects of 3D circuit integration.

Acknowledgments The authors wish to thank the following people for their contributions to this work: Roy R. Yu, Sampath Purushothaman, Kuan-neng Chen, Douglas C. La Tulipe, Narender Rana, Leathen Shi, Matthew R. Wordeman, Edmund J. Sprogis, Fei Liu, Steven Steen, Cornelia Tsang, Paul Andry, David Frank, Jyotica Patel, James Vichiconti, Deborah Neumayer, Robert Trzcinski, Latha Ramakrishnan, James Tornello, Michael Lofaro, Gil Singco, John Ott, David DiMilia, William Price, and Jesus Acevedo. The authors also acknowledge the support of the IBM Microelectronics Research Laboratory, and Central Scientific Services, as well as the staff at EV Group and Suss MicroTec. This project was funded in part by DARPA under SPAWAR contract numbers N66001-00-C-8003 and N66001-04-C-8032.

References

1. G. Moore, Cramming more components onto integrated circuits, *Electronics* 38, 114–117 (1965).
2. R. H. Dennard, F. H. Gaensslen, H. N. Yu, V. L. Rideout, E. Bassous, and A. R. LeBlanc, Design of ion-implanted MOSFETs with very small physical dimensions, *IEEE Journal of Solid-State Circuits*, SC-9, 256–268 (1974).
3. A. W. Topol, D. C. La Tulipe, L. Shi, D. J. Frank, K. Bernstein, S. E. Steen, A. Kumar, G. U. Singco, A. M. Young, K. W. Guarini, and M. Ieong, Three-dimensional integrated circuits, *IBM Journal of Research & Development*, 504: 91–506 (2006).
4. C. Narayan, S. Purushothaman, F. Doany, and A. Deutsch, Thin film transfer process for low cost MCM-D fabrication, *IEEE Transaction Transactions on Components, Packaging, and Manufacturing Technology*, B18: 42–46 (1995).
5. E. D. Perfecto, R. R. Shields, A. K. Malhotra, M. P. Jeanneret, D. C. McHerron, and G. A. Katopis, MCM-D/C packaging solution for IBM latest S/390 servers, *IEEE Transaction on Advance Packaging*, 23: 515–520 (2000).
6. S. E. Steen, D. C. La Tulipe, A. Topol, D. J. Frank, K. Belote, and D. Posillico, Wafer scale 3D integration: overlay as the key to drive potential, *Microelectronic Engineering*, 84: 1412–1415 (2007).
7. A. W. Topol, D. C. La Tulipe, L. Shi, S. M. Alam, A. M. Young, D. J. Frank, S. E. Steen, J. Vichiconti, D. Posillico, D. M. Canaperi, S. Medd, R. A. Conti, S. Goma, D. Dimilia, C. Wang, L. Deligianni, M. A. Cobb, K. Jenkins, A. Kumar, K. T. Kwietniak, M. Robson, G. W. Gibson, C. D'Emic, E. Nowak, R. Joshi, K. W. Guarini, and M. Ieong, Assembly technology for three dimensional integrated circuits, *Proceedings of 22nd International VLSI Multilevel Interconnection Conference (VMIC)*, pp. 83–88, Fremont CA, October 4–6, (2005).
8. K.-N. Chen, C. S. Tan, A. Fan, and R. Reif, Morphology and bond strength of copper wafer bonding, *Electrochemical and Solid-State Letters*, 7: G14–G16 (2004).

9. K.-N. Chen, C. K. Tsang, A. W. Topol, S. H. Lee, B. K. Furman, D. L. Rath, J.-Q. Lu, A. M. Young, S. Purushothaman, and W. Haensch, Improved manufacturability of Cu bond pads and implementation of seal design in 3D integrated circuits and packages, *23rd International VLSI Multilevel Interconnection (VMIC) Conference,* Fremont CA, September 25–28, 2006, VMIC Catalog No 06 IMIC-050, pp. 195–202 (2006).
10. K.-N. Chen. S. H. Lee, P. S. Andry, C. K. Tsang, A. W. Topol, Y. M. Lin, J. Q.Lu, A. M. Young, M. Ieong, and W. Haensch, Structure design and process control for Cu bonded interconnects in 3D integrated circuits, *IEDM Technical Digest*, 20–22 (2006).
11. H. B. Pogge, et al., Bridging the chip/package process divide, *Proceedings of AMC*, 129–136 (2001).
12. R. Yu, Wafer level 3D integration, *International VLSI Multilevel Interconnection (VMIC) Conference*, Fremont CA, September 24–27, (2007).
13. A. W. Topol, D. C. La Tulipe, L. Shi, S. M. Alam, D. J. Frank, S. E. Steen, J. Vichiconti, D. Posillico, M. Cobb, S. Medd, J. Patel, S. Goma, D. DiMilia, T. M. Robson, E. Duch, M. Farinelli, C. Wang, R. A. Conti, D. M. Canaperi, L. Deligianni, A. Kumar, K. T. Kwietniak, C. D'Emic, J. Ott, A. M. Young, K. W. Guarini, and M. Ieong, Enabling SOI based assembly technology for three-dimensional (3D) integrated circuits (ICs), *IEDM Technical Digest*, 363–366 (2005).
14. P. Leduc, F. de Crécy, M. Fayolle, B. Charlet, T. Enot, M. Zussy, B. Jones, J.-C. Barbe, N. Kernevez, N. Sillon, S. Maitrejean, D. Louis, and G. Passemard, Challenges for 3D IC integration: bonding quality and thermal management, *Proceedings of IEEE International Interconnect Technology Conference (IITC)*, pp. 210–212 (2007).
15. A. W. Topol, S. J. Koester, D. C. La Tulipe, and A. M. Young, 3D fabrication options for high performance CMOS technology, *Wafer Level 3D ICs Process Technology*, C. S. Tan, R. J. Gutmann, and L. R. Reif, Eds., Springer, New York; ISBN 978-0-387-76532-7 (2008).
16. K. W. Guarini, et al., Process technologies for three dimensional integration, *Proceedings of the 6th Annual International Conference on Microelectronics and Interfaces,*American Vacuum Society, pp. 212–214 (2005).
17. C. K. Tsang, P. S. Andry, E. J. Sprogis, C. S. Patel, B. C. Webb, D. G. Manzer, and J. U. Knickerbocker, CMOS-compatible through silicon vias for 3D process integration, *Proceedings of Material Research Society*, 970: 145–153 (2006).
18. P. S. Andry, C. Tsang, E. J. Sprogis, C. Patel, S. L. Wright, and B. C. Webb, A CMOS-compatible process for fabricating electrical through-vias in silicon, *Proceedings of the 56th Electronic Components and Technology Conference*, San Diego, CA, pp. 831–837 (2006).
19. C. S. Patel, C. K. Tsang, C. Schuster, F. E. Doany, H. Nyikal, C. W. Baks, R. Budd, L. P. Buchwalter, P. S. Andry, D. F. Canaperi, D. C. Edelstein, R. Horton, J. U. Knickerbocker, T. Krywanczyk, Y. H. Kwark, K. T. Kwietniak, J. H. Magerlein, J. Rosner, and E. J. Sprogis, Silicon carrier with deep through vias, fine pitch wiring and through cavity for parallel optical transceiver, *Proceedings of the 55th Electronic Components and Technology Conference*, pp. 1318–1324 (2005).

Chapter 3
Thermal and Power Delivery Challenges in 3D ICs

Pulkit Jain, Pingqiang Zhou, Chris H. Kim, and Sachin S. Sapatnekar

Abstract Compared to their 2D counterparts, 3D integrated circuits provide the potential for tremendously increased levels of integration per unit footprint. While this property is attractive for many applications, it also creates more stringent design bottlenecks in the areas of thermal management and power delivery. First, due to increased integration, the amount of heat per unit footprint increases, resulting in the potential for higher on-chip temperatures. The task of thermal management must necessarily be shared both by the heat sink, which transfers internally generated heat to the ambient, and by using thermally conscious design methods. Second, the power to be delivered to a 3D chip, per package pin, is tremendously increased, leading to significant complications in the task of reliable power delivery. This chapter presents an overview of both of these problems and outlines solution schemes to overcome the corresponding bottlenecks.

3.1 Introduction

One of the primary advantages of 3D chips stems from their ability to pack circuitry more densely than in 2D. However, this increased level of integration also results in side effects in the form of new limitations and challenges to the designer. Thermal and power delivery problems can both be traced to the fact that a k-tier 3D chip could use k times as much current as a single 2D chip of the same footprint while using substantially similar packaging technology. The implications of this are as follows:

- First, the 3D chip generates k times the power of the 2D chip, which implies that the corresponding heat generated must be sent out to the environment. If the

S.S. Sapatnekar (✉)
Department of Electrical and Computer Engineering, University of Minnesota, Minneapolis, MN 55455, USA
e-mail: sachin@umn.edu

Y. Xie et al. (eds.), *Three-Dimensional Integrated Circuit Design*, Integrated Circuits and Systems, DOI 10.1007/978-1-4419-0784-4_3,

design technique is thermally unaware and the package thermal characteristics for 2D and 3D circuits are similar, this implies that on-chip temperatures on 3D chips will be higher than for 2D chips. Elevated temperatures can hurt performance and reliability, in addition to introducing variabilities in the performance of the chip. Therefore, on-chip thermal management is a critical issue in 3D design.

- Second, the package must be capable of supplying k times the current through the power supply (V_{dd} and ground) pins as compared to the 2D chip. Moreover, the power delivery problem is worsened in 3D ICs as through-silicon vias (TSVs) contribute additional resistance to the supply network. Given that reliable power grid design is a major bottleneck even for 2D designs, this implies that significant resources have to be invested in building a bulletproof power grid for the 3D chip.

This chapter presents an overview of the issues related to thermal and power grid design. We first focus on thermal issues in Section 3.2, presenting an overview of methods for thermal analysis, followed by pointers to thermal optimization techniques, the details of which are addressed in several other places in this book. Next, in Section 3.3, we describe the challenges of power delivery in 3D systems and outline solutions to overcome them.

3.2 Thermal Issues in 3D ICs

3.2.1 The Thermal PDE

Full-chip thermal analysis involves the application of classical heat transfer theory. The differences lie in incorporating issues that are specific to the on-chip context: for example, on-chip geometries are strongly rectilinear in nature and involve rectangular geometric symmetries; the major sources of heat, the devices, lie in either a single layer in each 3D tier, and the points at which a user is typically interested in analyzing temperature are within the device layer(s).

Conventional heat transfer in a chip is described by Fourier's law of conduction [29], which states that the heat flux, q (in W/m^2), is proportional to the negative gradient of the temperature, T (in K), with the constant of proportionality corresponding to the thermal conductivity of the material, k_t (in W/(m K)), i.e.,

$$q = -k_t \nabla T \tag{3.1}$$

The divergence of q in a region is the difference between the power generated and the time rate of change of heat energy in the region. In other words,

$$\nabla \cdot q = -k_t \nabla \cdot \nabla T = -k_t \nabla^2 T = g(\mathbf{r},t) - \rho c_p \frac{\partial T(\mathbf{r},t)}{\partial t} \tag{3.2}$$

Here, $\mathbf{r}$ is the spatial coordinate of the point at which the temperature is being determined, t represents time (in s), g is the power density per unit volume (in W/m^3), c_p is the heat capacity of the chip material (in J/(kg K)), ρ is the density of the material (in kg/m^3). This may be rewritten as the following heat equation, which is a parabolic PDE:

$$\rho c_p \frac{\partial T(\mathbf{r},t)}{\partial t} = k_t \nabla^2 T(\mathbf{r},t) + g(\mathbf{r},t) \quad (3.3)$$

The thermal conductivity, k_t, in a uniform medium is isotropic, and thermal conductivities for silicon, silicon dioxide, and metals such as aluminum and copper are fundamental material properties whose values can be determined from standard tables. In practice, in early stages of analysis and for optimization purposes, integrated circuits may be assumed to be layer-wise uniform in terms of thermal conductivity. The bulk layer has the conductivity of bulk silicon, and the conductivity of the metal layers is often computed using an averaging approach: this region consists of a mix of silicon dioxide and metal, and depending on the metal density within the region, an effective thermal conductivity may be used for macroscale analysis.

The solution to Equation (3.3) corresponds to the transient thermal response. In the steady state, all derivatives with respect to time go to 0, and therefore, steady-state analysis corresponds to solving the PDE:

$$\nabla^2 T(\mathbf{r}) = -\frac{g(\mathbf{r})}{k_t} \quad (3.4)$$

This is the well-known Poisson's equation.

The time constants of heat transfer are of the order of milliseconds and are much longer than the subnanosecond clock periods in today's VLSI circuits. Therefore, if a circuit remains within the same power mode for an extended period of time and its power density distribution remains relatively constant, steady-state analysis can capture the thermal behavior of the circuit accurately. Even if this is not the case, steady-state analysis can be particularly useful for early and more approximate analysis, in the same spirit that steady-state analysis is used to analyze power grid networks early in the design cycle. On the other hand, when greater levels of detail about the inputs are available or when a circuit makes a number of changes between power modes at time intervals above the thermal time constant, transient analysis is possible and potentially useful.

To obtain a well-defined solution to Equation (3.3), a set of boundary conditions must be imposed. Typically, at the chip level, this involves building a package macromodel and assuming that this macromodel interacts with a constant ambient temperature.

3.2.2 Steady-State Thermal Analysis Algorithms

Next, we will describe steady-state analysis techniques based on the application of the finite difference method (FDM) and the finite element method (FEM). Both methods discretize the entire chip and form a system of linear equations relating the temperature distribution within the chip to the power density distribution. The major difference between the FDM and FEM is that while the FDM discretizes the differential operator, the FEM discretizes the temperature field. The primary advantage of the FDM and FEM is their capability of handling complicated material structures, particularly nonuniform interconnect distributions in a VLSI chip.

The FEM and FDM methods both lead to problem formulations that require the solution of large systems of linear equations. The matrices that describe these equations are typically sparse (more so for the FDM than the FEM, as can be seen from the individual element stamps) and positive definite. There are many different ways of solving these equations [9]. Direct methods typically use variants of Gaussian elimination, such as LU factorization, to first factor the matrices and then solve the system through forward and backward substitutions. The cost of LU factorization is $O(n^3)$ for a dense $n \times n$ matrix but is just slightly superlinear in practice for sparse systems. This step is followed by forward/backward substitution, which can be performed in $O(n)$ time for a sparse system where the number of entries per row is bounded by a constant. If a system is to be evaluated for a large number of right-hand side vectors, corresponding to different power vectors, LU factorization only needs to be performed once, and its cost may be amortized over the solution for multiple input vectors.

Iterative methods are seen to be very effective for large sparse positive definite matrices. This class of techniques includes more classical methods such as Gauss–Jacobi, Gauss–Seidel, and successive overrelaxation, as well as more contemporary approaches based on the conjugate gradient method or GMRES. The idea here is to begin with an initial guess solution and to successively refine it to achieve convergence. Under certain circumstances, it is possible to guarantee this convergence: in particular, FDM matrices have a structure that guarantees this property. For FDM methods, the similarity with the power grid analysis problem invites the use of similar solution techniques, including random walk methods [30, 31] and other methods such as multigrid approaches [23, 24].

3.2.2.1 Formulation of the FDM Equations

The steady-state Poisson's equation (3.4) can be discretized by writing the second spatial derivative of the temperature, T, as a finite difference in rectangular coordinates. The spatial region may be discretized into rectangles, each represented by a node, with sides along the x, y, and z directions with lengths Δx, Δy, and Δz , respectively. Let us assume that the region of interest is placed in the first octant, with a vertex at the origin. We will use $T_{i,j,k}$ to represent the steady-state temperature at node $(i\Delta x, j\Delta y, k\Delta z)$, and there is one equation associated with each node inside the chip.

This discretization can be used to write an approximation for the spatial partial derivatives. For example, in the x direction, we can write

$$\frac{\partial^2 T(\mathbf{r})}{\partial^2 x} \approx \frac{\frac{T_{i+1,j,k} - T_{i,j,k}}{\Delta x} - \frac{T_{i,j,k} - T_{i-1,j,k}}{\Delta x}}{\Delta x} \tag{3.5}$$

$$= \frac{T_{i+1,j,k} - 2T_{i,j,k} + T_{i-1,j,k}}{(\Delta x)^2} \tag{3.6}$$

Similar equations may be written in the y and z spatial directions.

Let us define the operators δ_x^2, δ_y^2, and δ_z^2 as

$$\begin{aligned} \delta_x^2 T_{i,j,k} &= T_{i+1,j,k} - 2T_{i,j,k} + T_{i-1,j,k} \\ \delta_x^2 T_{i,j,k} &= T_{i,j+1,k} - 2T_{i,j,k} + T_{i,j-1,k} \\ \delta_z^2 T_{i,j,k} &= T_{i,j,k+1} - 2T_{i,j,k} + T_{i,j,k-1} \end{aligned} \tag{3.7}$$

The FDM discretization of Poisson's equation using finite differences thus results in the following system of linear equations:

$$\frac{\delta_x^2 T_{i,j,k}}{(\Delta x)^2} + \frac{\delta_y^2 T_{i,j,k}}{(\Delta y)^2} + \frac{\delta_z^2 T_{i,j,k}}{(\Delta z)^2} = -\frac{g_{i,j,k}}{k_t} \tag{3.8}$$

A better visualization of the discretization process, particularly for an electrical engineering audience, employs another standard device in heat transfer theory that builds an equivalent *thermal circuit* through the so-called *thermal–electrical analogy*. Each node in the discretization corresponds to a node in the circuit. The steady-state equation corresponds to a network where "thermal resistors" are connected between nodes that correspond to spatially adjacent regions and "thermal current sources" that map on to power sources. The voltages at the nodes in this thermal circuit can then be computed by solving the circuit, and these yield the temperature at those nodes. Mathematically, this can be derived from Equation (3.4) by writing the discretized equation in a slightly different form from Equation (3.8). For example, in the x direction, the finite difference in Equation (3.5) can be rewritten as

$$\frac{\partial^2 T(\mathbf{r})}{\partial^2 x} \approx \left[\frac{T_{i+1,j,k} - T_{i,j,k}}{R_{i+1,j,k}} - \frac{T_{i,j,k} - T_{i-1,j,k}}{R_{i-1,j,k}} \right] \cdot \frac{1}{k_t A_x \Delta x} \tag{3.9}$$

where $R_{i-1,j,k} = \frac{\Delta x}{k_t A_x}$, and $A_x = \Delta y \Delta z$ is the cross-sectional area of the element when sliced along the x-axis, to obtain the following discretization:

$$\left[\frac{T_{i+1,j,k} - T_{i,j,k}}{R_{i+1,j,k}} + \frac{T_{i-1,j,k} + T_{i,j,k}}{R_{i-1,j,k}} \right] + \left[\frac{T_{i,j+1,k} - T_{i,j,k}}{R_{i,j+1,k}} + \frac{T_{i,j-1,k} - T_{i,j,k}}{R_{i,j-1,k}} \right] + \left[\frac{T_{i,j,k+1} - T_{i,j,k}}{R_{i,j,k+1}} + \frac{T_{i,j,k-1} - T_{i,j,k}}{R_{i,j,k-1}} \right] = -G_{i,j,k} \tag{3.10}$$

where $G_{i,j,k} = g_{i,j,k}\ \Delta V$ is the total power generated within the element and $\Delta V = A_x \Delta x = A_y \Delta y = A_z \Delta_z$.

Representation (3.10) can readily be recognized as being equivalent to the nodal equations at each node of an electrical circuit, where the node is connected to the nodes corresponding to its six adjacent elements through thermal resistors, as shown in Fig. 3.1. In other words, the solution to the thermal analysis problem using FDM amounts to the solution of a circuit of linear resistors and current sources.

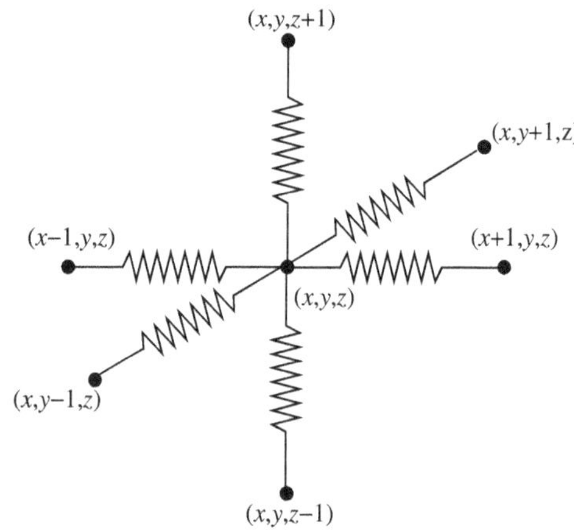

Fig. 3.1 Thermal resistances connected to a node (x,y,z) after FDM discretization

The ground node, or reference, for the circuit corresponds to a constant temperature node, which is typically the ambient temperature. If isothermal boundary conditions are to be used, this simply implies that the node(s) connected to the ambient correspond to the ground node. On the other hand, it is possible to use a more detailed thermal model for the package and heat spreader, consisting of an interconnection of thermal resistors and thermal capacitors, as used in HotSpot [1, 38], or another type of compact model such as a reduced–order model. In either case, one or more nodes of the package model will be connected to the ambient, which is taken to be the ground node. Such a model can be obtained by applying, for example, the FDM or FEM on the package, and extracting a (possibly sparsified) macromodel that preserves the ports that connect the package to the chip and to the ambient.

The overall equations for the circuit may be formulated using modified nodal analysis [7], and we may obtain a set of equations:

$$G\mathbf{T} = \mathbf{P} \tag{3.11}$$

Here G is an $n \times n$ matrix and $\mathbf{T}$, $\mathbf{P}$ are n-vectors, where n corresponds to the number of nodes in the circuit. It is easy to verify that the G matrix is a sparse conductance matrix that has a banded structure, is symmetric, and is diagonally dominant.

For transient thermal analysis, the time-dependent left-hand side term in Equation (3.3) is nonzero. Using a similar finite difference strategy as above, the equation may be discretized in the space domain as

$$\rho c_p \frac{\partial T_{i,j,k}}{\partial t} = k_t \left[\frac{\delta_x^2 T_{i,j,k}^{n+1} + \delta_x^2 T_{i,j,k}^{n}}{2(\Delta x)^2} + \frac{\delta_y^2 T_{i,j,k}^{n+1} + \delta_y^2 T_{i,j,k}^{n}}{2(\Delta y)^2} + \frac{\delta_z^2 T_{i,j,k}^{n+1} + \delta_z^2 T_{i,j,k}^{n}}{2(\Delta z)^2} \right] + \frac{g_{i,j,k}^{n}}{\rho c_p} \tag{3.12}$$

The time-independent terms on the right-hand side can again be considered to be the currents per unit volume through thermal resistors and thermal current sources. The left-hand side, on the other hand, represents a current source of value $\rho c_p \frac{\partial T_{i,j,k}}{\partial t}$. Recalling that in the thermal–electrical analogy, the temperatures correspond to voltages, it is easy to see that we can represent the left-hand side by a *thermal capacitor* of value ρc_p per unit volume.

Given this mapping, transient thermal analysis can be performed by creating the equivalent network consisting of resistors, current sources, and capacitors and using routine electrical techniques for transient analysis.

3.2.3 The Finite Element Method

The FEM provides another avenue to solve Poisson's equation described by Equation (3.4). While it is a generic, classical, and widely used technique for solving such PDEs, it is possible to use the properties of the on-chip problem, outlined at the beginning of Section 3.2.1, to compute the solution efficiently.

A succinct explanation of FEM, as applied to the on-chip case, is provided in [10]. In finite element analysis, the design space is first discretized or meshed into elements. Different element shapes can be used such as tetrahedra and hexahedra. For the on-chip problem, where all heat sources are modeled as being rectangular, a reasonable discretization [11] for the FEM divides the chip into 8-node rectangular hexahedral elements, as shown in Fig. 3.2. In the on-chip context, hexahedral elements also simplify the book-keeping and data management during FEM. The temperatures at the nodes of the elements constitute the unknowns that are computed during finite element analysis, and the temperature within an element is calculated using an interpolation function that approximates the solution to the heat equation within the elements, as shown below:

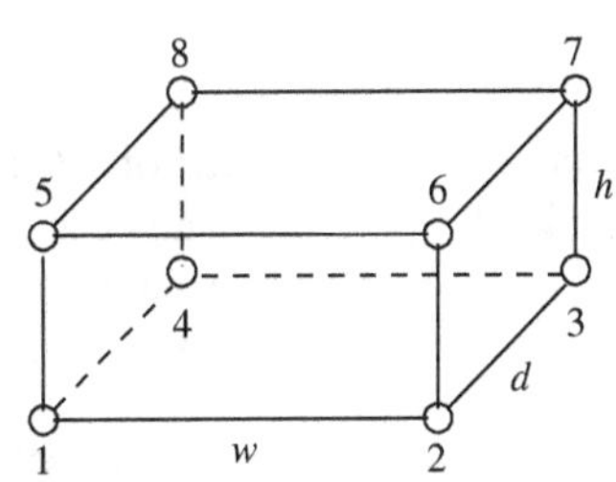

Fig. 3.2 An 8-node hexahedral element used in the FEM

$$T(x,y,z) = \sum_{i=1}^{8} N_i(x,y,z) T_i \tag{3.13}$$

where $N_i(x,y,z)$ is the shape function associated with node i and T_i is the temperature at node i. Let (x_c,y_c,z_c) be the center of the element and denote the width, depth, and height of the element by w, d, and h, respectively. The temperature at any point within the element is interpolated in FEM using a shape function, $N_i(x,y,z)$, which in this case is written as the trilinear function:

$$\begin{aligned} N_i(x,y,z) = {} & \left(\frac{1}{2} + \frac{2(x_i - x_c)}{w^2}(x - x_c)\right) \times \left(\frac{1}{2} + \frac{2(y_i - y_c)}{d^2}(y - y_c)\right) \\ & \times \left(\frac{1}{2} + \frac{2(z_i - z_c)}{h^2}(z - z_c)\right) \end{aligned} \tag{3.14}$$

The property of this function is that its value is 1 at vertex i and 0 at all other vertices, which satisfies the elementary requirement corresponding to a vertex temperature, as calculated in Equation (3.13).

From the shape functions, the thermal gradient, $\mathbf{g}$, can be found, using Equation (3.13), as follows:

$$g = \begin{bmatrix} \frac{\partial T}{\partial x} \\ \frac{\partial T}{\partial y} \\ \frac{\partial T}{\partial z} \end{bmatrix} = B\mathbf{T} \tag{3.15}$$

$$\text{where } B = \begin{bmatrix} \frac{\delta N_1}{\delta y} & \frac{\delta N_2}{\delta x} & \cdots & \frac{\delta N_8}{\delta x} \\ \frac{\delta N_1}{\delta y} & \frac{\delta N_2}{\delta y} & \cdots & \frac{\delta N_8}{\delta y} \\ \frac{\delta N_1}{\delta z} & \frac{\delta N_2}{\delta z} & \cdots & \frac{\delta N_8}{\delta z} \end{bmatrix} \tag{3.16}$$

As in the case of circuit simulation using the modified nodal formulation [7], stamps are created for each element and added to the global system of equations, given by

$$K_g \mathbf{T} = \mathbf{P} \tag{3.17}$$

where $\mathbf{T}$ is the vector of all the nodal temperatures. This system of equations is typically sparse and can be solved efficiently.

In the FEM, these stamps are called element stiffness matrices, K, and their values can be determined using techniques based on the calculus of variations. While a complete derivation of this theory is beyond the scope of this chapter and can be found in a standard text on FEM (such as [25]), it suffices to note that the end result yields the following stamps. For the case where only heat conduction comes into play, we have

$$K = \int_V B^T DBdV \tag{3.18}$$

where V is the volume of the element and $D = \mathrm{diag}(k_{t,x}, k_{t,y}, k_{t,z})$ is a 3×3 diagonal matrix in which $k_{t,i}, i \in \{x,y,z\}$ represents the thermal conductivity in each of the three coordinate directions, for the case where the region is anisotropic along the three coordinate directions; in many cases, $k_{t,x} = k_{t,y} = k_{t,z} = k_t$.

For our hexahedral element, the stamp for the conductive case is given by the 8×8 symmetric matrix whose entries depend only on w, h, and d and is given by

$$K = \begin{bmatrix} A\,B\,C\,D\,E\,F\,G\,H \\ B\,A\,D\,C\,F\,E\,H\,G \\ C\,D\,A\,B\,G\,H\,E\,F \\ D\,C\,B\,A\,H\,G\,F\,E \\ E\,F\,G\,H\,A\,B\,C\,D \\ F\,E\,H\,G\,B\,A\,D\,C \\ G\,H\,E\,F\,C\,D\,A\,B \\ H\,G\,F\,E\,D\,C\,B\,A \end{bmatrix} \tag{3.19}$$

where

$$A = \frac{k_{t,x}hd}{9w} + \frac{k_{t,y}wd}{9h} + \frac{k_{t,z}wh}{9d}, \quad B = -\frac{k_{t,x}hd}{9w} + \frac{k_{t,y}wd}{18h} + \frac{k_{t,z}wh}{18d}$$

$$C = -\frac{k_{t,x}hd}{18w} - \frac{k_{t,y}wd}{18h} + \frac{k_{t,z}wh}{36d}, \quad D = \frac{k_{t,x}hd}{18w} - \frac{k_{t,y}wd}{9h} + \frac{k_{t,z}wh}{18d}$$

$$E = \frac{k_{t,x}hd}{18w} + \frac{k_{t,y}wd}{18h} - \frac{k_{t,z}wh}{9d}, \quad F = -\frac{k_{t,x}hd}{18w} + \frac{k_{t,y}wd}{36h} - \frac{k_{t,z}wh}{18d}$$

$$G = -\frac{k_{t,x}hd}{36w} - \frac{k_{t,y}wd}{36h} - \frac{k_{t,z}wh}{36d}, \quad H = \frac{k_{t,x}hd}{36w} - \frac{k_{t,y}wd}{18h} - \frac{k_{t,z}wh}{18d}$$

The stamps from various elements, including separate conductive and convective stamps, if applicable, and the power dissipation vector may now be superposed to obtain the global stiffness matrix. The entire mesh consists of these hexahedral elements aligned in a grid, with each node being shared by at most eight different elements. The element stiffness matrices are stamped into a global stiffness matrix, K_g, by adding together the components of the element matrices corresponding to the same node. Each entry of the global power vector, P, contains the power dissipated or heat generated at the corresponding node, as well as possible additions from the convective element.

All of these stamps are incorporated into the global set of equations, given by (3.17). In the case of isothermal boundary conditions, or if a node is connected to the ambient, the corresponding temperature is set to the ambient. The number of equations and variables can be correspondingly reduced. For example, if $\mathbf{T}_1$ is the vector of unknown temperatures, and all nodes in the subvector $\mathbf{T}_2$ are connected to fixed temperatures, then the global stiffness matrix can be written in the form

$$\begin{bmatrix} K_{g,11} & K_{g,12} \\ K_{g,21} & K_{g,22} \end{bmatrix} \begin{bmatrix} \mathbf{T}_1 \\ \mathbf{T}_2 \end{bmatrix} = \begin{bmatrix} \mathbf{P}_1 \\ \mathbf{P}_2 \end{bmatrix} \tag{3.20}$$

The fixed values in $\mathbf{T}_2$ can be moved to the right-hand side to obtain the reduced set of equations:

$$K_{g,11}\mathbf{T}_1 = \mathbf{P}_1 - K_{g,12}\mathbf{T}_2 \tag{3.21}$$

3.2.4 Thermal Optimization of 3D Circuits

The illustration in Fig. 3.3 shows a simple thermal model for a 3D circuit and outlines techniques for overcoming thermal challenges in these structures. The figure shows a schematic of a 3D chip sitting atop a heat sink: this is modeled using a distributed power source feeding a distributed resistive network connected to a thermal resistance that models the heat sink. Although this is a coarse model, it suffices for illustrative purposes. By the thermal–electrical analogy, the voltage in this network represents the temperature on the chip. The temperature can therefore be reduced using the following schemes:

- *Through low-power design*: By reducing the power dissipation of the chip, the thermal current injected into the network can be reduced, controlling the IR drop, and therefore, the voltage.

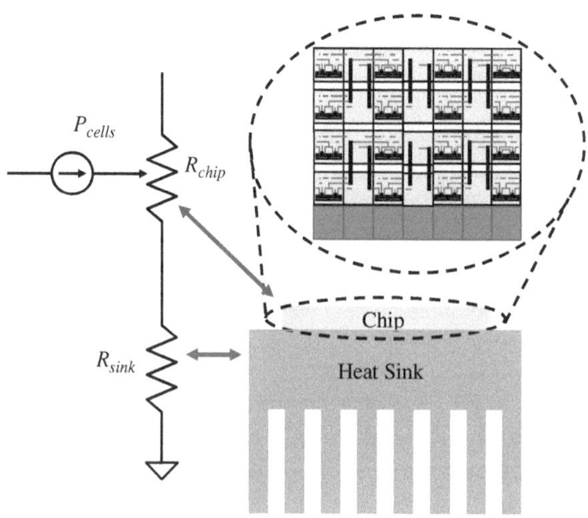

Fig. 3.3 A simple thermal model for a 3D chip

- *By rearranging the heat sources*: The locations of the heat sources can be altered through physical design (floorplanning and placement) to obtain improved temperatures. Coarsely speaking, this implies that high-power modules should be moved away from each other and closer to the heat sink.
- *By improving thermal conduits*: The temperature may also be reduced by improving the effective thermal conductivity of paths from the devices to the heat sink. An effective method for achieving this is through the insertion of *thermal vias*: thermal vias are structurally similar to electrical vias but serve no electrical purpose. Their primary function is to conduct heat through the 3D structure and convey it to the heat sink.
- *By improving the heat sink*: An improved heat sink results in an improvement in the value of R_{sink}, which can help reduce the temperature.

We will discuss several of these techniques elsewhere in this book, except for the last, which is beyond the scope of the book.

3.3 Power Delivery in 3D ICs

Despite the recent surge in 3D IC research, there has been very little work from the circuit design and automation community on power delivery issues for 3D ICs. On-chip power supply noise has worsened in modern systems because scaling of the power supply network (PSN) impedance has not kept up with the increase in device density and operating current due to the limited wire resources and constant RC per wire length [26], and as stated earlier, this situation is worsened in 3D ICs. The increased IR and Ldi/dt supply noise in 3D chips may cause a larger variation in operating speed leading to more timing violations. The supply noise overshoot due to inductive parasitics may aggravate reliability issues such as oxide breakdown, hot carrier injection (HCI), and negative bias temperature instability (NBTI) [2, 33] (which are also negatively affected by elevated temperatures). Consequently, on-chip power delivery will be a critical challenge for 3D ICs.

This section begins with a basic overview of power delivery issues in conventional high-performance 2D circuits in Section 3.3.1. Next, we examine the 3D IC power delivery problem, modeling techniques, and comparisons with conventional 2D ICs in Section 3.3.2. A few promising architectures that attempt to leverage the 3D IC topology to alleviate the specific power delivery problems are then described in Section 3.3.3, followed by a description of a few 3D-specific CAD techniques for power grid optimization.

3.3.1 The Basics of Power Delivery

According to scaling roadmaps, future high-performance ICs will need multiple, sub-1 V supply voltages, with total currents exceeding 100 A/cm^2 even for 2D chips

[4]. Conventional power delivery methods for high-performance ICs employ a DC–DC converter known as a voltage regulator module (VRM). The VRM is typically mounted on the motherboard, with external interconnects providing the power to the chip, as depicted in Fig. 3.4a. The intrachip power delivery network is shown in Fig. 3.4b, which shows a part of the modeled PSN of a microprocessor [28]. The package parasitics, contributed by the I/O pads and bonding wires, are modeled as an inductance and resistance in series. The decoupling capacitors (decaps) shown in the figure are intended to damp out transient noise and include the external decap as well as the capacitance due to the various circuit components such as the MOS gate capacitance.

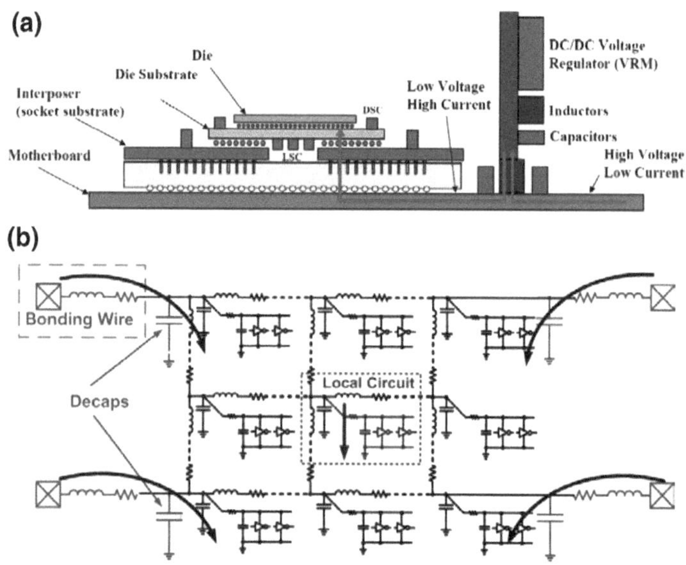

Fig. 3.4 (**a**) Conventional power delivery architecture [39] © 2007 IEEE. (**b**) On-chip power grid

The chip acts as a distributed noise source drawing current in different locations and at different frequencies, causing imperfections in the delivered supply. The supply that reaches the processor is affected by IR and Ldi/dt drop across the package constituting the supply noise: the package impedance has largely remained unaffected by technology scaling. Scaling does, however, result in some unwanted effects on-chip, namely, increased currents and faster transients from one technology node to the next. The former aggravate the IR drop, while the latter worsen the Ldi/dt drop. Over and above these effects is the issue of global resonant noise in which the supply impedance gets excited to produce large drops on supply at or near the resonant frequency. With these increased levels of noise and reduced noise margins, as Vdd levels scale down, reliable power delivery to power-hungry chips has become a major challenge.

The noise spectrum for a typical power grid is shown in Fig. 3.5a. The DC component of the noise is given by IR drop across the package and power grid. The first peak in the figure corresponds to the resonant frequency, given by $f_{res} = 1/(2\pi\sqrt{LC})$, which typically appears in the range of 100–300 MHz. An excitation at this frequency can be triggered during microprocessor loop operations or wakeup. Several other peaks are seen in the figure due to switching at clock frequency and its higher harmonics or due to local resonance: the corresponding noise is typically an order less in magnitude than the resonant peak. Figure 3.5b shows a measured supply impedance profile of a separate test structure, which validates the simulation model developed in Fig. 3.4. The noise at a particular frequency is estimated by multiplying the impedance with the current component at that frequency [15, 41].

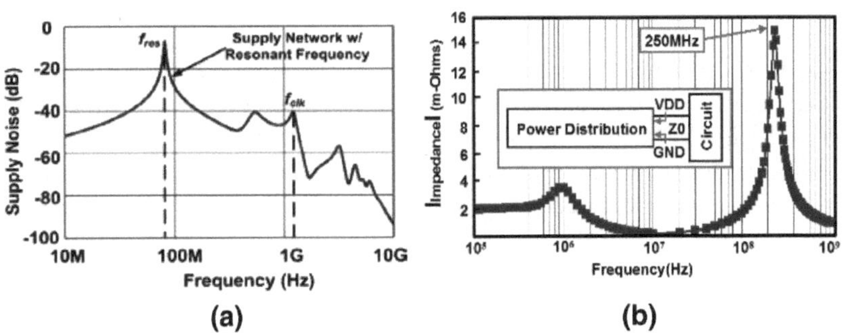

Fig. 3.5 (**a**) Simulation of the supply noise spectrum. (**b**) Measurement results for supply noise [15] © 2006 IEEE

3.3.2 Three-Dimensional IC Power Delivery: Modeling and Challenges

In this section, we focus on the power delivery problem for 3D ICs and analyze the PSN noise problem in this regime.

A model for 3D ICs, based on distributed models of the on-chip and package power supply structures, is shown in Fig. 3.6 [19]. Power is fed from the package through power I/O bumps distributed over the bottom-most tier and travels to the upper tiers using TSVs. The footprint of the chip can be divided into cells, which are identical square regions between a pair of adjacent power and ground pads, as shown in Fig. 3.6a. The cells are connected in Fig. 3.6b in the form of a grid formed by several subcells between adjacent TSVs. Electrically, each TSV is modeled as a series combination of resistance and inductance. The planar square cells use a lumped model, where R_{si}, J_i, and C_{di} represent, respectively, the grid resistance, effective current density, and chip decap on a per-unit basis. Since each pad is shared by four independent cells, the package parameters are normalized by a factor of four. The subcell can be then repeated multiple times to realize the complete 3D IC functional block.

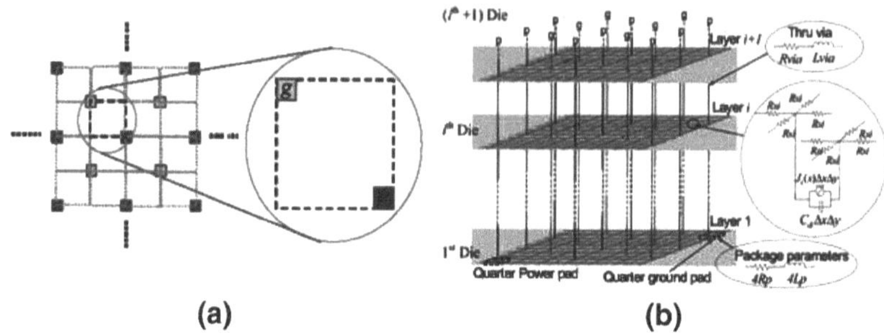

Fig. 3.6 Distributed model for 3D IC [19]. (**a**) Division of the power grid into independent cells. (**b**) A model for one such cell. © IEEE 2007

The power grid model must necessarily be tied to a real 3D process. Figure 3.7a depicts a 3D IC cross-sectional model of a production level 0.18 μm 3D process from MIT Lincoln Laboratory [5]. This process has three tiers. The bonding pads are on the top tier, while the heat sink is typically below the bottom tier. Processors or other power intensive circuits would ideally be placed on the bottom tier in close proximity with the heat sink.

The tiers are interconnected through TSVs for electrical and thermal conduction. Figure 3.7b shows the cross-sectional scanning electron microscope (SEM) photograph of a stacked TSV connecting the back metal of the top tier with the top level metal of the bottom tier. A simplified resistance model is superimposed. Based on actual parameter extraction [5], each stacked cone-shaped TSV has a resistance of 1 Ω in this process. The top and middle tiers are aligned face-to-back, while the

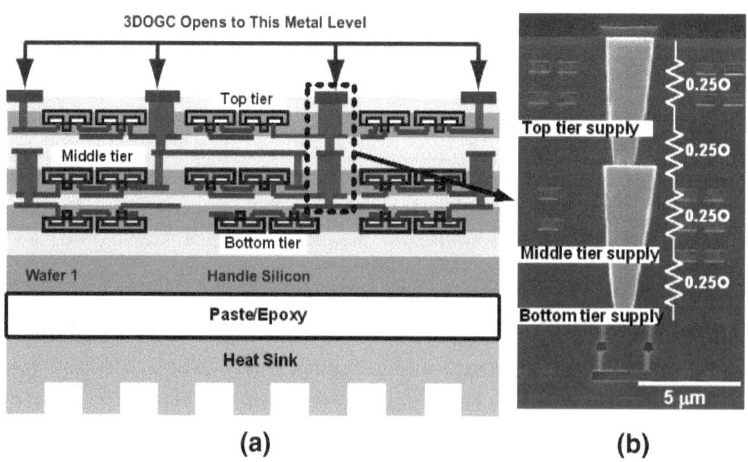

Fig. 3.7 (**a**) Cross section of 3D FD-SOI process. (**b**) Simplified via resistance model aligned with a cross-sectional SEM photograph [21]. © IEEE 2007

middle and bottom tiers one face-to-face, making the path from the top to middle tier longer and more resistive. This configuration can be modeled by breaking up the total 1Ω-stacked via resistance into chunks of 0.25 Ω, 0.5 Ω, and 0.25 Ω, as shown in Fig. 3.7b. The values of TSV inductance and capacitance can be ignored as their values, found experimentally, are fairly small [45].

The TSV resistance in the supply path potentially imposes new challenges in 3D power delivery vis-à-vis the conventional 2D case [20]. First, the lower tiers experience worsened PSN noise due to the increased resistance in the PSN. Moreover, power intensive circuits have to be placed at the bottom tier, which makes reliable power delivery further difficult.

In 3D, there are two significant points of departure, in comparison with models for conventional 2D chips. First, for the same circuitry, the reduced footprint of the 3D die effectively increases the package parasitics: since ratios of the number of supply pins and bonding wires to the supply current are reduced, the role of the package resistance and inductance is increased. Second, the noise characteristics in each tier are affected by the additional TSV resistance in the supply path.

Figure 3.8a shows the circuit models developed to compare the 3D and 2D cases. The models are based on curve fits with the impedance profile of a distributed supply network model, along with typical decap and package parasitic values. In 3D, we see that the supply path would be dominated by the TSVs. The overall chip capacitance (3nF in the 2D case) within an equal footprint is assumed to be split equally in the 3D IC between its three tiers. Moreover, due to the reduced footprint of the 3D die, the number of power pins is assumed to be a third of the 2D case, leading to 3× increase in package parasitic inductance and resistance values.

Since the noise at the bottom tier is predictably worst, we compare the impedance response of this tier with the 2D case. The normalized impedance comparison is shown in Fig. 3.8b, which illustrates the following:

- *Low-frequency impedance*: At low frequencies, the capacitors and inductors are open and short circuited, respectively. Therefore, the 2D model has an impedance of 2(0.01+0.03)=0.08 Ω, while the 3D model has an impedance of 2(0.03+0.05+0.1+0.05)=0.46 Ω. This indicates that for the same amount of current, the 3D chip will have 0.46/0.08=5.75× more 2R drop compared to 2D.
- *Resonant peak impedance*: The resonant peak is determined by the amount of damping and the value of inductance. Here, the increased role of inductance in 3D is counteracted by the increased damping provided by the larger resistance drop to the bottom tier, and the peaks show comparable values.
- *Resonant frequencies*: Two-dimensional circuits typically have a resonant frequency of around 50–300 MHz, given by $f_{res} = 1/(2\pi\sqrt{LC})$. If the equivalent capacitance in 3D is same as in our model, due to the increased L, the peak is shifted to a lower frequency as seen in Fig. 3.8c.
- *High-frequency impedance*: At high frequencies, 2D and 3D impedances become comparable, and this is attributed to the shielding effect of the bottom tier capacitance, due to the fact that the capacitance becomes virtually a short circuit at high frequencies.

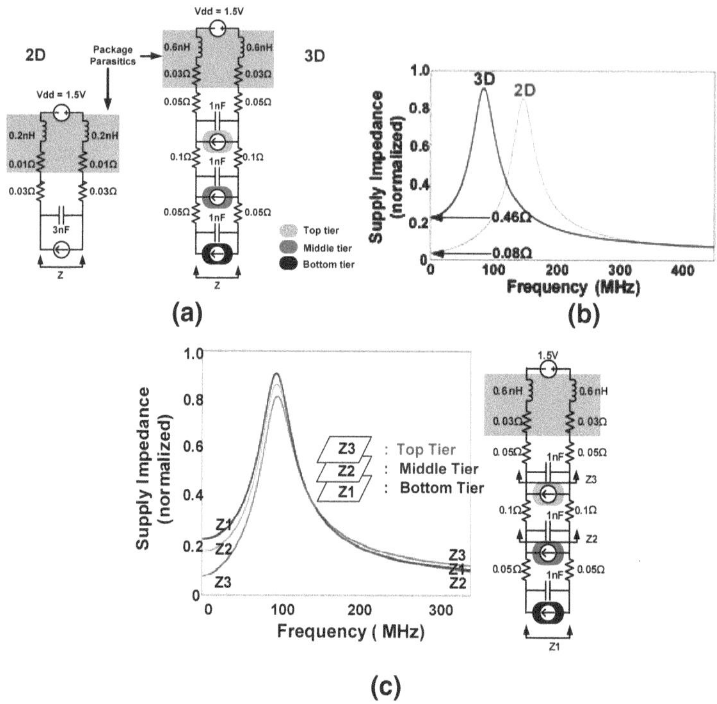

Fig. 3.8 (**a**) Simplified PSN models for comparing impedance response in 2D and 3D. (**b**) Impedance response comparison between 2D and 3D. (**c**) Impedance response of the three tiers in a 3D IC [20] © IEEE 2007

Clearly, it can be seen that DC supply noise becomes a greater concern in 3D designs as compared to its 2D counterpart.

To understand the supply noise behavior in different tiers, we analyze the impedance spectrum (see Fig. 3.8c) across different tiers obtained by simulating the 3D IC model. The key results are as follows:

- *Low-frequency impedance*: As expected, the DC- and low-frequency impedances, which are governed by the TSV resistances, show a worsening trend for the lower level tiers.
- *High-frequency impedance*: At high frequencies, the top tier has the largest impedance, while the middle tier has the minimum AC impedance. Although this seems to be counter intuitive, it can be explained by the shielding/decap effect of the adjacent tier capacitances, which causes the effective damping resistances to be the largest for the middle tier and smallest for the top tier. The above trend is more noticeable at high frequencies beyond the resonance peak.
- *Resonant behavior*: Since the shielding effect mentioned above is not significant at mid-frequencies, the resonance peak follows the low-frequency trend, with the bottom tier being the worst case. However, there is a reduced noise offset as noted

from the simulated curves. Moreover, since the effect of capacitance is the same for all tiers, the resonant frequencies are almost identical.

In summary, the AC impedance is worst for the bottom tier until the resonant frequency, while beyond this point, the top tier has a slightly larger impedance value. Since thermal constraints dictate that the bottom tier is likely to contain circuit blocks with large current consumption, the supply noise in the bottom tier (i.e., the product of current and impedance) will become a significant concern for 3D implementations.

The aim of the above discussion was to provide some quantitative understanding of power delivery in 3D ICs. It should be pointed out that these numbers are tied to a specific process and will change depending on the process. For example, if the technology allows TSVs with much lower resistance or area, then the impedance bottleneck in a path may be due to the supply pads, and the PSN models should account for that. However, regardless of this, it remains likely that PSN will be a key problem in 3D designs.

3.3.3 Design Techniques for Controlling PSN Noise

The presence of severe power delivery bottlenecks necessitates a look at entirely novel power delivery schemes for 3D chips. In this section, we introduce several possible approaches for this purpose.

3.3.3.1 On-Chip Voltage Regulation

One way of dealing with the power delivery problem in 3D ICs (and also in conventional 2D ICs) is to bring the DC–DC converter module closer to the processor, conceptually shown in Fig. 3.9 [37]. Boosting the external voltage and locally down-converting it ensures that the current through external package, I_{ext}, is small, and relaxes the scaling requirement on external package impedance. Moreover, this point of load (PoL) regulation isolates the load from global resonant noise from external package and decap.

Traditionally, the efficiency of monolithic DC–DC converters has been limited by the small physical inductors allowed on-chip. Typical off-chip DC–DC conversion requires high-Q inductors of the order of 1-100 μH [17], which are difficult to implement on-chip due to their area requirements. With growing power delivery problems, the focus has been on building compact inductors through technologies

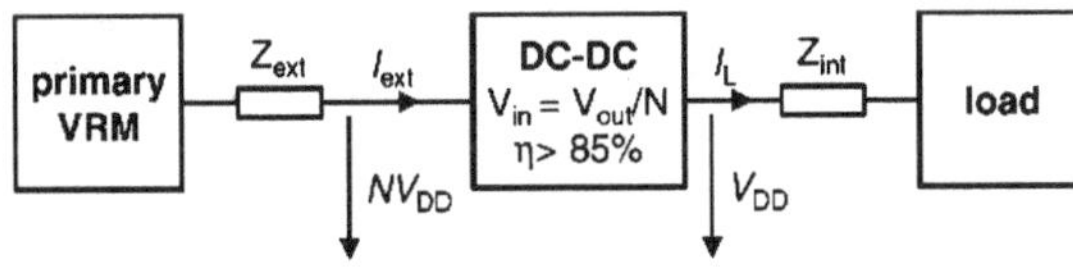

Fig. 3.9 Insertion of a DC–DC converter near the load [37]. © 2004 IEEE

like thin film inductors [22] or on more efficient, but costly, DC–DC converters through multiphase/interleaving topologies [42, 44]. Clearly, there is an onus to incorporate these on-chip, which calls for a different process altogether. The possibility to stack different wafers with heterogeneous technologies, as offered by three-dimensional wafer-level stacking in 3D ICs, is thus the natural solution for realizing on-chip switching converters.

3.3.3.2 *Z*-axis Power Delivery

Z-axis or 3D power delivery [6, 16], in which the PSN is vertically integrated with the processor in a 3D stack, promises an attractive solution for on-chip DC–DC conversion. Figure 3.10 shows the schematic visualization [39] of such a *Z*-axis power delivery technique using wafer–wafer integration. This still requires that all passives, including the inductors and output capacitors, must be monolithically integrated with the power switches and control circuitry. The idea is gaining traction in research, and implementation of such a structure, using two interleaved buck converter cells each operating at 200 MHz switching frequency and delivering 500 mA output current has been reported [39]. In the future, we may see a 3D IC with several tiers, with one whole tier dedicated to voltage regulation, incorporating various passives and other circuitry.

One main issue with *Z*-axis power delivery is the area overhead in dedicating a tier to an on-chip DC–DC converter, whose footprint should be at par with the processor in a wafer–wafer 3D process. Moreover, high-efficiency switching regulators for DC–DC conversion require monolithic realization of bulky passive components. On the other hand, typical linear regulators, though less bulky, suffer from efficiency loss.

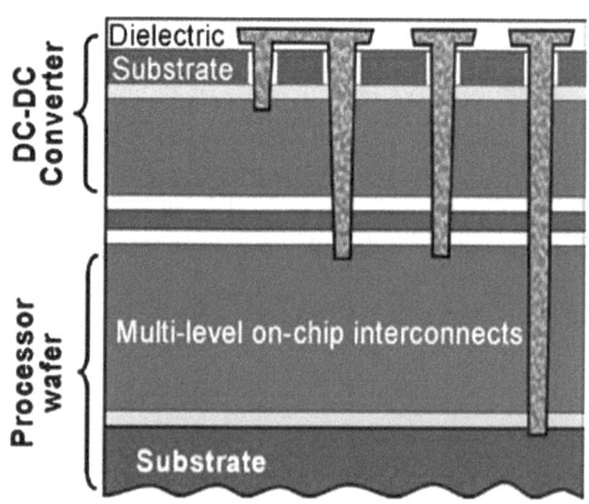

Fig. 3.10 *Z*-axis power delivery based on monolithic power conversion and wafer–wafer bonding [39]. © 2007 IEEE

3.3.3.3 Multistory Power Delivery

A promising technique for achieving high-efficiency on-chip DC–DC conversion and supply noise reduction is the multistory power delivery (MSPD) scheme [14, 32]. It has been demonstrated in [20] that the idea becomes particularly attractive for 3D IC structures involving stacked processors and memories.

Figure 3.11 demonstrates the basic concept of MSPD. A schematic of a conventional supply network is shown in Fig. 3.11a, where all circuits draw current from a single power source. Figure 3.11b shows the multistory supply network, with subcircuits operating between two supply stories. The concept of a "story" is merely an abstraction to illustrate the nature of the power delivery scheme, as opposed to the 3D IC architecture, where circuits are physically stacked in tiers. In this scheme, current consumed in the "$2\,V_{dd}$–V_{dd} story" is subsequently recycled in the "V_{dd}–Gnd story." Due to this internal recycling, half as much current is drawn compared to the conventional scheme, with almost the same total power consumption. A reduced current is beneficial since it cuts down the supply noise. Thus, in the best case, if the currents in the two subcircuits are completely balanced, the middle supply path will sink zero current. This results in minimal noise on that rail, as also illustrated in Fig. 3.11.

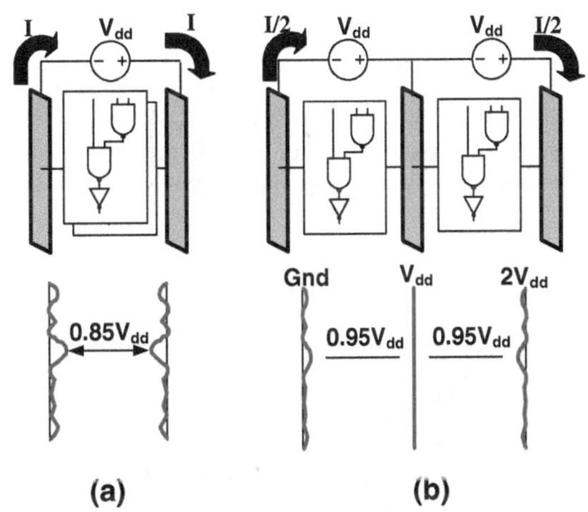

Fig. 3.11 The (**a**) conventional and (**b**) multistory power delivery schemes [14]. © 2005 IEEE

The main issue with this technique is the requirement of separate body islands. This may be difficult in typical bulk processes. However, if we consider 3D ICs, the tiers are inherently separated electrically, which makes MSPD particularly attractive.

Figure 3.12 is a simplistic 3D IC model for a memory (M), memory (M), processor (P) stacked configuration. To model the difference between M and P blocks, the latter is assumed to draw twice the current of the former. We denote the two currents

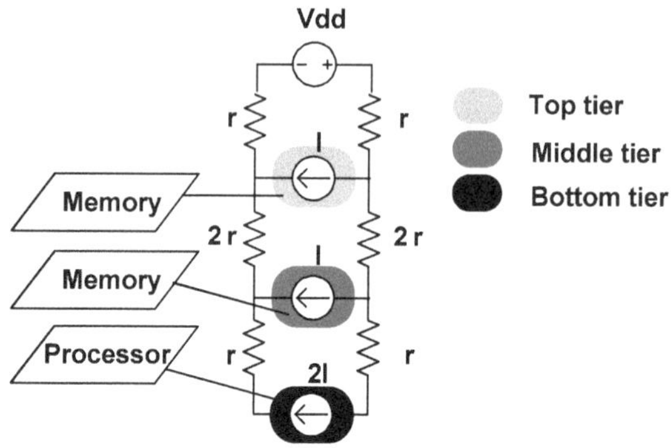

Fig. 3.12 Benchmark model for a power delivery network in a 3D IC. The processor is assumed to draw twice as much current as the memory [20]. © 2008 IEEE

by I and $2I$, respectively. The tier–tier path impedance, constituted by the TSVs, is denoted by r. Note that r is inversely proportional to the number of parallel TSVs in the path.

Considering the benchmark model for a 3D IC in Fig. 3.12, the application of MSPD can lead to a variety of different electrically equivalent architectures, depicted in Fig. 3.13. Here, the tier–tier per-path impedance is denoted by R. Note that MSPD requires another supply rail, implying that the number of supply rails has increased by a factor of 3/2. If we assume that all structures in Fig. 3.12 and 3.13 are normalized to a fixed number of supply path vias, each supply rail in the latter will have two-thirds the number of dedicated vias. This will correspond to a proportional 3/2 times increase in impedance, R, i.e., $R = 1.5\text{r}$.

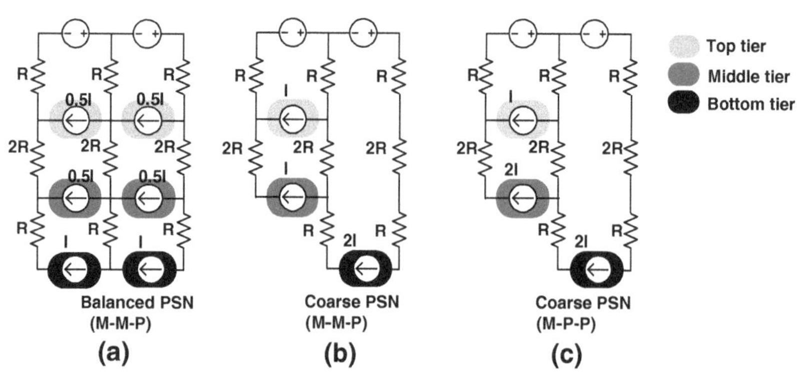

Fig. 3.13 Application of MSPD in 3D ICs [20]. (**a**) Balanced PSN in Memory–Memory–Processor. (**b**) Coarse PSN in Memory–Memory–Processor. (**c**) Coarse PSN with identical tiers. Here, M and P denote memory and processor blocks, respectively, and $R = 1.5r$ for a fixed TSV count. © 2008 IEEE

A fine-grained application of MSPD to each tier in a 3D IC can yield the balanced PSN configuration of Fig. 3.13a. Here, the power supply domain of each tier has been split into two equal stories, with the current from one story being recycled to the other, within and across the different tiers. This scheme leverages the inherent balanced topology to obtain maximal reductions in supply noise levels. However, there are obvious implementation challenges, especially related to body voltage separation in the case of a bulk process. Figure 3.13b is a coarse-grained application of MSPD exploiting the readily segregated tiers in a 3D IC. Here, in spite of maintaining current recycling from a higher supply story to a lower one, each tier has only one dedicated story and single body. Note that the operating current is recycled between the processor in the bottom tier and the memories in the other two tiers. This scheme promises easier implementation, but may not be effective in mitigating the supply noise as there may be a substantial difference between the processor and memory switching currents, which might negate the balancing effect. Making MSPD effective in such a scenario would require a redistribution of TSVs across different supply paths. On the contrary, as shown in Fig. 3.13c, if there is a 3D IC with identical tiers, such as a dual-processor stack, we may exploit the balance in the switching currents between two processor blocks. This could provide a more effective implementation of the coarse-grained multistory PSN idea and can potentially cut down DC supply noise while being easier to implement. However, thermal issues are likely to be an obvious concern here, as the middle tier is isolated from the heat sink at the bottom.

Table 3.1 presents an overview of the above discussion, demonstrating that MSPD can promise substantial PSN noise reduction if the implementation issues are satisfactorily addressed and are achievable. From the last column we see that the power dissipation in the PSN itself can be reduced as well due to current recycling, which is an added benefit of MSPD.

Table 3.1 Overview of various MSPD schemes

Scheme	Architecture (M=memory, P=processor)	Issues in implementation	Noise reduction	Power reduction
Balanced PSN	M–M–P	Easy in SOI, difficult in bulk	Best	Best
Coarse PSN	M–M–P	TSV/supply pad redistribution required	Good	Good
Coarse PSN	M–P–P	Easy in bulk and SOI thermal issues for the middle tier	Better	Better

Figure 3.14 presents a view of the overall effectiveness of MSPD and a comparison of the various discussed schemes in terms of DC noise at different leakage values, and PSN power reductions obtained [20]. The values are normalized with the corresponding nominal non-MSPD 3D IC model shown in black. Clearly, the MSPD technique promises a DC noise reduction of 20–40% depending on the topology and

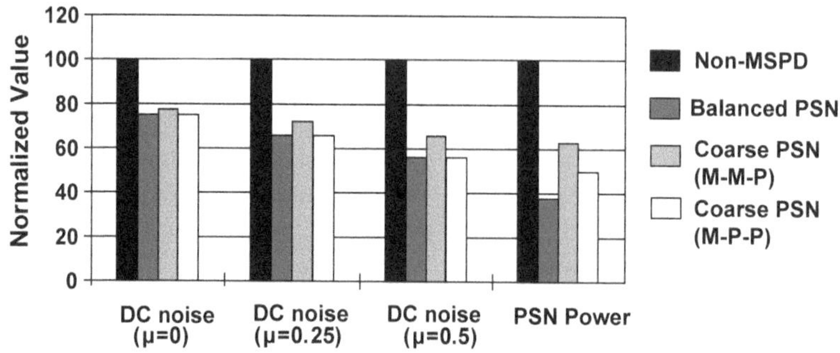

Fig. 3.14 DC noise and PSN power for different schemes [20]. Here, μ represents the fraction of the total operating current attributable to leakage. © 2008 IEEE

the amount of leakage current. It is interesting to note that DC noise is reduced with leakage in an MSPD scenario.

3.3.4 CAD Techniques for Controlling PSN Noise

3.3.4.1 Decap Allocation

Several techniques are available to increase the reliability of power grids and to control power grid noise, such as wire widening, grid topology optimization, and decoupling capacitor (decap) insertion [36]. Of all these techniques, the decaps are arguably the most powerful method for reducing transient noise. Decaps serve as local current reservoirs and can be used to satisfy sudden surges in current demand by the functional blocks/cells while keeping supply voltage levels relatively stable. Active/passive damping methods for resonant noise using decaps [12, 13, 46] have also been proposed.

Conventional technologies for implementing decaps are based on SiO_2-based structures that are widely used in robust power delivery network design. Three-dimensional power grid optimization has been studied in [19, 27, 43,]. Unlike the 2D case, new considerations come into play while optimizing a 3D power grid using CMOS decaps:

- Since CMOS decaps are usually fabricated using white space on the device layer, they must compete for area with TSVs, or with the landing pads of 3D vias, for the limited white space. This leads to a new resource contention problem. One way to resolve this contention problem is to increase the chip size in order to make room for CMOS decaps. However, one of the advantages of 3D circuits over 2D implementations is that they result in a reduced chip footprint: increasing the chip size may counteract this benefit.
- Leakage power is an important issue in 3D circuit design. The CMOS decaps added to the 3D circuit will consume extra leakage power and make things worse.

While new high-k dielectrics have been proposed, they are not in widespread use yet and even when they are deployed, they will provide temporary relief to the gate leakage problem.

The approach in [49] presents an approach for decap allocation in 3D power grids, using both conventional CMOS decaps and metal–insulator–metal (MIM) decaps. Unlike CMOS capacitors that are built in the device layer, MIM capacitors are fabricated between metal layers. These structures have high capacitance density and low leakage current density [3, 18, 34, 35, 40, 50].

Figure 3.15 shows the positions of CMOS and MIM decaps in a 3D circuit. MIM decaps are usually fabricated between the top two metal layers in each 2D tier. A significant advantage of MIM decaps lies in their extremely low leakage: in [34], the leakage current for the 250 nF MIM decaps is reported to be about 1.0×10^{-8} A (with leakage density of $3.2 \times 10^{-}8$ A/cm^2), while the leakage current for a 25 nF CMOS decap in parallel with MIM is approximately 3.2×10^{-6} A (with leakage density of 1.45×10^{-4} A/cm^2).

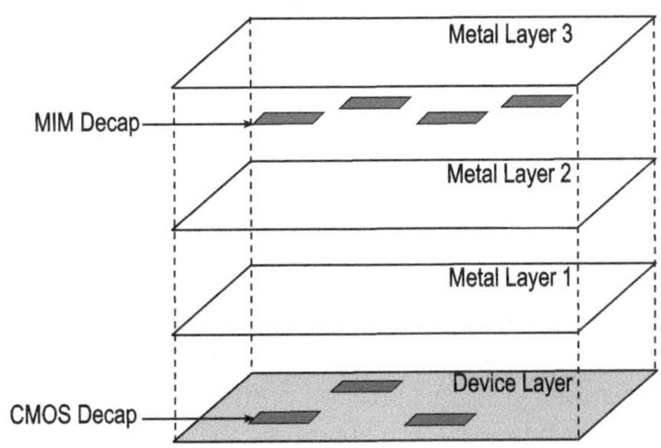

Fig. 3.15 MIM and CMOS decaps in one 2D tier with three metal layers [49]. © 2009 IEEE

However, MIM decaps cannot be used unconditionally to replace CMOS decaps, since their use incurs a cost: they present routing blockages to nets that attempt to cross them. In [49], the decap budgeting problem, using both CMOS and MIM decaps, is formulated as a linear programming (LP) problem, and an efficient congestion-aware algorithm is proposed to optimize the power supply noise while trying to find a balance between the routing congestion deterioration and leakage power increase.

An iterative flow is used to solve the decap allocation problem. In each iteration a relatively small amount of decap is allocated to the current circuit for two reasons. First, the decap allocation problem is highly nonlinear, and this iterative approach permits the optimization process to be controlled by solving a sequence of linear programs, one in each iteration. In order to enable the formulation of these linear programs, it is necessary to model the noise violation and the congestion using models that are linear in the decap value. The second reason is related to this approximation: it avoids the excessive allocation of decaps that could invalidate the approximate linear model of congestion and noise violation used in the algorithm; these models are predicated on the assumption of small perturbations.

Experimental results demonstrate that the use of CMOS decaps alone is insufficient to overcome the violations; the use of MIM decaps results in high levels of congestion; and the optimal mix of the two meets both congestion and noise constraints with low leakage.

3.3.4.2 Automated MSPD Assignments for 3D ICs

The MSPD idea described in Section 3.3.3.3 can be automated by solving an optimization problem [47, 48], which formulates the two-story problem as one of module assignment between the two stories. An important consideration in the design of an MSPD circuit is to locally maintain the current balance between logic blocks operating in different Vdd domains because otherwise, the imbalance current will flow through voltage regulators and be wasted. Another important issue that has to be considered is the design stage at which the circuit should be partitioned into different Vdd domains. Note that a level shifter is required at the output of a logic block if it is used to drive another logic block operating in a different Vdd domain. Level shifters occupy silicon area and cause extra delays in the circuit.

The module assignment problem is addressed at the floorplanning level where the number of modules is usually not very large, and their area is largely ignored. It is assumed that K voltage regulators are distributed across the chip: these regulators are well designed and output stable voltage levels at V_{dd}.

Each regulator is represented by the point at which it taps into the V_{dd} grid. As shown in Fig. 3.16a, the chip is divided into K regions accordingly, such that there is one regulator in each region and the *ith* region contains all the points on-chip that primarily draw(sink) currents from(to) the *ith* regulator. The division of the chip into these nonoverlapping regions can be achieved by meshing the die area using a fine grid and determining which grid cell each block belongs to, e.g., each cell can be said to belong to the region controlled by the nearest voltage regulator.

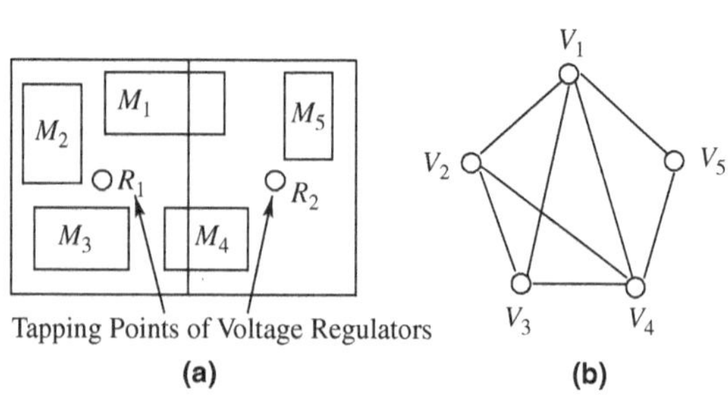

Fig. 3.16 Graph construction: (**a**) partitioning the chip into disjoint regions each of which is controlled by a voltage regulator and (**b**) constructing a graph where node V_i corresponds to module M_i [48] © 2007 IEEE

Once the chip is partitioned into disjoint regions, it is assumed that any "imbalanced" current, i.e., current that is not recycled to the next story in a particular region, goes through the regulator in the same region and is wasted. If a module is located at the boundary between multiple regions, it will be decomposed into several submodules, with one submodule in each region it overlaps with, and with the constraint that all submodules must be assigned to the same Vdd domain.

Let us focus on a particular region corresponding to a particular voltage regulator. Assume the modules located in this region are $M_1, M_2, \ldots, M_n$, where the current flowing through module M_i as a function of time t is given by $I_i(t)$. Because voltage regulators can only respond to the low-to mid-frequency components of the imbalance currents, while the high-frequency components are usually handled by on-chip decaps, we preprocess the input current traces obtained through cycle-accurate power simulations to smooth out the high-frequency components in the current signals. Therefore, $I_i(t)$ should be understood as containing only the low-to mid-frequency-components of the current flowing through module M_i.

If we associate a 0/1 integer variable x_i with module M_i defined as

$$x_i = \begin{cases} 0 \text{ if } M_i \text{ operates between the } 2V_{dd} \text{ and } V_{dd} \text{ rails} \\ 1 \text{ if } M_i \text{ operates between the } V_{dd} \text{ and } GND \text{ rails} \end{cases} \tag{3.22}$$

then the total current flowing through the voltage regulator at time t will be approximated by

$$I_R(t) = \left| \sum_{i=1}^{n} I_i(t) \cdot (1 - x_i) - \sum_{i=1}^{n} I_i(t) \cdot x_i \right| = \left| \sum_{i=1}^{n} I_i(t) \cdot (1 - 2x_i) \right| \tag{3.23}$$

This problem can be shown to map on to one of graph partitioning to maximize the cut size between the partitions, where the weights of the edges are given by

$$w(V_i, V_j) = \left[\sum_{k=1}^{K} \frac{S_{ik} S_{jk}}{S_i S_j} \right] \overline{I_i(t) I_j(t)} \tag{3.24}$$

Here, S_i represents the area of the *ith* module, and the overlap area between the *ith* modules and the *kth* region is denoted by S_{ik}.

The intuition behind (3.24) is that for any pair of modules, only the portions that are located in the same region over the chip count toward the calculation of the correlation between them. If modules M_i and M_j are completely separated into two disjoint regions, the weight $w(V_i, V_j)$ will be 0, and therefore, the corresponding edge can be removed from the graph. An example of graph construction is shown in Fig. 3.16b. A Fidducia–Mattheyses-like approach [8] is used to speedily find the partition that maximizes the cut.

Experimental results in [47] demonstrate that the method is effective in building partitions for multistory power grids in both 2D and 3D chips under an SOI-based process, where blocks from multiple stories may coexist on the same tier. It is shown that the partitioning-based method is successful in recycling a large amount of power through the system, and the quality of results of the partitioning-based method compares favorably with an annealing approach.

3.4 Conclusion

In this chapter, we have extensively analyzed the thermal and power delivery issues in future 3D ICs. The two issues share a common origin, in that they are caused by the increased current per unit footprint for a 3D IC and cause significant reliability problems, as well as potential logic incorrectness. Thermal issues will be dealt with in greater depth elsewhere in this book, but we have provided an overview of solutions to overcoming the power delivery problem through design and CAD approaches.

References

1. HotSpot. available at http://lava.cs.virginia.edu/HotSpot/index.htm.
2. M. A. Alam, B. E. Weir, and P. J. Silverman. A study of soft and hard breakdown - part II: Principles of area, thickness, and voltage scaling. *IEEE Transactions on Electron Devices*, 49(2):239–246, February 2002.
3. M. Armacost, A. Augustin, P. Felsner, Y. Feng, G. Friese, J. Heidenreich, G. Hueckel, O. Prigge, and K. Stein. A high reliability metal insulator metal capacitor for 0.18 μm copper technology. In *Proceedings of the IEEE International Electronic Devices Meeting*, pp. 157–160, 2000.
4. Semiconductor Industry Association. International technology roadmap for semiconductors (ITRS), 2007. http://public.itrs.net/.
5. J. A. Burns, B. F. Aull, C. K. Chen, C. L. Keast, J. M. Knecht, V. Suntharalingam, K. Warner, P. W. Wyatt, and D. Yost. A wafer-scale 3-D circuit integration technology. *IEEE Transactions on Electron Devices*, 53(10):2507–2516, October 2006.
6. S. Chandrasekaran, J. Sun, and V. Mehrotra. Vertically packaged switched-mode power converter, 2006. US Patent #7012414.
7. L. O. Chua and P.-M. Lin. *Computed-Aided Analysis of Electronic Circuits: Algorithms and Computational Techniques*. Prentice-Hall, Englewood Cliffs, NJ, 1975.
8. C. M. Fiduccia and R. M. Mattheyses. A linear-time heuristic for improving network partitions. In *Proceedings of the ACM/IEEE Design Automation Conference*, pp. 175–181, 1982.
9. G. Golub and C. F. Van Loan. *Matrix Computations*. John Hopkins University Press, Baltimore, MD, 3rd edition, 1996.
10. B. Goplen. *Advanced Placement Techniques for Future VLSI Circuits*. PhD thesis, University of Minnesota, Minneapolis, MN, 2006.
11. B. Goplen and S. S. Sapatnekar. Efficient thermal placement of standard cells in 3D ICs using a force directed approach. In *Proceedings of the IEEE/ACM International Conference on Computer-Aided Design*, pp. 86–89, 2003.
12. J. Gu, H. Eom, and C. H. Kim. Sleep transistor sizing and control for resonant supply noise damping. In *Proceedings of the ACM International Symposium on Low Power Electronics and Design*, pp. 80–85, 2007.

13. J. Gu, H. Eom, and C. H. Kim. A switched decoupling capacitor circuit for on-chip supply resonance damping. In *Proceedings of the IEEE International Symposium on VLSI Circuits*, pp. 126–127, 2007.
14. J. Gu and C. H. Kim. Multi-story power delivery for supply noise reduction and low voltage operation. In *Proceedings of the ACM International Symposium on Low Power Electronics and Design*, pp. 192–197, 2005.
15. E. Hailu, D. Boerstler, K. Miki, J. Qi, M. Wang, and M. Riley. A circuit for reducing large transient current effects on processor power grids. In *Proceedings of the IEEE International Solid-State Circuits Conference*, pp. 2238–2245, 2006.
16. J. A. Harrison and E. R. Stanford. Z-axis processor power delivery system, 2003. US Patent #6523253.
17. P. Hazucha, G. Schrom, J. Hahn, B. A. Bloechel, P. Hack, G. E. Dermer, S. Narendra, D. Gardner, T. Karnik, V. De, and S. Borkar. A 233-MHz 80%–87% efficient four-phase DC-DC converter utilizing air-core inductors on package. *IEEE Journal of Solid-State Circuits*, 40(4):838–845, April 2005.
18. H. Hu, S.-J. Ding, H. F. Lim, C. Zhu, M. F. Li, S. J. Kim, X. F. Yu, J. H. Chen, Y. F. Yong, B. J. Cho, D.S.H. Chan, S. C. Rustagi, M. B. Yu, C. H. Tung, A. Du, D. My, P. D. Foot, A. Chin, and D.-L. Kwong. High performance ALD HfO_2-Al_2O_3 laminate MIM capacitors for RF and mixed signal IC applications. In *Proceedings of the IEEE International Electronic Devices Meeting*, pp. 15.6.1–15.6.4, 2003.
19. G. Huang, M. Bakir, A. Naeemi, H. Chen, and J. D. Meindl. Power delivery for 3D chip stacks: Physical modeling and design implication. In *Proceedings of the IEEE Electrical Performance of Electronic Packaging Meeting*, pp. 205–208, 2007.
20. P. Jain, T. Kim, J. Keane, and C. H. Kim. A multi-story power delivery technique for 3D integrated circuits. In *Proceedings of the ACM International Symposium on Low Power Electronics and Design*, pp. 57–62, 2008.
21. C. Keast, B. Aull, J. Burns, N. Checka, C.-L. Chen, C. Chen, M. Fritze, J. Kedzierski, J. Knecht, B. Tyrrell, K. Warner, B. Wheeler, D. Shaver, V. Suntharlingam, and D. Yost. 3D integration for integrated circuits and advanced focal planes, 2007. Available at http://vmsstreamer1.fnal.gov/VMS_Site_03/Lectures/Colloquium/070228Keast/index.htm .
22. K. H. Kim, J. Kim, H. J. Kim, S. H. Han, and H. J. Kim. A megahertz switching DC/DC converter using FeBN thin film inductor. *IEEE Transactions on Magnetics*, 38(5):3162–3164, September 2002.
23. P. Li, L. T. Pileggi, M. Asheghi, and R. Chandra. Efficient full-chip thermal modeling and analysis. In *Proceedings of the IEEE/ACM International Conference on Computer-Aided Design*, pp. 319–326, 2004.
24. P. Li, L. T. Pileggi, M. Asheghi, and R. Chandra. IC thermal simulation and modeling via efficient multigrid-based approaches. *IEEE Transactions on Computer-Aided Design of Integrated Circuits and Systems*, 25(9):1763–1776, September 2006.
25. D. L. Logan. *A First Course in the Finite Element Method*. Brooks/Cole Publishing Company, Pacific Grove, CA, 3rd edition, 2002.
26. R. Mahajan, R. Nair, V. Wakharkar, J. Swan, J. Tang, and G. Vandentop. Emerging directions for packaging technologies. *Intel Technology Journal*, 6(2):62–75, May 2002.
27. J. R. Minz, S. K. Lim, and C.-K. Koh. 3D module placement for congestion and power noise reduction. In *Proceedings of the Great Lakes Symposium on VLSI*, pp. 458–461, 2005.
28. N. Na, T. Budell, C. Chiu, E. Tremble, and I. Wernple. The effects of on-chip and package decoupling capacitors and an efficient ASIC decoupling methodology. In *Proceedings of the IEEE Electronic Components and Technology Conference*, pp. 556–567, 2004.
29. M. N. Özişik. *Heat Transfer: A Basic Approach*. McGraw-Hill, New York, NY, 1985.
30. H. Qian, S. R. Nassif, and S. S. Sapatnekar. Random walks in a supply network. In *Proceedings of the ACM/IEEE Design Automation Conference*, pp. 93–98, 2003.
31. H. Qian and S. S. Sapatnekar. Hierarchical random-walk algorithms for power grid analysis. In *Proceedings of the Asia-South Pacific Design Automation Conference*, pp. 499–504, 2004.

32. S. Rajapandian, K. Shepard, P. Hazucha, and T. Karnik. High-tension power delivery: Operating 0.18 μm CMOS digital logic at 5.4 V. *Proceedings of the IEEE International Solid-State Circuits Conference*, pp. 298–599, February 2005.
33. V. Reddy, A. T. Krishnan, A. Marshall, J. Rodriguez, S. Natarajan, T. Rost, and S. Krishnan. Impact of negative bias temperature instability on digital circuit reliability. In *Proceedings of the IEEE International Reliability Physics Symposium*, pp. 248–254, 2002.
34. D. Roberts, W. Johnstone, H. Sanchez, O. Mandhana, D. Spilo, J. Hayden, E. Travis, B. Melnick, M. Celik, B. W. Min, J. Edgerton, M. Raymond, E. Luckowski, C. Happ, A. Martinez, B. Wilson, P. Leung, T. Garnett, D. Goedeke, T. Remmel, K. Ramakrishna, and B.E. Jr. White. Application of on-chip MIM decoupling capacitor for 90 nm SOI microprocessor. In *Proceedings of the IEEE International Electronic Devices Meeting*, pp. 72–75, 2005.
35. H. Sanchez, B. Johnstone, D. Roberts, O. Mandhana, B. Melnick, M.Celik, M. Baker, J. Hayden, B. Min, J. Edgerton, and B. White. Increasing microprocessor speed by massive application of on-die high-k MIM decoupling capacitors. In *Proceedings of the IEEE International Solid-State Circuits Conference*, pp. 2190–2199, 2006.
36. S. S. Sapatnekar and H. Su. Analysis and optimization of power grids. *IEEE Design & Test*, 20(3):7–15, 2003.
37. G. Schrom, P. Hazucha, J. Hahn, V. Kursun, D. Gardner, S. Narendra, T. Karnik, and V. De. Feasibility of monolithic and 3D-stacked DC-DC converters for microprocessors in 90*nm* technology generation. In *Proceedings of the ACM International Symposium on Low Power Electronics and Design*, pp. 263–268, 2004.
38. K. Skadron, M. R. Stan, W. Huang, S. Velusamy, K. Sankaranarayan, and D. Tarjan. Temperature-aware microarchitecture. In *Proceedings of the ACM International Symposium on Computer Architecture*, pp. 2–13, 2003.
39. J. Sun, J. Lu, D. Giuliano, T. P. Chow, and R. J. Gutmann. 3D power delivery for microprocessors and high-performance ASICs. In *Proceedings of IEEE Applied Power Electronics Conference*, pp. 127–133, 2007.
40. Y. L. Tu, H. L. Lin, L. L. Chao, D. Wu, C. S. Tsai, C. Wang, C. F. Huang, C. H. Lin, and J. Sun. Characterization and comparison of high-k metal-insulator-metal (MiM) capacitors in 0.13 μm cu BEOL for mixed-mode and RF applications. In *Proceedings of the IEEE International Symposium on VLSI Circuits*, pp. 79–80, 2003.
41. A. Waizman. CPU power supply impedance profile measurement using FFT and clock gating. In *Proceedings of the IEEE Electrical Performance of Electronic Packaging Meeting*, pp. 29–32, 2003.
42. J. Wibben and R. Harjani. A high efficiency DC-DC converter using 2 nH on-chip inductors. In *Proceedings of the IEEE International Symposium on VLSI Circuits*, pp. 22–23, 2007.
43. E. Wong, J. Minz, and S. K. Lim. Decoupling capacitor planning and sizing for noise and leakage reduction. In *Proceedings of the IEEE/ACM International Conference on Computer-Aided Design*, pp. 395–400, 2006.
44. P. Wong, P. Xu, P. Yang, and F. C. Lee. Performance improvements of interleaving VRMs with coupling inductors. *IEEE Transactions on Power Electronics*, 16(4):499–507, July 2001.
45. J. H. Wu. *Through-Substrate Interconnects for 3-D Integration and RF Systems*. PhD thesis, Department. of EECS, Massachusetts Institute of Technology, October 2006.
46. J. Xu, P. Hazucha, M. Huang, P. Aseron, F. Paillet, G. Schrom, J. Tschanz, and C. Zhao. On-die supply-resonance suppression using band-limited active damping. In *Proceedings of the IEEE International Solid-State Circuits Conference*, pp. 286–603, 2007.
47. Y. Zhan and S. S. Sapatnekar. Automated module assignment in stacked-Vdd designs for high-efficiency power delivery. *ACM Journal on Emerging Technologies in Computing Systems*, 4(4):1–20, 2008.
48. Y. Zhan, T. Zhang, and S. S. Sapatnekar. Module assignment for pin-limited designs under the stacked-Vdd paradigm. In *Proceedings of the IEEE/ACM International Conference on Computer-Aided Design*, pp. 656–659, 2007.

49. P. Zhou, K. Sridharan, and S. S. Sapatnekar. Congestion-aware power grid optimization for 3D circuits using MIM and CMOS decoupling capacitors. In *Proceedings of the Asia-South Pacific Design Automation Conference*, pp. 179–184, 2009.
50. P. Zurcher, P. Alluri, P. Chu, A. Duvallet, C. Happ, R. Henderson, J. Mendonca, M. Kim, M. Petras, M. Raymond, T. Remmel, D. Roberts, B. Steimle, J. Stipanuk, S. Straub, T. Sparks, M. Tarabbia, H. Thibieroz, and M. Miller. Integration of thin film MIM capacitors and resistors into copper metallization based RF-CMOS and Bi-CMOS technologies. In *Proceedings of the IEEE International Electronic Devices Meeting*, pp. 153–156, 2000.

Chapter 4
Thermal-Aware 3D Floorplan

Jason Cong and Yuchun Ma

Abstract Three-dimensional integration makes floorplanning a much more difficult problem because the multiple device layers dramatically enlarge the solution space and the increased power density accentuates the thermal problem. This chapter introduces the algorithms for 3D floorplanning with both 2D blocks and 3D blocks. In addition to stochastic optimizations based on various representations that are briefly introduced, the analytical approach is also introduced. The effects of various 3D floorplanning techniques on wirelength, area, and temperature are demonstrated by experimental results.

4.1 Introduction

Three-dimensional IC design provides another dimension for topological arrangement of logic blocks. Therefore, physical design tools play an important role in the adoption of 3D technologies. As the critical step in the process of physical design, floorplanning influences the performance of the final design greatly. Three-dimensional integration makes floorplanning a much more difficult problem because

---

J. Cong (✉)
Department of Computer Science, University of California, Los Angeles, CA 90095, USA
e-mail: cong@cs.ucla.edu

This chapter includes portions reprinted with permission from the following publications, (a) J. Cong, J. Wei, and Y. Zhang, A thermal-driven floorplanning algorithm for 3D ICs, *Proceedings of ICCAD*, pp. 306–313, 2004, © 2004 IEEE. (b) Y. Liu, Y. Ma, E. Kursun, J. Cong, and G. Reinman, Fine grain 3D integration for microarchitecture design through cube packing exploration, *Proceedings of IEEE ICCD*, pp. 259–266, October 2007, © 2007 IEEE. (c) Y. Ma, X. Hong, S. Dong, Y. Cai, C. K. Cheng, and J. Gu, Floorplanning with abutment constraints and L-shaped/T-shaped blocks based on Corner Block List, *Proceedings of DAC*, pp. 770–775, 2001, © 2001 IEEE. (d) Y. Ma, X. Hong, S. Dong, and C.K.Cheng, 3D CBL: an efficient algorithm for general 3-Dimensional packing problems, *Proceedings of the 48th MWSCAS*, Vol. 2, pp. 1079–1082, 2005, © 2005 IEEE. (e) P. Zhou, Y. Ma, Z. Li, R.P. Dick, L. Shang, H. Zhou, X. Hong, and Q. Zhou, 3D-STAF: scalable temperature and leakage aware floorplanning for three-dimensional integrated circuits, *Proceedings of ICCAD*, pp. 590–597, 2007, © 2007 IEEE.

Y. Xie et al. (eds.), *Three-Dimensional Integrated Circuit Design*,
Integrated Circuits and Systems, DOI 10.1007/978-1-4419-0784-4_4,

the multiple layers dramatically enlarge the solution space and the increased power density accentuates the thermal problem. Therefore, moving to 3D designs increases the problem complexity greatly:

1. The design space of 3D IC floorplanning increases exponentially with the number of active layers.
2. The addition of a temperature constraint or temperature minimization objective complicates optimization, requiring trade-offs among area, wirelength, and thermal characteristics. And with the high temperature in 3D chips, it is necessary to account for the closed temperature/leakage power feedback loop to accurately estimate or optimize either one.
3. Multi-layer stacking offers a reduction in inter-block latency. It can also be used to help the intra-block wire latency when the block is implemented in multiple layers. Use of multi-layer blocks requires a novel physical design infrastructure to explore three-dimensional design space.

Therefore, it is imperative to develop thermally aware floorplanning tools that consider 3D design constraints. The goal of 3D floorplanning is to pack blocks on multiple layers with no overlaps by optimizing some objectives without violating some design constraints. According to the block representation, we can classify the 3D floorplanning problem into two types. The first type is a 3D floorplan with 2D blocks in which each block is a 2D rectangle and the packing on each layer can be treated as a 2D floorplan. A 3D floorplan with 2D blocks can be represented by an array of 2D representations (2D array), each representing all blocks located on one device layer. The second type of 3D floorplanning involves 3D blocks where each block is treated as a cubic block with non-zero height in the *Z*-dimension. In this case, the existing 2D representations no longer apply, and we need new representations.

This chapter discusses these two kinds of approaches and the applications with 3D IC designs. Section 4.2 formulates the problem of 3D floorplanning; Sections 4.3 and 4.4 introduce representations used for the optimization of the 3D floorplan with 2D blocks and 3D blocks, respectively. In Section 4.5, we introduce the optimization techniques for 3D floorplanning. Besides the commonly used simulated annealing optimization approaches, we also introduce analytical approaches, such as the force-directed approach. The experimental results for various techniques are provided in Section 4.6. Finally, we arrive at a conclusion in Section 4.7.

4.2 Problem Formulation

Similar to the traditional 2D floorplanning, 3D floorplanning also aims at a small packing area, short wirelength, low power consumption, and high performance. As shown in Chapter 3, although 3D integration has many potential benefits, thermal distribution becomes a critical issue during every stage of 3D design. Therefore, 3D floorplanning distributes blocks on a certain number of layers without overlapping each other so that the design metrics, such as the chip area, wirelength, inter-layer via number, and maximal on-chip temperature, are optimized or meet some design constraints.

With the additional Z-direction, not only can the 2D blocks be spread among multiple layers, but some individual components can be folded into the designs of a multi-layer block so that the intra-block wire latency can be reduced, as well as the power consumption. Recent studies lead to various 3D architectural structures, including 3D cache [30, 9, 28], 3D register files [31], 3D arithmetic units [25], and 3D instruction scheduler [26]. The 3D components with different layer numbers can be treated as cubic blocks to be packed in the 3D space. The dimension in the Z-direction represents the layer information. Therefore, in 3D floorplanning the blocks to be packed can be 2D blocks or 3D blocks. Figure 4.1a shows the two-layer packing for Alpha 21264 in which all blocks are 2D blocks, and Fig. 4.1b shows the packing with some 3D blocks. The implementation for each 3D component may have multiple choices with different area-delay-power trade-offs. As shown in Fig. 4.1b, it is possible that an optimal floorplan has a subset of the architectural units occupying a single device layer, while others are implemented on multiple strata with potentially different heights in the Z-dimension. According to the block representation, we classify the 3D floorplanning problem into two types: 3D floorplan with 2D blocks only and 3D floorplan with possible 3D blocks.

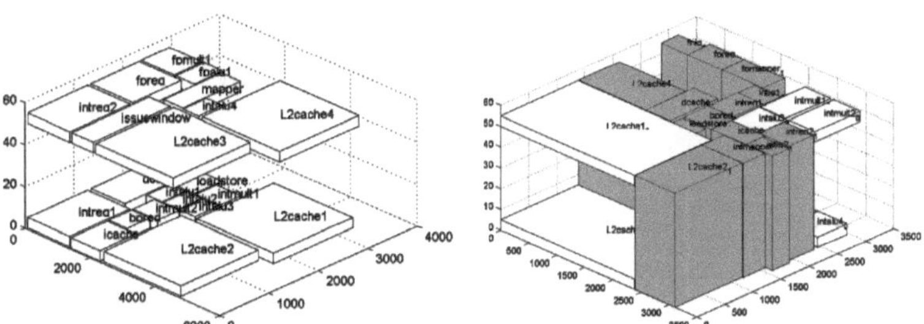

(a) Packing with 2D blocks (b) Packing with 3D blocks

Fig. 4.1 Three-dimensional floorplanning

4.2.1 3D Floorplanning with 2D Blocks

Though 3D packing with 2D blocks can be treated as multiple stacked 2D packings, the additional concern at the chip level relates to the large number of active devices that are packed into a much smaller area, so that the power density is much higher than in a corresponding 2D circuit. As a result, in addition to the common objectives of packing area and wirelength, thermal issues are given primacy among the set of design objectives. Hence, we can formulate a 3D floorplan with 2D blocks as follows.

An instance of the 3D floorplanning problem with 2D blocks is composed of a set of blocks $\{m_1, m_2, \ldots, m_n\}$. A block m_i is a $W_i \times H_i$ rectangle with area A_i, aspect ratio H_i/W_i, and power density PD_i. Each block is free to rotate. There are a fixed number of layers L. Let the tuple (x_i, y_i, l_i) denote the coordinates of the bottom-left corner of block m_i, where $1 \leq l_i \leq L$. A 3D floorplan F is an assignment of (x_i, y_i, l_i) for each block m_i such that no two blocks overlap. The common objectives of 3D floorplanning algorithms are to minimize (1) chip peak temperature $T_{\max}$, (2) total wirelength (or total power), and (3) chip area. Chip area is the product of the maximum height and width over all layers. Wirelength is the half perimeter wirelength estimation. In addition, some other design objectives, such as noise, performance, the number of inter-layer vias, can be considered at the same time. Also, some design constraints can be included, such as pre-packed blocks (the positions of the constrained blocks are pre-defined), alignment constraints (some specific blocks are constrained to be aligned in X-, Y-, or Z-direction).

Since 3D floorplanning with 2D blocks can be represented with an array of 2D representations, the 2D floorplanning algorithm can be extended to handle multi-layer designs by introducing new operations in optimization techniques. Though floorplanning for 2D design is a well-studied problem, with the additional layer of information, the design space of 3D IC floorplanning increases exponentially. Li et al. showed that, given a floorplanning problem with n blocks, the solution space of 3D floorplanning with L layers increases by $n^{L-1}/(L-1)!$ times compared to the 2D case [11]. Though the multi-layer design can be represented by an array of 2D packings, the specific optimization techniques are still needed for efficient exploration. Thermal-aware optimization is especially critical in 3D designs.

4.2.2 3D Floorplanning with 3D Blocks

Fine-grain three-dimensional integration provides reduced intra-block wire delay as well as improved power consumption (more details are provided in the Appendix). The implementation for each component may have multiple choices due to various configurations. Therefore, the components might be implemented on multiple layers, such as a four-layer or two-layer cache, by different stacking techniques. But locally, the best implementation of an individual unit may not necessarily lead to the best design for the entire multi-layered chip. To obtain the trade-off between multiple objectives, it is possible to have cubic blocks, which have different heights in the Z-direction, in the packing design. Therefore, a cube-packing algorithm should be developed to arrange the given circuit components in a rectangular box of the minimum volume without overlapping each other. With the various implementations for each critical component, the block implementation is partially defined. Without the physical information, it is impossible to obtain the optimal implementations for components for the final chip. Thus, 3D floorplanning with 3D blocks should not only determine the coordinates of the blocks, but also be able to choose the configurations for components, such as the number of layers and the partitioning approaches. Therefore, we can formulate the 3D packing with 3D blocks as follows.

Given a list of 3D blocks: Suppose for block i, there are k different implementations that are recorded in a candidate list as $\{c^i{}_1, c^i{}_2, \ldots c^i{}_k\}$. And each candidate $c^i{}_j$ has the width ($w^i{}_j$), height ($h^i{}_j$), layer number ($z^i{}_j$), delay ($d^i{}_j$), and power ($p^i{}_j$) (assume each layer has the same power consumption). The objective is to generate a floorplan that optimizes for the die area, maximum on-chip temperature, etc. At the same time, the number of layers is normally fixed, which means, given the layer number constraints as Z_{con}, the blocks should not exceed the layer number constraint. In this chapter, we will provide several typical representations which can represent cubic packings in section 4.4.

4.3 Representations for 3D Floorplanning with 2D Blocks

Since 3D floorplanning with 2D blocks can be represented with an array of 2D representations, the 2D floorplanning algorithm can be extended to handle multi-layer designs by introducing new operations in optimization techniques. Before we discuss the detailed 3D-related operations, we introduce the basic 2D representations briefly; these are the fundamental techniques for 3D floorplanning optimization.

4.3.1 Basic Representations for 2D Packing

The geometrical relationship among the blocks is commonly specified by a rectangular dissection of the floorplan region. In order to restrict the size of the solution space, three different ways of dissection are proposed. The corresponding floorplanning structures are called slicing [34, 23], mosaic [15, 7, 42], and general floorplan [20, 18]. The slicing floorplan is a special case of the mosaic floorplan, and the mosaic floorplan is a special case of the general floorplan. The relationship among the solution spaces of slicing, mosaic, and general floorplans is illustrated in Fig. 4.2.

A slicing floorplan can be obtained by recursively cutting a rectangle into two parts using a vertical or horizontal line. In a mosaic floorplan, the floorplan region is dissected into exact rooms so that each room is occupied by one and only one block. The general floorplan is similar to the mosaic floorplan in that non-slicing structures are allowed. However, the floorplan region can be dissected into a number of rooms that are larger than the number of blocks, such that some rooms are not occupied by any block. Besides these three categories, a compact packing is also introduced which establishes that a packing is *L-compact* (*B-compact*) if no block can be moved left (down) while other blocks are fixed. As shown in Fig. 4.2, the packing of three blocks, a, b, and c, is slicing and mosaic, but is not a compact one.

Fig. 4.2 Relationship among the solution spaces of slicing, mosaic, and general floorplans

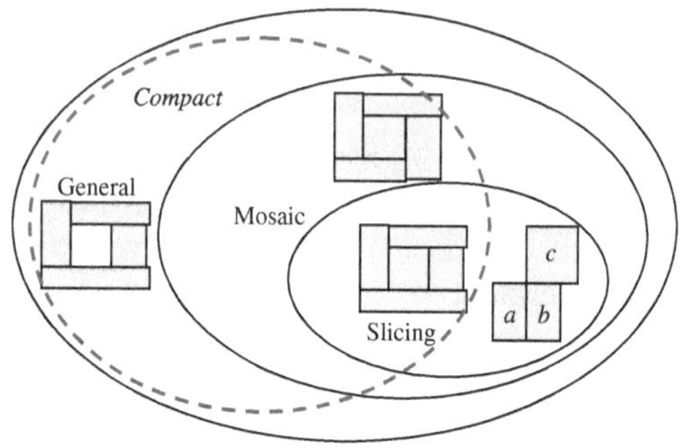

Therefore, compact packing is a special kind of the packings whose solution space intersects with the solution spaces for slicing and mosaic, but does not fully cover them.

Since the 2D and 3D rectangular packing problems are NP-hard, most floorplanning algorithms are based on stochastic combinatorial optimization techniques such as simulated annealing. During the optimization, topological representations are commonly used since they guarantee that all encoded packings are overlap-free and each topological representation encodes relative positions (such as left to, right to, above, below) among blocks in a way that is amenable to perturbation. The slicing structure can be encoded by slicing trees [22] or Polish expressions [34]. After a sequence pair [18] was proposed as the first representation for general floorplans, various kinds of floorplan representations such as BSG [19], O-tree [6], B*-tree [1], CBL [7], and TCG [13, 14] were proposed. These representations may have different degrees of redundancy and efficiency for solution exploitations. In the following, we briefly describe several typical representations.

4.3.1.1 Slicing Structure

Since the slicing structure can be recursively partitioned by vertical or horizontal lines, an oriented rooted binary tree called a slicing tree [22, 23, 34] is used to represent the partitioning between blocks (Fig. 4.3). Each internal node of the tree is labeled by a "$*$" or a "+" operator, corresponding to a vertical or a horizontal cut, respectively. Each leaf corresponds to a basic block and is labeled by a number from 1 to n (n is the number of the blocks).

However, for a given slicing floorplan, there may be more than one slicing tree representation. In order to obtain a nonredundant representation of all slicing floorplans, Wong and Liu [34] proposed a special kind of slicing tree named Skewed Slicing Tree (SST). A SST is a slicing tree in which no node and its right child have the same label (Fig. 4.3). They used the post-order traversal of SST called the normalized Polish expression (NPE) as the floorplanning representation. A Polish expression is said to be normalized if there is no consecutive "$*$"s or "+"s in the sequence. It was proved in [34] that there is a 1:1 correspondence between the set of

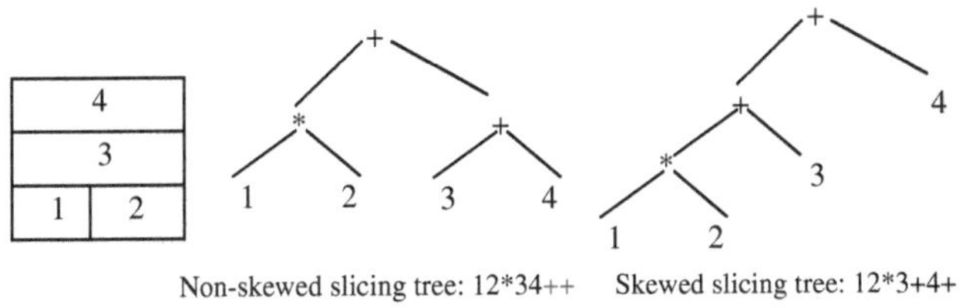

Fig. 4.3 Slicing tree representation and polish expression representation of a slicing floorplan

normalized Polish expressions of length $2n-1$ and the set of slicing floorplans with n blocks. The packing can be constructed by scanning the slicing tree or normalized Polish expressions in linear time.

4.3.1.2 Mosaic Structure

Mosaic floorplanning was first introduced in [7] and it has the following characteristics:

1. There is no empty space within the floorplan, i.e., each rectangle is assigned to one and only one block. As shown in Fig. 4.4, some structures with empty rooms cannot be represented in the mosaic floorplan.
2. The topology is equivalent before and after the non-crossing segment slides to adjust the block sizes.
3. There is no degenerate case where two segments meet at the same point. If a degenerate situation happens, the segments can be separated and slid apart by a small distance (for further discussions on the degeneration case, see Section 4.3.2.2).

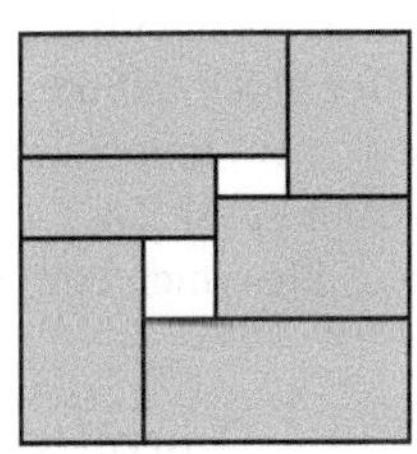

Fig. 4.4 A packing with empty rooms

The corner block list [7, 15, 40], twin-binary sequence [39], Q-sequence [42], etc., are proposed to represent mosaic floorplans. In the following discussion, we will briefly introduce the corner block list.

The corner block list (CBL) uses a triple list of (S, L, T), where S records the sequence of the blocks' IDs. L records the orientation for each block: $L_i = 0$ denotes that block i covers other blocks from the top; $L_i = 1$ denotes that block i covers other blocks from the right (as shown in Fig. 4.5). Binary list T records how many blocks are covered by a block when it is packed. That is, in list T, the length of each sub-list, which has a number of successive "1"s and an ending "0," corresponds to the number of the blocks covered by block i. Figure 4.6 is an example of a non-slicing floorplan and its corresponding CBL. In Fig. 4.6, block g covers two blocks $\{f, e\}$. Therefore, the sub-list {10} is used for block g in list T.

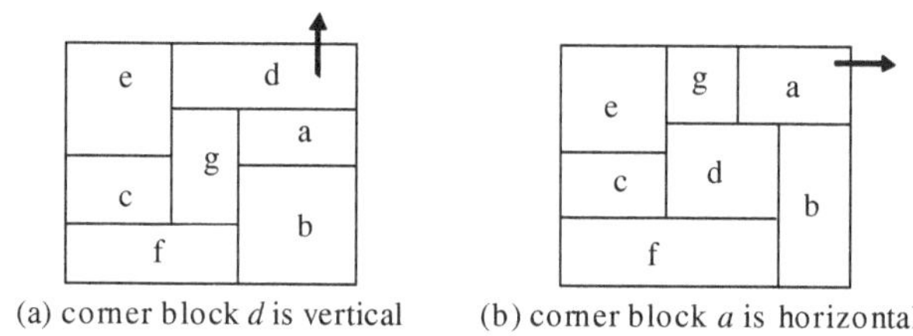

Fig. 4.5 The orientation of the corner block

Fig. 4.6 A non-slicing floorplan and its corresponding CBL list

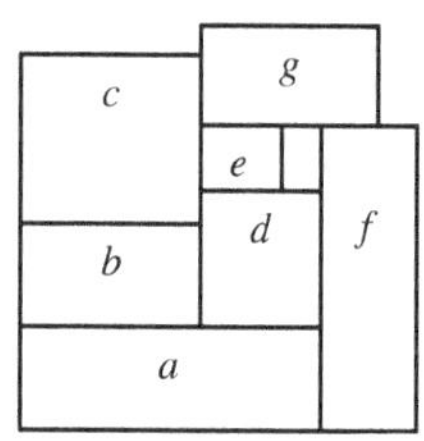

Corner Block List:
S = (a b c d e f g)
L = (001010)
T =(0 0 10 0 110 10)

Note that in a CBL list, it may happen that the number of successive "1"s in sub-list T_i is greater than the number of the available uncovered blocks in the corresponding direction. To amend this, an ending "0" can be automatically/virtually inserted, indicating that the block will cover all the available blocks in the corresponding direction. Suppose the number of available blocks is t. When scanning list T for block i, if there are t successive "1"s, then a "0" can be automatically/virtually inserted immediately after those successive "1"s to end this sub-list T_i. Then the scanning of list T can proceed for the next block. Therefore, an arbitrary CBL is feasible which corresponds to a mosaic floorplan.

4.3.1.3 General Structure

The general floorplan is similar to the mosaic floorplan in that non-slicing structures are allowed. However, the packing region can be dissected into more than n rooms such that some rooms are not occupied by any blocks. There was no efficient topological representation for the general floorplan until the sequence pair (SP) [18] and the bounded-sliceline grid (BSG) [19] appeared in the mid-1990s. The sequence pair is an elegant representation for the general floorplan and has been widely used. Later, TCG/TCG-S [13, 14] was proposed to represent the general structure. In this subsection, we briefly introduce sequence pair and TCG.

Sequence pair: A sequence pair is a pair of sequences of n elements representing a list of n blocks. Two permutations ($\Gamma+$, $\Gamma-$) capture geometric relations between each of two blocks. In general, the sequence pair imposes the relationship between each pair of blocks as follows:

$$(<\ldots,a,\ldots,b,\ldots>;<\ldots,a,\ldots,b,\ldots>) \rightarrow a \text{ is left to } b$$
$$(<\ldots,a,\ldots,b,\ldots,>;<\ldots,b,\ldots,a,\ldots>) \rightarrow a \text{ is above } b$$

Every two blocks constrain each other in either vertical or horizontal directions, and only these constraints are recorded. Therefore, the positions of blocks are pushed to low-left as much as possible, while satisfying the topological relations encoded in the sequence pair. Figure 4.7 is an example of a sequence pair.

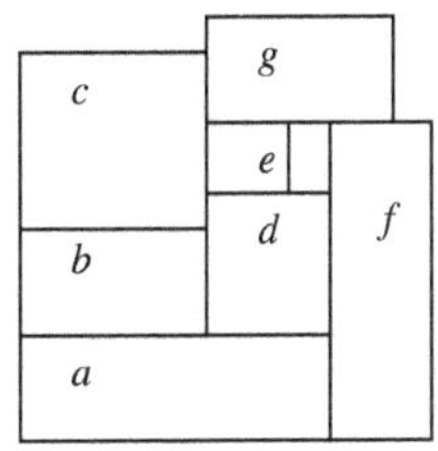

Fig. 4.7 Sequence pair for a packing: (*c b g e d a f, a b c d e f g*)

The original $O(n^2)$-time evaluation algorithm from [18] has been considerably improved in [32]. The algorithm in [32] runs in time $O(n \log(n))$ which is to evaluate a sequence pair based on computing the longest common subsequence in a pair of weighted sequences. Later work in [33] improves the algorithm in [32] and reduces runtime to $O(n \log(\log(n)))$ without affecting the resulting block locations.

TCG: TCG describes the geometric relations between blocks based on two graphs – namely, a *horizontal transitive closure graph* C_h and a *vertical transitive closure graph* C_v, in which a node n_i represents a block b_i and an edge (n_i, n_j) in C_h (C_v) denotes that block b_i is left of (below) block b_j.

Figure 4.8 shows a placement with five blocks, *a*, *b*, *c*, *d*, and *e*, and the corresponding TCG graphs. The value associated with a node in C_h (C_v) is the width (height) of the corresponding block, and the edge (n_i, n_j) in C_h (C_v) denotes the horizontal (vertical) relation of b_i and b_j. Here, *S* and *T* are the dummy nodes representing the source node and target node. For clarity, we omit the transitive edges connecting the dummy nodes in Fig. 4.8. Since there exists an edge (n_b, n_d) in C_h, block *b* is left of *d*. Similarly, *a* is below *b* since there exists an edge (n_a,n_b) in C_v. Therefore, by traversing the constraint graphs finding the longest path, the positions for each block are determined.

4.3.1.4 Compact Structure

The huge solution spaces of general floorplan representations restrict the applicability of these representations in the large floorplan problem. The O-tree [6] and

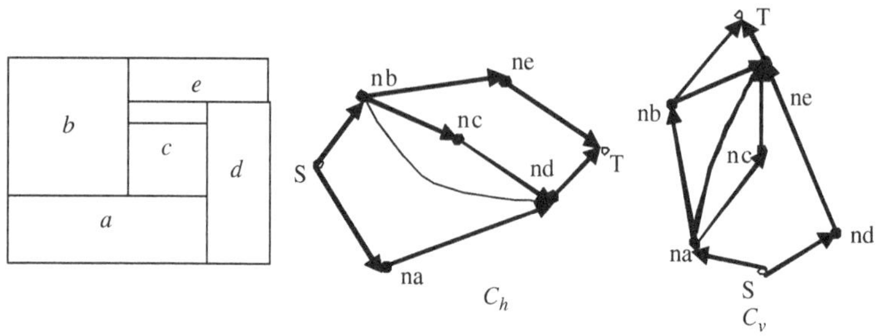

Fig. 4.8 A packing and the corresponding TCG

B*-tree [1] were proposed to represent a compacted version of the general floorplan. Compared to SP and TCG, these two representations have a much smaller solution space. However, they represent only partial topological information, and the dimensions of all blocks are required in order to describe an exact floorplan. In addition, not all possible rectangular dissections can be represented by the O-tree and B*-tree. For instance, the packing in Fig. 4.9a is not compacted since block A is not pushed to the most left as it is in Fig. 4.9b. But if block A has many connections with block B, the packing in (a) will have better wirelength than the packing in (b). Therefore, some packings represented by a sequence pair or CBL cannot be captured by the B*-tree or O-tree. Since the structure of the O-tree and B*-tree are somewhat similar, we briefly introduce the B*-tree in the following.

Fig. 4.9 Packing examples: (**a**) non-compact packing; (**b**) compact packing

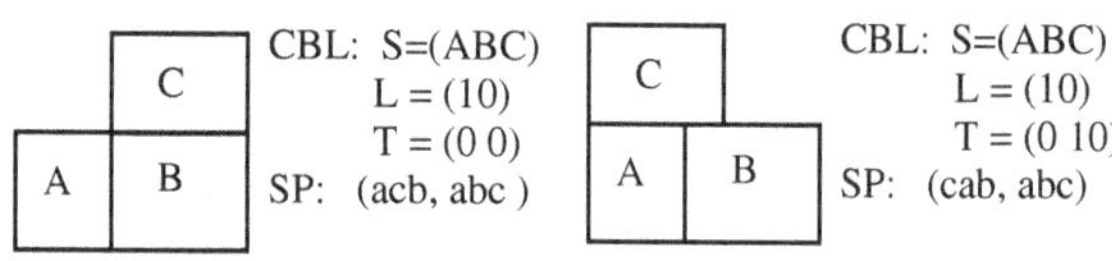

B-tree*: B*-tree represents a compact packing by a binary tree, in which each node corresponds to a block (see Fig. 4.10). The root node represents the bottom-left block. For example, *B*3 in Fig. 4.10 is the bottom-left block. A left child is the lowest-right neighbor of its parent and a right child is the lowest block above its parent that shares the same *x*-coordinate with its parent. In Fig. 4.10, *B*5 and *B*2 are the left and right children of *B*3, respectively. Given a B*-tree, block locations can be found by a depth-first traversal of the tree. After block *A* is placed at $(x_A;y_A)$, we consider its left child *B* and set $x_B = x_A + w_A$, where w_A is the width of *A*. Then y_B is the smallest non-negative value that avoids overlaps with previously placed blocks. After returning from recursion at block *B*, we consider the right child *C* of *A*: $x_C = x_B$, and y_C is the smallest possible so as to avoid overlaps. This algorithm can be implemented in $O(n)$ time with the *contour* data structure. The contour of a packing defines its upper outline (jagged) and can be implemented as a doubly linked list of line segments (Fig. 4.10). When a new block is put on top of the contour at a certain x-coordinate, it takes amortized $O(1)$ time to determine its y-coordinate. All packings represented by B*-trees are necessarily compacted so that no single block can move down without creating overlaps. Therefore, the B*-tree may not be able to represent min-wirelength packing. As shown in Fig. 4.9, if block C has tight connections with block B, then the packing in (a) has a shorter wirelength than the packing in (b), but (a) is not a compact packing so it cannot be represented by the B*-tree.

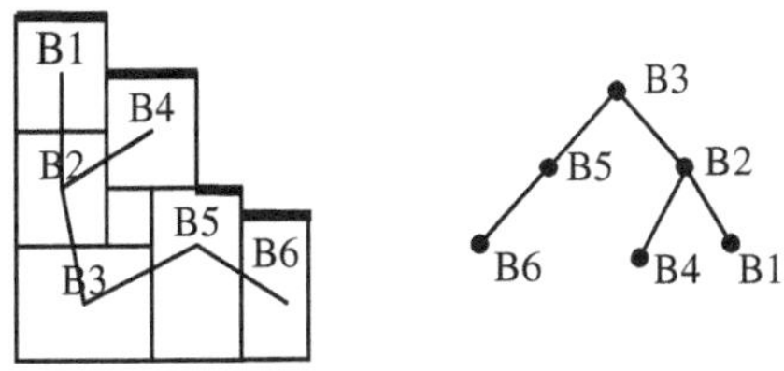

Fig. 4.10 A packing and its B*-tree representation: the contour of the packing is shown in *thick lines*

4.3.2 Analysis of Different Representations

In the previous section we described several typical representations for 2D floorplanning. Each of them can be extended to solve the 3D floorplanning with 2D blocks by keeping an array of the representation for each layer and allowing the swapping of blocks between different layers. But the solution spaces of those representations may be quite different. Some of the packing may not be captured in some representations. Some representations may actually capture the exact same set of floorplans. Therefore, we analyze the representations from several points of view.

4.3.2.1 Complexity

Based on the representations, we need a scanning procedure to construct the packing of blocks; we call this procedure a floorplan construction. Floorplan representations are always justified by the algorithmic complexity of floorplan construction and the total number of encoded configurations. Mathematical properties of various floorplan representations are discussed in [38, 29] and are briefly summarized here. It has been shown that the exact number of mosaic floorplans is given by the Baxter number [38], which can be represented as

$$B(n) = \binom{n+1}{1}^{-1}\binom{n+1}{2}^{-1}\sum_{k=1}^{n}\binom{n+1}{k-1}\binom{n+1}{k}\binom{n+1}{k+1}$$

The exact number of slicing floorplans is twice that of the Super Catalan number when the number of blocks is larger than 1. The Super Catalan number can be expressed as

$$A_0 = 1; A_1 = 1;$$
$$A_n = (2(2n-3)A_{n-1} - (n-3)A_{n-2})/n$$

Figure 4.11 shows the exact number of combinations for different structures. We can see that the number of combinations for SP/TCG increases very fast with the number of blocks; O-tree/B*-tree, which represent the compact structures, have the least number of combinations.

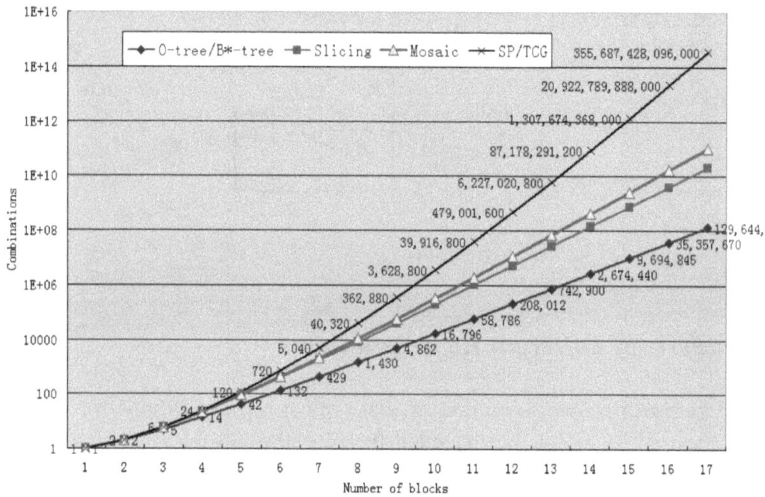

Fig. 4.11 The exact number of combinations for different structures. (Note that the logarithmic coordinates are used for the number of combinations in Y-axis)

Table 4.1 shows the comparisons between representations in terms of solution space, packing time, and packing category.

Some of the representations are equivalent in solution space though they may have different representations. The solution space defines an intrinsic bound on

Table 4.1 Comparisons between various representations

Representation	Solution space	Complexity of floorplan construction	Move	Packing category
NPE (SST)	$O(n!2^{3n-3}/n^{1.5})$	$O(n)$	$O(1)$	Slicing
SP	$n!^2$	$O(n \log\log n) - O(n^2)$	$O(1)$	General
BSG	$n!C(n^2,n)$	$O(n^2)$	$O(1)$	General
O-tree	$O(n!2^{2n}/n^{1.5})$	$O(n)$	$O(1)$	Compact
B*-tree	$O(n!2^{2n}/n^{1.5})$	$O(n)$	$O(1)$	Compact
CBL	$O(n!2^{3n-3}/n^{1.5})$	$O(n)$	$O(1)$	Mosaic
TCG	$n!^2$	$O(n^2)$	$O(n)$	General

the expressiveness of a representation, and this bound may directly impact solution quality. If the representations share the same solution space, one can equalize the move set, and thus the differences will only affect runtime (some moves may be faster or slower). Sequence pair and TCG are shown to be equivalent in [14] in the sense that they share the same $(n!)^2$ solution space and capture the exact same set of floorplans. Each sequence pair corresponds to one TCG and vice versa. Although TCG and SP are equivalent, their properties and induced operations are significantly different. Both SP and TCG are considered very flexible representations and construct constraint graphs to evaluate their packing cost. However, like most existing representations, the geometric relations among blocks are not transparent to the operations of SP (i.e., the effect of an operation on the change of module relation is not clear before packing); thus we need to construct constraint graphs from scratch after each perturbation to evaluate the packing cost. This deficiency makes SP harder to converge to a desired solution and to handle placement with constraints (e.g., boundary modules, pre-placed modules). In contrast to SP, the geometric relations among blocks are transparent to TCG, as well as its operations, thus facilitating the convergence to a desired solution. Further, TCG supports incremental update during operations and keeps the information of boundary modules as well as the shapes and the relative positions of modules in the representation.

Both the O-tree and B*-tree use a single tree to represent a horizontally compact packing, but differ in the bit-level implementation of the tree. The O-tree uses a rooted ordered tree with arbitrary vertex degrees, while the B*-tree uses a binary tree. Therefore, they share the same solution space of size $O(n!2^{2n}/n^{1.5})$ and capture the same set of floorplans. Compared to other representations, the B*-tree and O-tree have a much smaller solution space. However, they represent only partial topological information, and the dimensions of all blocks are required in order to describe an exact floorplan.

4.3.2.2 Redundancy

Redundancy means that there is more than one representation that can represent a certain floorplan. The redundancy in representation can waste steps in various search procedures. Actually, if we consider a degenerate case (as shown in Fig. 4.12), most of the representations will have at least two representations for this packing. Take NPE as an example, since the partitioning has two choices and the resultant slicing tree is still skewed no matter which partition is taken first. But most of the work that

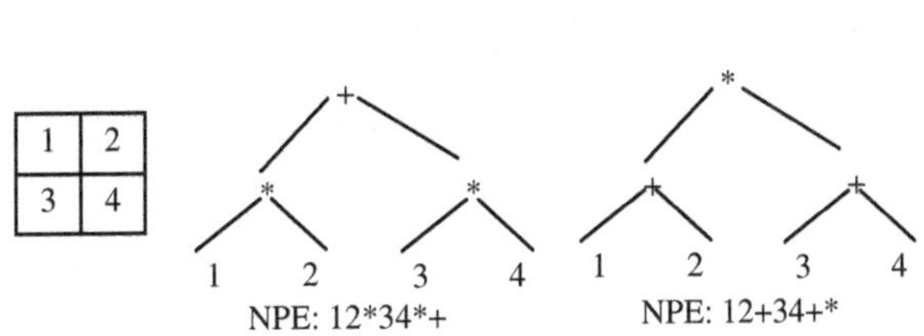

Fig. 4.12 Degenerate case with two corresponding NPE lists

has been done [7, 38, 40, 39] takes the degenerate cases as special cases and assumes the crossing segments are separated by a small distance so that the topological relations between blocks can be settled. Therefore, we do not consider the multiple representations for the degenerate case as redundant representations. But even without the degenerate case, there still exists redundancies in some representations – some are amendable, but some are inevitable.

For a corner block list, the arbitrary binary list T with another two lists S and L can represent a mosaic packing. List T is composed of a binary list whose length is no more than $2n-3$. The length of T is dynamically changed with the packing structure; most of the packings do not need to use the full length of $2n-3$ in list T. But if we allocate the fixed length for list T in the representation, some CBL lists that have the same S and L, but differ in list T in the tail, may represent the same packing. To remedy this, we can record the valid length for list T while packing, so that during the optimization, we can control the possibility of redundant moves. As shown in Fig. 4.13, if we fix the length of T to be $2n-3=5$, both of the two lists in Fig. 4.13 represent the same packing, since the valid list for T is only {0 0 0} which means block 2, block 3, and block 4 cover only one block. Therefore, the valid length for T should be 3; however, if we take this information into consideration, those two lists are the same.

Fig. 4.13 Redundancy in CBL representation: the valid length for T is 3

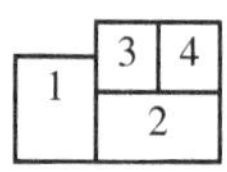

S = (1 2 3 4)
L = (0 1 0)
T = (0 0 0 ***0 0***)

S = (1 2 3 4)
L = (0 1 0)
T = (0 0 0 ***1 1***)

In SP, two sequences, $\Gamma+$ and $\Gamma-$, sort all the blocks from top-left to bottom-right and from bottom-left to top-right. When the relative positions of the two blocks are both above and right-to, their relative positions in $\Gamma+$ have multiple choices (as blocks D and E shown in Fig. 4.14). Similarly, if the relative positions of the two blocks are both below and right-to, their relative positions in $\Gamma-$ have multiple choices. This redundant representation causes a one-to-many mapping from a floorplan to the representations.

Fig. 4.14 Redundant SP representations for the same packing

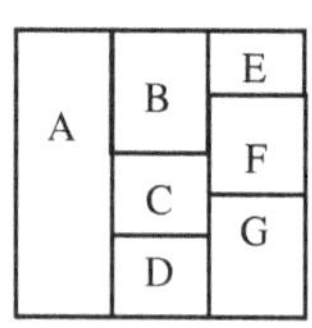

SP1 = (ABECFDG, ADCGBFE)

SP2 = (ABCDEFG, ADCBGFE)

Regarding floorplan representations, NPE [34] is a nonredundant representation for the slicing floorplan. TBS [39] and Q-sequence [42] are two nonredundant representations for the mosaic floorplan. However, there is no nonredundant representation for the general floorplan. Although all general floorplans can be produced by inserting empty rooms into TBSs, the information describing which empty room to insert is not uniform. Hence, TBS cannot be easily extended to a succinct representation that describes a general floorplan completely.

4.3.2.3 Suitability for 3D Design

To extend the 2D floorplan representations to handle the 3D floorplan with 2D blocks, an array of 2D representations (2D array) can be constructed, each representing all blocks located on one device layer using any kind of 2D representations. There are two ways to implement the layer assignment: (1) assign the blocks to layers before the packing optimization with some inter-layer constraints or objectives considered, and then keep the layer assignment unchanged during the packing optimization; (2) initialize the layer information and then swap blocks between layers during the packing optimization. The first method might limit the solution space and lose the optimality of the final results, but it simplifies the problem and makes the inter-layer constraints more easily satisfied. The second method is more flexible and will get a possibly better trade-off between multiple objectives. Therefore, in this chapter, we assume to use the second approach, by which the layer assignments and the floorplans of each layer are determined simultaneously.

Compared to 2D floorplanning, 3D floorplanning with 2D blocks needs to take more issues into consideration, such as thermal distribution, vertical relative position constraints, and thermal via insertion. Since the B*-tree and O-tree represent only compact packings, they may not capture min-wirelength solutions and optimal temperature. In consideration of the thermal distribution, the packings are not necessarily compacted since the whitespace between blocks is useful to separate hot blocks and might be used for thermal via insertion. And to handle the physical relation constraints, such as alignment constraints, the geometric relations among blocks are very useful. We compare the typical 2D representations to show the pros and cons for 3D floorplanning with only 2D blocks.

Both SP and TCG can represent general packing with $O(n^2)$ complexity. The redundancy is typically viewed as a limitation of the sequence pair and TCG. The sequence pair representation is simpler and the moves take less time to evaluate, but the geometric relations among blocks are less clear than those in TCG, so it is easier to extend the TCG to handle physical constraints.

The room-based representations such as CBL are also a good choice since the blocks can be moved within the rooms while the representation and topological relations are unchanged; thus the local incremental improvement might be easier. The CBL representation can be evaluated in linear time with smaller solution space compared to SP and TCG, but it can only represent mosaic packing.

Hence, with different complexity and flexibility trade-off, multiple representations are possible for 3D floorplanning with 2D blocks. In section 4.5.2, we take the TCG as the representation, and a bucket structure is posed to encode the Z-axis neighboring information – the so-called combined bucket and 2D array (CBA) [3].

4.4 Representations for the 3D Floorplan with 3D Blocks

Similar to the 2D packings, 3D cube packings can also be classified into two main categories: slicing and general non-slicing. Among the general 3D packings, there is also a subset called 3D mosaic packings which includes all slicing structures and part of the non-slicing structures. Therefore, in the following section, we describe several typical representations: the 3D slicing tree [2], the 3D CBL [16], and the sequence triple, and sequence quintuple [35].

4.4.1 3D Slicing Tree

To get a slicing structure, we can recursively cut the 3D block by planes that are perpendicular to the *x*-, *y*-, or *z*-axis. (It is assumed that the faces of the 3D block are perpendicular to the *x*, *y*, and *z* axes.) A slicing floorplan can be represented by an oriented rooted binary tree called a slicing tree (see Fig. 4.13). Each internal node of the tree is labeled by *X*, *Y*, or *Z*. The label *X* means that the corresponding super module is cut by a plane that is perpendicular to the *x*-axis. The same is true for label *Y* and label *Z* in terms of the *y*-axis and *z*-axis, respectively. Each leaf corresponds to a basic 3D block and is labeled by the name of the block. Similar to 2D slicing representations, the skewed 3D slicing tree can also be used to avoid redundancy. In the 3D skewed slicing tree, no node and its right child have the same label (Fig. 4.15).

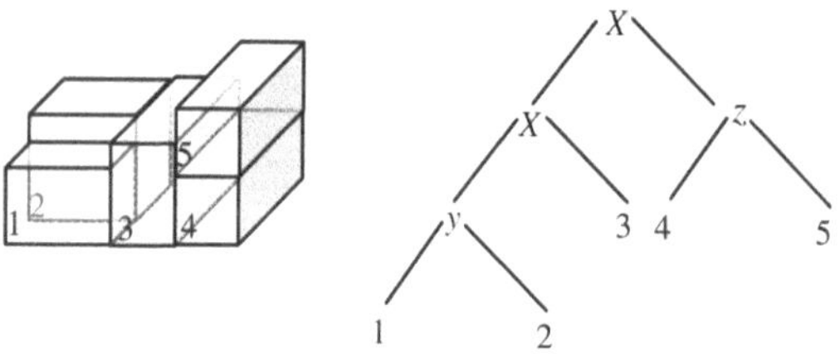

Fig. 4.15 Three-dimensional slicing floorplan with skewed slicing tree representation

4.4.2 3D Cbl

The topology of 3D packing is a system of relative relations between pairs of 3D blocks in such a way that block "*a*" is said to be left-of block "*b*" when any point of "*a*" is left-of any point of "*b*." Relations of "right-of," "above," "below," "front-of," and "rear-of" are analogously defined. Similar to the mosaic structure in 2D packing, the 3D floorplan divides the total packing region into cubic rooms with sides. Each cubic room holds one cubic block.

Therefore, to represent the topological relations in the 3D mosaic floorplan, each cubic block is represented by a cubic room and rooms cover each other in the *x*-,

y-, or z-direction. During the packing process from the bottom-left-front corner to top-right-rear corner, if the room of block A covers the room of block B, the room of block A is totally covered by a side and the side extension of the room of block B. As shown in Fig. 4.16, the direction for each room is defined by the direction in which it covers other rooms. Therefore, if a new block 4 is going to be inserted into the packing in Fig. 4.16b, the room of block 4 can cover those packed blocks {1,2,3} in x-, y-, or z-direction. The new inserted block will be located at the top-right-rear corner so that it is defined as the *corner cubic block*. For each direction, not all the rooms of packed blocks are available to be covered since some of them may have already been covered by the previous packed rooms. As shown in Fig. 4.16b, block 1 has already been covered by block 2 in x-direction; the new room of block 4 can only cover the rooms for block 2, block 3, or both in x-direction. Therefore, the uncovered block list in a packing sequence can be defined for each direction, which records the current available blocks to be covered. In Fig. 4.16b, before block 4 is inserted, the uncovered block list in z-direction is {1, 2, 3}, the uncovered block list in y-direction is {1, 3}, and the uncovered block list in x-direction is {2, 3}.

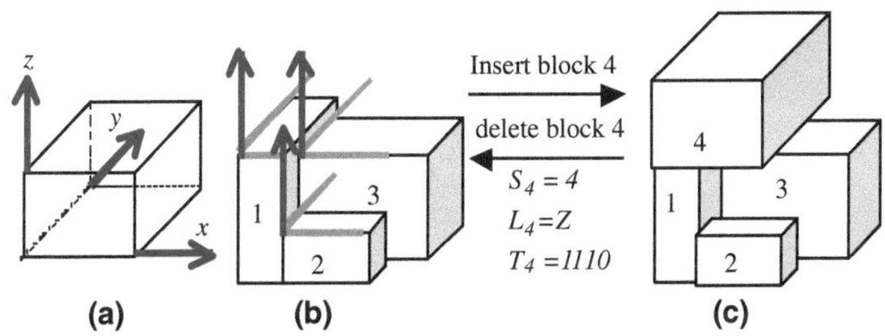

Fig. 4.16 The process of corner cubic block: (**a**) x-, y-, z-directions; (**b**) corner cubic block is 3 and uncovered block list in z-direction is {1, 2, 3}; (**c**) corner cubic block is 4 which covers 3 blocks in {1, 2, 3} since $T_4 = 1110$; uncovered block list in z-direction becomes {4} and the corresponding 3D CBL: $S = \{1, 2, 3, 4\}$ $L = \{X, Y, Z\}$ $T = \{10\ 10\ 1110\}$

With the covering direction and the uncovered block list in the corresponding direction, information is still needed concerning which block or blocks should be covered to determine the position of the inserted block. Suppose that the uncovered block list in a direction in a packing sequence is $\{B_1, B_2, \ldots B_k\}$. As shown in Fig. 4.16c, the uncovered block list in z-direction is {1, 2, 3}. If the room of block 4 covers the room of block 1, the room of block 4 will cover the rooms for blocks 2 and 3 at the same time. Therefore, to determine the position of inserted blocks, the number of blocks covered by the room of this block is recorded in the uncovered block list. With the new inserted block, the uncovered block list should be updated dynamically. The last m blocks $\{B_{k-m+1}, \ldots, B_k\}$ are no longer available to be covered in this direction, Hence, the updated uncovered block list after block B is inserted should be $\{B_1, \ldots, B_{k-m}, B\}$.

Therefore, the information related to the packing process of the inserted corner block B should include the following: the block's name, the covering direction, and the number of blocks covered by B in the uncovered block list. To favor the generation of new solutions during the optimization process, a binary sequence T_i is used to record the number of blocks covered in the uncovered block list, in which the number of 1s corresponds to the number of covered blocks. Each string of 1s is ended with a 0 to separate it from the record of the next block. Given a 3D packing, we can have a sequence S of block names, a list L of orientations, and a list $\{T_2, T_3, \ldots, T_n\}$ of covering information. The three-element triple (S,L,T) composes a 3D CBL (as shown in Fig. 4.16c). Figure 4.17 shows a packing example step by step.

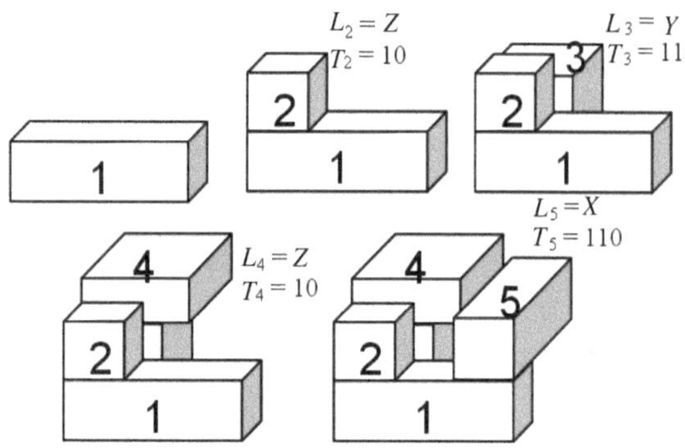

Fig. 4.17 The packing process: $S = \{1\ 2\ 3\ 4\ 5\}$; $L = (Z, Y, Z, X)$; $T = (10,110,10,1110)$

4.4.3 Sequence Triple

The sequence triple (ST) [35] is the system of ordered three-sequence block labels, which is extended from the sequence pair used for 2D packings. A sequence triple (ST) is denoted as ST(Γ_1, Γ_2, Γ_3). Similar to the SP, ST also has its decoding rules to represent the topological relationships between blocks.

$$(\ldots a \ldots b \ldots, \ldots a \ldots b \ldots, \ldots a \ldots b \ldots) \rightarrow b \text{ is rear-of } a$$

$$(\ldots a \ldots b \ldots, \ldots a \ldots b \ldots, \ldots b \ldots a \ldots) \rightarrow b \text{ is left-of } a$$

$$(\ldots a \ldots b \ldots, \ldots b \ldots a \ldots, \ldots a \ldots b \ldots) \rightarrow b \text{ is right-of } a$$

$$(\ldots a \ldots b \ldots, \ldots b \ldots a \ldots, \ldots b \ldots a \ldots) \rightarrow b \text{ is below } a$$

$$(\ldots b \ldots a \ldots, \ldots b \ldots a \ldots, \ldots b \ldots a \ldots) \rightarrow b \text{ is front-of } a$$

$$(\ldots b \ldots a \ldots, \ldots b \ldots a \ldots, \ldots a \ldots b \ldots) \rightarrow b \text{ is right-of } a$$

$$(\ldots b \ldots a \ldots, \ldots a \ldots b \ldots, \ldots b \ldots a \ldots) \rightarrow b \text{ is left-of } a$$

$$(\ldots b \ldots a \ldots, \ldots a \ldots b \ldots, \ldots a \ldots b \ldots) \rightarrow b \text{ is above } a$$

Given an ST, the realization of the 3D packing is as follows: decode the representation to the system of RL-, FR-, and AB-topology. Then, construct three constraint graphs G_{RL}, G_{FR}, G_{AB}, all analogously as in 2D packing. Then, the longest path length to each vertex locates the corresponding box, i.e., the coordinate (x, y, z) of the left-front-bottom corner.

Figure 4.18 is an example of a packing with three blocks and its corresponding ST. Since there is an empty hole between the three blocks, the packing shown in Fig. 4.18 is not 3D mosaic packing and cannot be represented by 3D CBL.

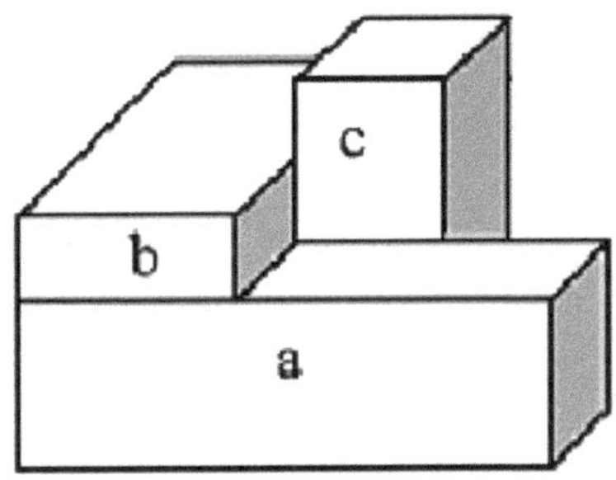

Fig. 4.18 A packing with an empty room which can be represented by sequence triple as (*bac, acb, abc*)

Similar to 3D CBL, the relative relations between pairs of 3D blocks are left-of, right-of, above, below, front-of, or rear-of. Both 3D CBL and ST give each box pair exactly one direct relation constraint so that the constraint holds transitiveness. In other words, an indirect relation constraint on a pair, if existing, is not different from the direct relation constraint. As shown in Fig. 4.19, *a* is below *b*. Similarly, *b* must be constrained to be below *d*. Although *a* need not be constrained to be below *d* directly, *a* is indirectly constrained to be below *d* through *c*. Similarly, *a* is indirectly constrained to be front-of *d* through *c*. Consequently, *a* is indirectly constrained to be below and front-of *d*, and the pair of *a* and *d* has two indirect relative relation constraints. There are 3D packings which should contain two or three indirect relation constraints on a block pair. These 3D packings are called "β-type." It is known that 3D packings of β-type cannot be represented by ST or 3D CBL [16].

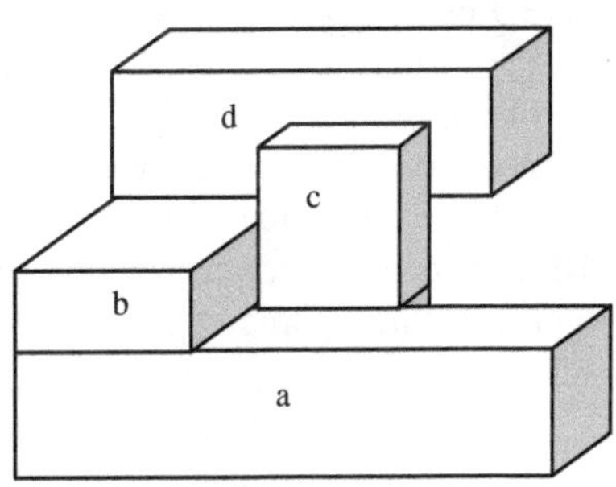

Fig. 4.19 β-type 3D packing in which blocks *a* and *d* have two indirect relations

Therefore, the system of five sequences, denoted as $Q=(\Gamma_1,\Gamma_2,\Gamma_3,\Gamma_4,\Gamma_5)$ and called sequence quintuple (Squin), is proposed to represent all 3D packings. The algorithm to construct packing from sequence quintuple is as follows:

- Step 1: Construct the right–left constraint graph G_{RL} to represent the RL-topology by Γ_1,Γ_2 following the rule similar to the sequence pair, but only restricted to the right–left relation that

$$(\Gamma_1: < \dots, a, \dots, b, \dots > ; \Gamma_2 < \dots, a, \dots, b, \dots >) \rightarrow a \text{ is front-of } b$$

The front–rear constraint graph G_{FR} is constructed from (Γ_3, Γ_4) by the rule that

$$(\Gamma_3: < \dots, a, \dots, b, \dots > ; \Gamma_4 < \dots, a, \dots, b, \dots >) \rightarrow a \text{ is front-of } b$$

- Step 2: Determine the longest path in G_{RL} and G_{FR} so that every block has been located at its $x{-}y$ coordinate. Two blocks are said to be $x{-}y$ overlapping if they overlap in the projected $x{-}y$ plane.
- Step 3: Construct the above–below constraint graph G_{AB} as follows. For each pair of blocks, an edge from a to b is added if and only if (1) a and b are $x{-}y$ overlapping and (2) Γ_5: $<\dots, a, \dots, b, \dots>$.
- Step 4: Determine the z-coordinate by the longest path in G_{AB}.

  It is proven that the sequence quintuple can represent all the 3D packings. The complexity of the algorithm to construct packing from the sequence quintuple is $O(n^2)$.

4.4.4 Analysis of Various Representations

The 3D packing problem is more complicated than that of 2D packings. We analyze the various representations from two views: complexity, and flexibility for 3D floorplanning with 3D blocks.

Complexity: Table 4.2 shows the features for several 3D packing representations. The slicing tree can represent fewer 3D packings than 3D CBL, and 3D CBL can represent fewer packings than ST. ST can represent fewer packings than Squin (3D slicing tree $\subset$ 3D CBL $\subset$ ST $\subset$ Squin). Squin can represent any 3D packing.

If some partitioning sides meet at the same line, we call this case a degenerated topology. If we treat degenerated topology as a special case, than we can separate two sides by sliding one side by a small distance so that the topology between the blocks is unique. With this assumption, the skewed 3D slicing tree can be a

Table 4.2 Features for several 3D packing representations

Representation	Complexity of floorplan construction	Move complexity	Packing category	Solution space
ST	$O(n^2)$	O(1)	General but not all	$n!^3$
Squin	$O(n^2)$	O(1)	All Slicing	$n!^5$
3D Slicing tree	$O(n)$	O(1)		$O(n!3^{n-1}2^{2n-2}/n^{1.5})$
3D-subTCG	$O(n^2)$	$O(n^2)$	General but not all	$n!^3$
3D-CBL	$O(n)$	O(1)	Mosaic	$O(n!\,3^{n-1}2^{4n-4})$

nonredundant representation. With the dynamically updated T_list information, 3D CBL can also represent the 3D mosaic packings with no redundancy. But both ST and Squin have redundancy since they have transitive properties. When the relative positions of two blocks are both up and right or both down and right, etc., their relative positions in list have multiple choices. This redundant representation causes a one-to-many mapping from a floorplan to the representations.

Flexibility for 3D floorplan with 3D blocks: The problem of 3D floorplanning with 3D blocks is that we need to select the best configuration for each block from a pool of the candidates. And some design constraints, such as Z-height constraints, are imposed during the optimization. Therefore, it would be better if the representations for 3D floorplanning with 3D blocks are flexible to handle this requirement. Based on the implementation, the transformation from lists to packing in 3D CBL representation can be incrementally processed from bottom-left to top-right in linear time. Compared to a graph-based representation or a sequence family, it is much easier to handle constraints by fixing the violations dynamically. In the following section, a heuristic based on CBL representation can be used to fix violations of a Z-height constraint while doing the packing. This approach guarantees the feasibility of the final results and improves the converge process. The construction of the cubic floorplan based on 3D CBL is $O(n)$ time, where n is the number of blocks. But the major shortage of 3D CBL is that the packing solution space is much less than the ST and Squin. Therefore, any 3D packing representation can be used for the 3D floorplan.

4.5 Optimization Techniques

Since the 2D and 3D rectangular packing problems are NP-hard, most floorplanning algorithms are based on stochastic combinatorial optimization techniques such as simulated annealing and genetic algorithm. But some of the recent research focuses on deterministic approaches so that the analytical algorithms are proposed for handling 3D floorplanning.

4.5.1 Simulated Annealing

Stochastic optimization methods are now being used in a multitude of applications when enumerative methods are too costly. The objective of 3D floorplanning is to minimize a given cost function by searching the solution space represented by a specific representation. Normally, the cost function describes the combination of chip area, wirelength, maximal on-chip temperature, or other factors.

In this section we introduce the simulated annealing approach with the application of 3D floorplanning. Generally, simulated annealing is a generalization of a Monte Carlo method for examining the equations of state and frozen states of *n*-body systems [20, 8]. As one of the most popular stochastic optimization methods, simulated annealing has been applied successfully to many optimization problems in the area of VLSI layout. The algorithm simulates the annealing temperature near melting point, and then slowly cools it down so that it will crystallize into a highly ordered state. The time spent at each temperature should be sufficiently long enough to allow a thermal equilibrium to be approached. Figure 4.20 shows the optimization flow based on the simulated annealing approach.

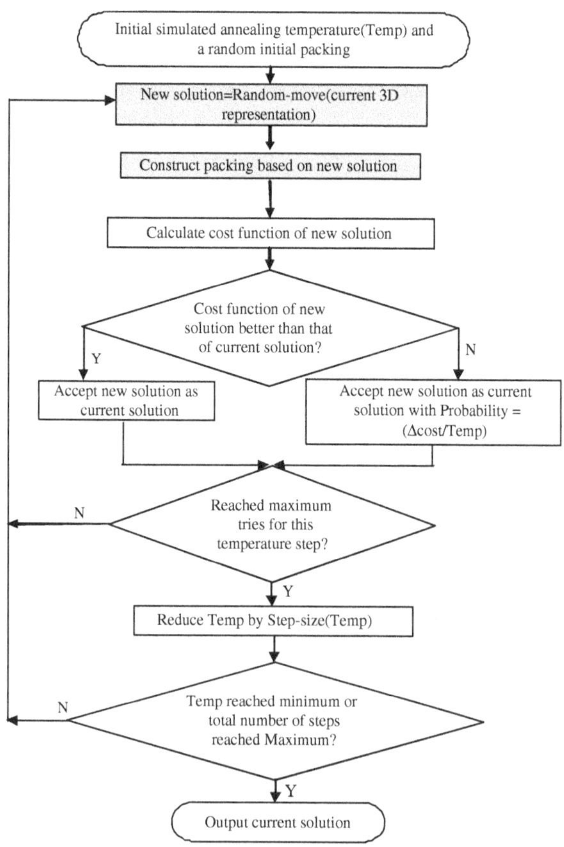

Fig. 4.20 The flow of the simulated annealing approach

4.5.2 SA-Based 3D Floorplanning with 2D Blocks

With the additional Z-direction, the stacking structure dramatically enlarges the solution space. Therefore, some of the 3D floorplanning approaches based on SA [10, 36] proposed the hierarchical framework, in which the layer assignment and the floorplanning are performed successively. The layer number for each block is fixed during the simulated annealing process. Though these approaches reduce the complexity of the problem, they may lose the optimality by limiting the layer assignment during optimization. Here we introduce a flat design framework [3] in which the layer assignments and the floorplans of each layer are determined simultaneously. Therefore, the block can be moved from one layer to another during the searching process. With the representations introduced in the previous section, we can apply the SA optimization scheme to 3D floorplanning with 2D blocks. To design an efficient SA scheme, there are several issues which are critical.

1. *Representation of the solution:* Since the packing on each layer can be represented by a 2D representation, the multi-layer packing can be represented by an array of 2D representations and blocks are moved inside each layer or swapped between layers for solution perturbations. But to overcome limitations caused by the lack of relative position information of blocks on different layers, we may encode the Z-direction neighboring information using an additional bucket structure. In each bucket i, indexes of the blocks that intersect with the bucket are stored; the index set is referred to as $IB(i)$, no matter which layer the block is on. In the meantime, each block j stores indexes to all buckets that overlap with the block; the index set is referred to as $IBT(j)$. Therefore, the combined bucket and 2D array (CBA) is proposed; this is composed of two parts – a 2D floorplan representation used to represent each layer, and a bucket structure to store the vertical relationship between blocks. In this chapter we choose TCG to represent the 2D packing on each layer.
2. *Cooling schedule:* The whole cooling schedule includes the set up of the initial temperature, cooling function, and end temperature. It depends on the size of the problem and the property of the problem.
3. *Solution perturbation:* We take the CBA representation as an example. There are seven kinds of operations on CBA, as shown in the following:
   - Rotation, which rotates a block
   - Swap, which swaps two blocks in one layer
   - Reverse, which exchanges the relative position of two blocks in one layer
   - Move, which moves a block from one side (such as top) of a block to another side (such as left)
   - Inter-layer swap, which swaps two blocks at different layers
   - z-neighbor swap, which swaps two blocks at different layers but close to each other
   - z-neighbor move, which moves a block to a position at another layer close to the current position

4. *Cost function:* Every time a block configuration is generated, a weighted cost of the optimization objectives and the constraints will be evaluated. The cost function can be written as

$$\text{Cost} = \alpha \text{WL} + \beta \text{Area} + \gamma N_{\text{via}} + \theta T$$

where WL is the wirelength estimation using a half perimeter model, Area is the product of the maximal height and width over all layers, N_{via} is the number of inter-layer vias, and T is the maximal temperature. Here in 3D designs, the on-chip temperature is so high that it is necessary to account for the closed temperature/leakage power feedback loop to accurately estimate or optimize either one.

4.5.3 SA-Based 3D Floorplanning with 3D Blocks

The cubic packing process with 3D blocks based on simulated annealing is similar to 3D floorplanning with 2D blocks. But the 3D floorplanning problem with 3D blocks that we investigate here considers not only the positions for blocks, but also the configurations of blocks. Different than the previous simulated annealing-based floorplanning approach, the choosing of the blocks' configurations is integrated dynamically while doing the packing.

Since representation is the key issue in the simulated annealing approach, we choose 3D CBL as an example to explain the process of the SA-based approach. With the block candidates varying in dimensions, delay, power consumption, and layer numbers according to different partitioning approaches, the block configuration can be chosen during the optimization. Therefore, to choose the best feasible configuration for blocks, the new operation "*Alternative_Selection*" is defined to create a new solution.

Operation: Alternative_Selection:

1. Randomly choose a block i with multiple candidates
2. Randomly choose a feasible candidate from the candidate list
3. Update block i with the dimension of the chosen candidate

The move used to generate a neighboring solution is based on any one of the following operations:

1. Randomly exchange the order of the blocks in S
2. Randomly choose a position in L and change the orientation
3. Randomly choose a position in T, change "1" to "0" or change "0" to "1"
4. Alternative_Selection

The various candidates of components greatly enlarge the solution space. Especially with some layer number constraints, parts of the solutions are infeasible. Therefore, heuristic methods are devised to speed up the searching process.

The cost function also uses a weighted combination of area, temperature, and wirelength, which can be represented by

$$\text{Cost} = wl^{*}\text{Area} + w2^{*}\text{Temp} + w3^{*}\text{Wire}$$

With that floorplan of the blocks, Area is the total area of the floorplan. Temp corresponds to the maximum on-chip temperature based on the temperature simulator. The coefficients of $w1$, $w2$, and $w3$ are used to control the different weights for each component.

In 3D microarchitecture design, the number of chip layers is often given as a constraint. To handle the layer number constraints, the traditional method is to penalize the violations in the cost function. However, this method does not guarantee the feasibility of the final results and may slow down the convergence of the optimization. With 3D CBL representation, the blocks are packed in sequence. Therefore, the blocks or CBL list can be dynamically changed during the packing. If some block exceeds the layer number constraint, the violation can be fixed by either lowering the block or changing the direction of the block. We take the following steps to fix the violation:

1. To maintain the topology as much as possible, first try to change the implementation of the block by choosing a candidate with a lower z-dimension.
2. If the violation cannot be fixed by changing the candidate, try to modify the 3D CBL list to achieve a feasible packing. If block B covers previous blocks in the z-direction, which means block B will be placed on top of packed blocks, and if block B exceeds the layer number constraint, we can change the covering direction to x or y so that block B can be placed on the right side or behind previous blocks. But if the z-position of block B is still too high, we can dynamically move block B to the lower position by increasing the number of "1"s in T_B. Since T_B means the number of blocks covered by B in the direction L_B, block B will be moved to lower blocks when we increase the number of "1"s in T_B. This process will be continued until block B satisfies the layer number constraint.

Given the number of layers of the design Z_{con}, the CBL list is scanned to pack the blocks from bottom-left-front corner to top-right-rear corner. The coordinates of the bottom-left-front corner for packed block B are (x_B, y_B, z_B) with the corresponding implementation $c^B{}_j$. Hence, the process can be described as follows :

Algorithm Fix_Violation
Input: block B which exceeds the layer number constraint: $z_B + z^B{}_j > Z_{con}$;
3D_CBL and the candidate list for block B.
Output: New 3D_CBL with the new candidate selection c^B for B;
If $z_B < Z_{con}$

For candidate $c^B{}_j$ in candidate list of B
If $z_B + z^B{}_j \le Z_{con}$
choose this candidate $c^B = c^B{}_j$ and update the positions of B;
return; // violation be fixed by changing candidate
EndIf;
EndFor;
choose the candidate with the lowest z-height and update the information of B;
If $L_B = Z$ // cover previous block from z-direction
Change L_B to Xor Y and update position of B;
EndIf
While $(z_B + z^B{}_j > Z_{con})$
Increase the number of "1"s in $T^{LB}{}_B$ which means the number of blocks covered by B in the direction L_B is increased.
Update the position of B;
EndWhile;
EndIf.

The extreme case is that block B is moved to the bottom (z_B=0). The candidate list should be constructed with the constraint that every block's z-height should be less than Z_{con}. Block B will not exceed the layer number constraint if $z_B = 0$. Therefore, our algorithm guarantees the feasibility of the results.

4.5.4 Analytical Approach

Most of the floorplanning algorithms are based on simulated annealing techniques. But stochastic optimization approaches generally have long run times that scale poorly with problem size. Therefore, the analytical approaches provide relatively stable and scalable techniques for 3D floorplanning optimization. The analytical approach has been widely explored in placement algorithms for standard cells [4, 5, 21] (which will be introduced in Chapter 5 in detail). But in floorplanning with macro blocks, the heterogeneity in block sizes and shapes complicated the problem. A small change during optimization can cause large displacements in the final legalized packing. In stochastic optimization approaches, the topological relations between blocks are described by the representations so that non-overlapping between blocks is guaranteed. It is hard for the mathematics computation to formulate the non-overlapping constraints between blocks in linear way; therefore, the legalization which removes overlapping between blocks is necessary in most of the analytical approaches. In this section we briefly introduce the force-directed approach to handle the 3D temperature-aware floorplanning with 2D blocks that was proposed in [41]. The floorplanning with 3D blocks is similar and the approach introduced in the following can be easily extended to handle 3D blocks.

Starting from a placement solution requires translation from continuous space to a discrete, layer-assigned, and legalized solution. Therefore, the analytical approach has three phases: global placement, layer assignment, and legalization (as shown in Fig. 4.21).

1. *Global placement:* There are many mathematical methods to optimize the cell locations in a continuous region (this will be introduced in Chapter 5 in detail).

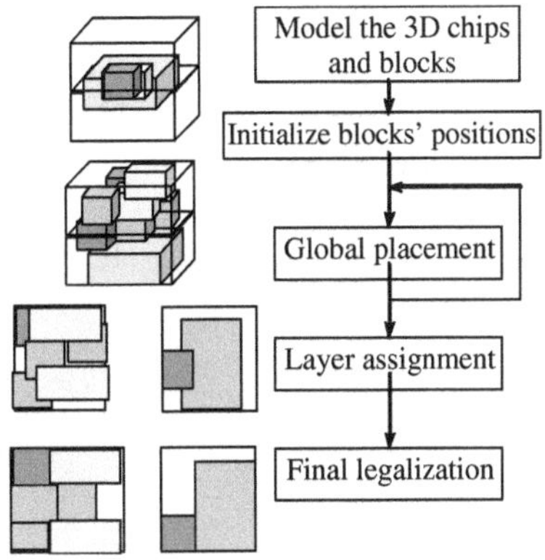

Fig. 4.21 Three-dimensional force-directed floorplanning flow

Here, we take the basic force-directed approach as an example. The force-directed algorithm simulates the mechanics problem in which particles are attached to springs and their movement obeys Hooke's law. A homogeneous cubic bin structure is overlayed on the 3D space to simplify computation of forces. Based on this bin structure, two kinds of forces, filling forces and thermal forces in 3D space, are introduced to eliminate overlaps and reduce placement peak temperature.

- *Filling Force*: Filling force is used to eliminate overlap between blocks and distribute them evenly over the 3D placement region. It drives the placement to remove overlap by pushing blocks away from regions of high density and pulling blocks toward regions of low density in 3D space. The bin density is defined as the sum of the block areas covering the bin. Each bin's filling force is equal to its bin density. A block receives a filling force equal to the sum of the prorated filling forces of the bins the cell covers.
- *Thermal Force*: The thermal model (described in Chapter 3) obtains thermal gradient for a placement. We would like to move blocks (which produce heat) away from regions of high temperature. This goal is achieved by using the thermal gradient to determine directions and magnitudes of the thermal forces on blocks. The filling force and thermal force for a given block are calculated by summing the individual forces upon the bins that the block occupies in each level of the tree. Forces from a bin and from its nearest neighbors are considered. Large blocks span numerous bins. As a consequence, they receive greater forces than small blocks.

2. *Layer Assignment:* After optimizing the placement in continuous 3D space, blocks must be assigned to discrete IC layers. In the above approach, each block is modeled as a 3D rectangle that can be moved freely in continuous 3D space. Layer assignment moves blocks from continuous space to discrete space, forcing each block to occupy exactly one IC layer. The force-directed approach tries

to gradually distribute the blocks evenly in space. Layer assignment is based on block positions on the z-axis, derived from the current placement obtained by the force-directed approach. Figure 4.22 illustrates the process of layer assignment for three blocks.

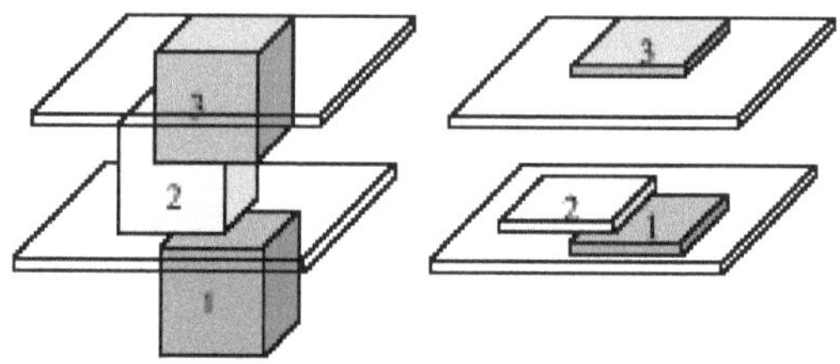

Fig. 4.22 Layer assignment

3. *Final Legalization:* After the global placement described in the previous sections, we arrive at a multi-layer packing solution with little residual overlap. To obtain a feasible placement, the legalization strategy perturbs the solution slightly to produce an overlap-free packing while attempting to maintain the original topological relationships among blocks. The legalization problem definition can be described in this way: construct the topological relations between overlapping blocks so that the displacements of blocks are minimized. The blocks are sorted according to their positions from the bottom-left to top-right corners of the chip to get a rough topological sequence. As shown in Fig. 4.23, block a is before block b in the sequence, and they overlap each other. We must determine whether block b should be right-of or above block a and choose the best orientation. The 2D representations introduced in the previous section can be used to represent the topological relation between blocks. In addition, the blocks can be rotated during the legalization process, while it can help to control the displacement caused by overlap removal. Since the topological relations between blocks are settled by some heuristic rules, the straightforward legalization may generate large displacement to the original placement. It is very natural to design a post-process to further improve the legalized results. The stochastic approach can be used to shuffle the packing locally so that the packing can be further optimized.

This force-directed analytical approach is effective in terms of wirelength optimization compared to simulated annealing, as shown in the next section. However, it

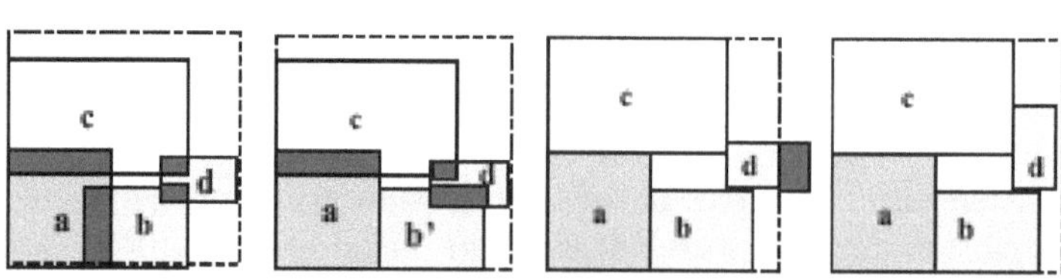

Fig. 4.23 Legalization process

faces two problems: (i) Satisfying the density constraints for the bins does not necessarily lead to legal 3D placement solution in terms of layer assignment. We may consider using the recent force-directed 3D placement formulated (to be presented in Section 5.4 and also published in [4]), which introduced density constraints on pseudo device layers to guarantee the legality of the placement into the discrete 3D layers. (ii) The force-directed analytical approach presented here has only been applied to 2D blocks. Extensions to handle 3D blocks require more study.

4.6 Effects of Various 3D Floorplanning Techniques

In this section we summarize the experimental results reported by various 3D floorplanners for both 2D blocks and 3D blocks.

4.6.1 Effects of 3D Floorplanning with 2D Blocks

Although a significant amount of work has been done on 3D floorplanning with 2D blocks, here we summarize the results of two representative algorithms which used a common set of examples: 3D floorplanner using simulated annealing based on CBA [3] and the force-directed 3D floorplanner [41]. All algorithms are tested on MCNC benchmarks and GSRC benchmarks. Four device layers are used for all circuits. We first compare the results of 3D floorplanning algorithms with 2D blocks using various representations without thermal awareness, as shown in Table 4.3. The wirelength is estimated using half parameter wirelength estimation (HPWL). Compared to the 3D floorplanner using simulated annealing based on CBA [3], the force-directed approach degrades area by 4%, improves wirelength by 12%, and completes execution in 69% of the time required by CBA, when averaged over all benchmarks.

Table 4.4 shows the comparison between CBA and the force-directed approach when optimizing area, wire length, and temperature. Here, the power density for each block is assigned between 10^5 and 10^7 W/m^2 [3]. An extended version of a spatially adaptive 3D multi-layer chip-package thermal analysis software package

Table 4.3 Area and wirelength optimization for two 3D floorplanner with 2D blocks

	CBA-based [3]			Force-directed [41]		
Circuit	Area (mm^2)	HPWL (mm)	Time (s)	Area (mm^2)	HPWL (mm)	Time (s)
Ami33	35.3	22.5	23	37.9	22	52
Ami49	1490	446.8	86	1349.1	437.5	57
N100	5.29	100.5	313	5.9	91.3	68
N200	5.77	210.3	1994	5.9	168.6	397
N300	8.90	315.0	3480	9.7	237.9	392
	1	1	1	+4%	−12%	−31%

Table 4.4 Comparison between CBA and force-directed approach when optimizing area, wire length, and temperature

	CBA-based [3]				Force-directed [41]			
Circuit	Area (mm^2)	HPWL (mm)	Temp (°C)	Time (s)	Area (mm^2)	HPWL (mm)	Temp (°C)	Time (s)
Ami33	43.2	23.9	212.4	486	41.5	24.2	201.3	227
Ami49	1672.6	516.4	225.1	620	1539.4	457.3	230.2	336
N100	6.6	122.9	172.7	4535	6.6	91.5	156.8	341
N200	6.6	203.7	174.7	6724	6.2	167.8	164.6	643
N300	10.4	324.9	190.8	18475	9.3	236.7	168.2	1394
	1	1		1	−16%	−12%		−75%

[37] is used as the thermal model to evaluate the thermal distribution. Here, the leakage power consumption is assumed to be fixed. But the temperature-dependent leakage power model can be applied to figure out the leakage–temperature feedback. Readers may refer to [41] for more information.

Figure 4.24 shows a four-layer packing obtained by the force-directed approach, with the corresponding power distribution and thermal profile. The blocks with the

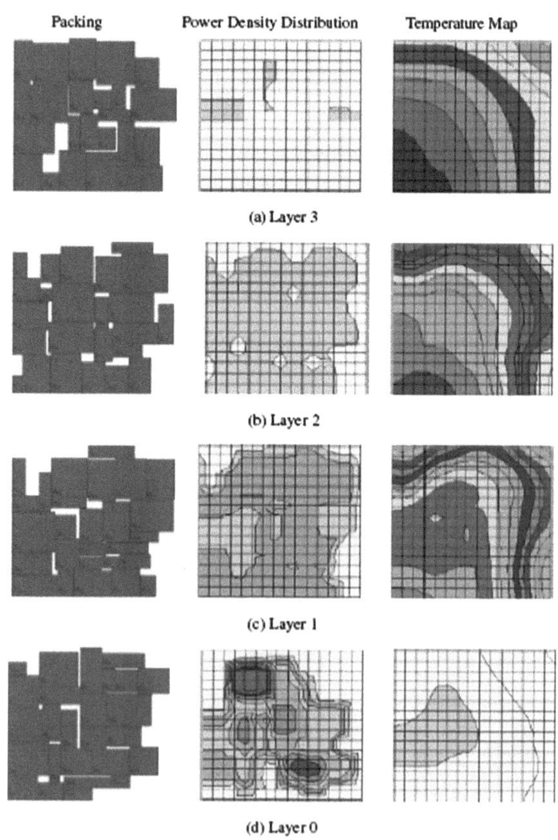

Fig. 4.24 A four-layer packing obtained by the force-directed approach, with the corresponding power distribution and thermal profile

high power density are assigned to the bottom layer to reduce peak temperature. Compared with SA-based approaches, the analytical approach is more stable, and it can obtain better results in shorter time. But the SA-based approach is more flexible for handling additional objectives and constraints.

4.6.2 *Effects of 3D Floorplanning with 3D Blocks*

Most of the published 3D floorplanning algorithms with 3D blocks tested their algorithms on the benchmark which is for knapsack problems in [17]. The weight factor is assumed as the third dimension, and therefore it is changed to a three-dimensional rectangle packing problem. Table 4.5 shows the comparison of results for three algorithms: 3D CBL, ST, and 3D-TCG. From the results, we can see that 3D CBL runs faster than the other two algorithms since 3D CBL has linear time complexity when constructing floorplanning from lists. But due to the limitation of solution space, the packing results of 3D CBL are not as good as 3D-sub TCG, especially in larger cases.

To show the effects of the thermal-aware 3D floorplanning algorithm with 3D blocks, the evaluation results are presented for a high-performance superscalar processor [12, 24]. Table 4.6 shows the baseline processor parameters used. Since each critical component has different implementations which can be represented as 3D blocks, the packing engine can pack the blocks successfully and choose the best implementation for each, with the layer number constraints. Figure 4.25a displays a 3D view of the floorplan for two-layer packing with 3D blocks. The area is 3.6 × 3.6 mm^2. The packing engine selects between single-layer or two-layer

Table 4.5 Comparison of results for three algorithms: 3D CBL, ST, and 3D-TCG

			ST		3D-sub TCG		3D CBL	
Test	No. of blocks	Sum of volume	Dead space (%)	Run time (S)	Dead space (%)	Run time (S)	Dead space (%)	Run Time (S)
beasley1	10	6218	28.6	7.7	17.1	8.5	23.5	6
beasley2	17	11497	21.5	45.2	7.2	28.5	17.0	7
beasley3	21	10362	35.3	44.1	18.0	18.0	17.0	12
beasley5	14	16734	26.4	18.2	11.5	16.0	13.5	12
beasley6	15	11040	26.3	27.9	16.3	24.8	15.4	20
beasley7	8	17168	30.1	3.8	16.5	2.3	24.6	4
beasley10	13	493746	25.2	13.0	14.2	10.8	15.2	10
beasley11	15	383391	24.8	17.5	12.6	9.8	13.2	10
beasley12	22	646158	29.9	100.0	21.5	58.5	21.2	40
okp1	50	$1.24*10^8$	42.6	1607.2	28.4	387.3	29.1	202
okp2	30	$8.54*10^7$	33.2	285.3	22.3	73.8	27.0	57
okp3	30	$1.23*10^8$	33.1	280.7	23.0	70.6	26.3	56
okp4	61	$2.38*10^8$	42.8	791.3	27.3	501.9	28.6	320
okp5	97	$1.89*10^8$	57.7	607.8	35.8	565.9	36.2	340

Table 4.6 Architectural parameters for the design driver

Processor width	6-way out-of-order superscalar, two integer execution clusters
Register files	128 entry integer (two replicated files), 128 entry FP
Data cache	8 KB 4-way set associative, 64B block size
Instruction cache	8 KB 2-way set associative, 32B block size
L2 cache	4 banks, each 128 KB 8-way set associative, 128B block size
Branch predictor	8 K entry gshare and a 1 K entry, 4-way BTB
Functional units	2 IntALU+1 Int MULT/DIV in each of two clusters; 1 FPALU and 1MULT/DIV

Fig. 4.25 A two-layer packing obtained by 3D floorplanner with 3D blocks based on 3D CBL representation

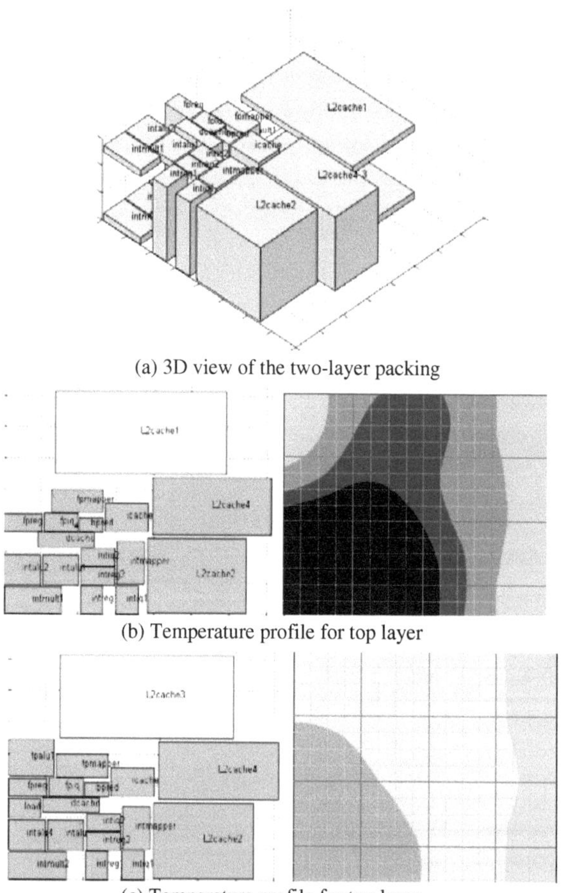

(a) 3D view of the two-layer packing

(b) Temperature profile for top layer

(c) Temperature profile for top layer

block architectures. For blocks such as ALU, MUL, and L2 cache units, single-layer implementation was selected. The rest of the blocks were implemented in two layers. (We use cubic blocks to represent multi-layer blocks.)

Figure 4.25 also shows the temperature profiles of the layers in the two-layer design, where the top layer is significantly hotter than the bottom layer with a hotspot temperature of 90°C. The bottom layer, which is in contact with the heat spreader and the heat sink, is cooler compared to the top layer. Although the bottom layer has a higher power density compared to the top layer, the thermal resistance from the top layer to the heat sink is higher. Even though silicon is considered to be a good thermal conductor, vertical heat conduction is negatively affected by the combination of metal layers, bonding materials, and the increased distances. Thermal vias that enable improved vertical heat conduction from the top layer to heat sink can be used to keep the hotspot temperatures below the given thermal thresholds.

Figure 4.26 illustrates the temperature comparison of the 2D and 3D architectural block technologies. The x-axis shows the different configurations with different silicon layers in the 3–6 GHz frequency range. The y-axis has the temperature in °C for 3D and 2D block technologies and the results of thermal via insertion. The ambient temperature is assumed to be 27°C. As the analysis in [12] shows, multi-layer 3D blocks can save about 10–30% power consumption over single-layer blocks. But temperature heavily relies on the layout. To relieve the hotspots, it is often necessary to keep potential hotspots away from one another. Even though single-layer blocks may seem to have advantages over multi-layer blocks in this, the 3D packing engine overcomes this issue by its intelligent layer selection for blocks depending on their thermal profile. Therefore, we can see that for two-layer and three-layer designs, the temperatures can be reduced due to the power reduction of multi-layer blocks and alternative selection.

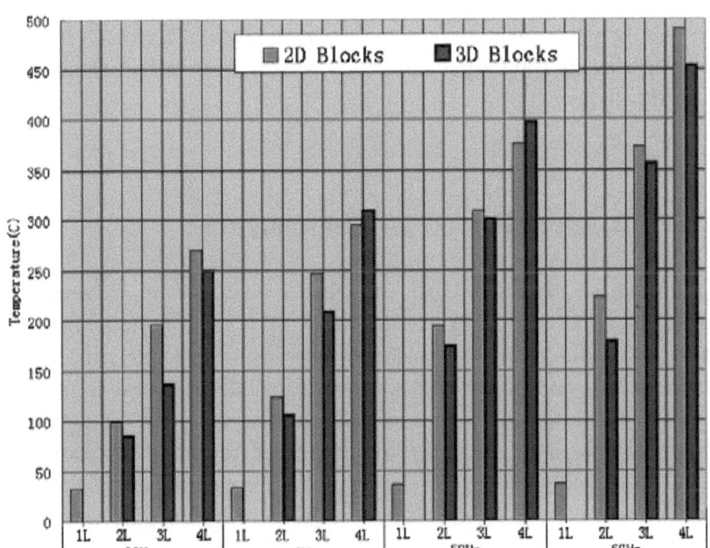

Fig. 4.26 Temperature comparison of 2D and 3D for 3–6 GHz and 1–4 layer cases

4.7 Summary and Conclusion

As a new integrated circuit (IC) design technology, the physical design of three-dimensional (3D) integration is now challenged by design methodologies and optimization orientations arising from the multiple device layer structure; this is in addition to those arising from the design complexities in deep submicron technology. In this chapter we introduce the algorithms for 3D floorplanning with both 2D blocks and 3D blocks.

According to the block representation, the 3D floorplanning problem can be classified into two types: 3D floorplan with 2D blocks and 3D floorplan with 3D blocks. As described in Section 4.2, these two types of 3D floorplanning need different representation and optimization techniques. Therefore, in Sections 4.3 and 4.4, we provide the introduction of representations for 2D blocks and 3D blocks, respectively.

Since the 3D floorplan with 2D blocks can be represented with an array of 2D representations, the 2D floorplanning algorithm can be extended to handle multi-layer designs by introducing new operations in optimization techniques. In Section 4.3, several basic 2D representations are introduced briefly; these are the fundamental techniques for 3D floorplanning optimization. The analysis on different representations shows the pros and cons of these representations. As described in Section 4.4, there are several typical representations: the 3D slicing tree, 3D CBL, sequence triple, and sequence quintuple are introduced to represent 3D packing with 3D blocks.

In Section 4.5, in addition to a brief introduction to the stochastic optimization based on various representations, the analytical approach is also introduced. We introduce simulated annealing as the typical optimization approach for 3D floorplanning with 2D/3D blocks. And the thermal-aware analytical approach introduced in this section applies the force-directed method which is normally used in placement with standard cells.

Appendix: Design of Folded 3D Components

Recent studies have provided block models for various architectural structures including 3D cache [30, 9, 28], 3D register files [31], 3D arithmetic units [25], and 3D instruction scheduler [26]. To construct multi-layer blocks to reduce intra-block interconnect latency and power consumption in architecture design, there are two main strategies for designing blocks in multiple silicon layers: *block folding (BF)* and *port partitioning (PP)*. Block folding implies a folding of blocks in the X- or Y-direction – potentially shortening the wirelength in one direction. Port partitioning places the access ports of a structure in different layers. The intuition here is that the additional hardware needed for replicated access to a single block entry (i.e., a multi-ported cache) can be distributed in different layers, which can greatly reduce the length of interconnect within each layer. As an example, the use of these strategies for cache-like blocks is briefly described. For all the other components, such as issue queue, register files, a similar analysis can be performed accordingly.

Caches are commonly found in architectural blocks with regular structures. They are composed of a number of tag and data arrays. Figure 4.27 shows a single cell for a three-ported structure. Each port contains bit, bitbar lines, a wordline, and two transistors per bit. The four transistors that make up the storage cell take much less space than that allocated for ports. The wire pitch is typically five times the feature size. For each extra port, the wirelength in both X- and Y-directions is increased by twice the wire pitch. On the other hand, the storage, which consists of four transistors, is twice the wire pitch in height and has a width equal to the wire pitch. Therefore, the more ports a component has, the larger the portion of the silicon area that is allocated to ports. A three-ported structure would have a port area to cell area ratio of approximately 18:1.

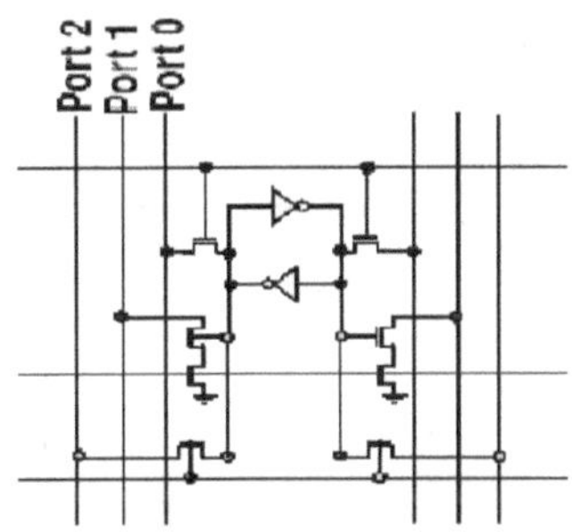

Fig. 4.27 Three-ported SRAM cell

Figure 4.28a demonstrates a high-level view of a number of cache tag and data arrays connected via address and data buses. Each vertical and horizontal line represents a 32-bit bus. It can be assumed that there are two ports on this cache, and therefore the lines are paired. The components of caches can easily be broken down into subarrays. CACTI [27, 30] can be used to explore the design space of different subdivisions and find an optimal point for performance, power, and area.

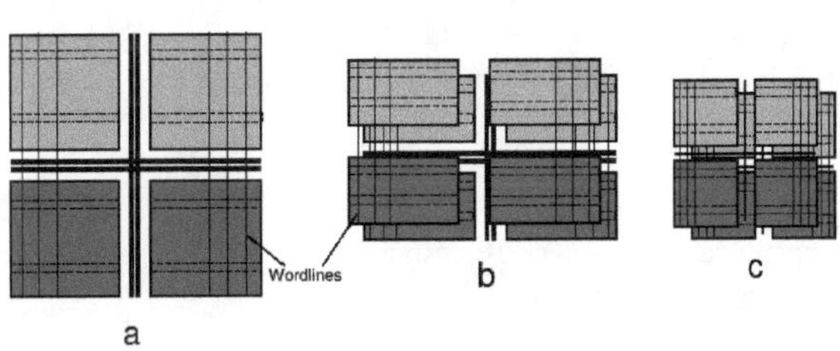

Fig. 4.28 Three-dimensional block alternatives for a cache: (**a**) 2D two-ported cache: the two *lines* denote the input/output wires of two ports; (**b**) Wordline folding: only Y-direction is reduced. Input/output of the ports is duplicated; (**c**) Port partitioning: ports are placed in two layers. Length in both X- and Y-directions is reduced

Block Folding (BF): For block folding, there are two folding options: wordline folding and bitline folding. In the former, the wordlines in a cache subarray are divided and placed onto different silicon layers. The wordline driver is also duplicated. The gain from wordline folding comes from the shortened routing distance from predecoder to decoder and from output drivers to the edge of the cache. Similarly, bitline folding places bitlines into different layers but needs to duplicate the pass transistor. Our investigation shows that wordline folding has a better access time and lower power dissipation in most cases compared to a realistic implementation of bitline folding. Here, the results using wordline folding are presented in Fig. 4.29.

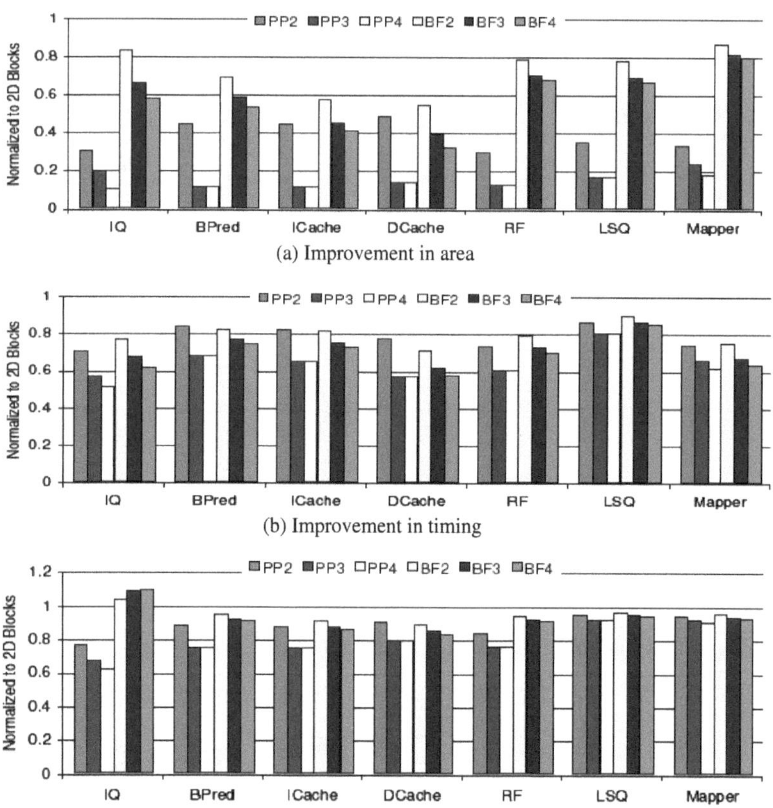

Fig. 4.29 Improvements for multi-layer F2B design (PP2 means port partition for 2 layer design, BF2 means block folding for 2 layer design)

Port Partitioning (PP): There is a significant advantage to partitioning the ports and placing them onto different layers, as shown in Fig. 4.28c. In a two-layer design, we can place two ports on one layer and one port and the SRAM cells on the other layer. The width and height are both approximately reduced by a factor of two and the area by a factor of four. Port partitioning allows reductions in both vertical and horizontal wirelengths. This reduces the total wirelength and capacitance, which translates into a savings in access time and power consumption. Port partitioning requires vias to connect the memory cell to ports in other layers. Depending on the technology, the via pitch can impact size as well. In our design, a space of 0.7 μm × 0.7 μm is allocated for each via needed. The same model as [30] is used to obtain via capacitance and resistance.

Figure 4.29 shows the effects of different partitioning strategies on different components. To summarize the effects, we can see

- Port partitioning is more effective for area reduction consistently, over all structures. This is because port partitioning reduces lengths in both x- and y-directions.
- For caches, port partitioning does not provide extra improvement in power or timing as the number of layers increases. This is because there are not as many ports on these caches. At the same time, the transistor layer must accommodate the size of vias. On the other hand, with wordline folding, the trend continues with consistent improvement for more numbers of layers.
- On an average, port partitioning performs better than block folding for area. Block folding is more effective in reducing the block delay, especially for the components with fewer ports. Port partitioning also performs better in reducing power.
- Though multi-layer blocks have reduced delay and power consumption compared to single-layer blocks, the worst-case power density may increase substantially by stacking circuits. Therefore, the power reduction in individual blocks alone cannot guarantee the elimination of hotspots. The thermal effect depends not only on the configuration of each block, but also on the physical information of the layout.

The diversity in benefit from these two approaches demonstrates the need for a tool to flexibly choose the appropriate implementation based on the constraints of an individual floorplan. With wire pipelining considered, the process of choosing the appropriate implementation should consider the physical information. The best 3D configuration of each component may not lead to the best 3D implementation for the whole system. In some cases, such as in a four-layer chip, if a component is chosen as a four-layer block, other blocks cannot be placed on top of it and the neighboring positions. Additionally, this block may not be enough for all the other highly connected blocks. Therefore, the inter-block wire latency may be increased and some extra cycles may be generated. On the other hand, if a two-layer implementation is chosen for this component, though the intra-block delay is not the best, the inter-block wire latency may be favored since other blocks that are heavily connected with this component can be placed immediately on top of the component, and the vertical interconnects are much shorter. Therefore, the packing with a two-layer implementation may perform better than the packing with a four-layer implementation of this component. Furthermore, to favor the thermal effect, the reduction in delay of 3D blocks may provide the latency slack to allow the trade-off between timing and power. But this optimization should also depend on the timing information which comes from physical packing results.

Therefore, to utilize 3D blocks, the decision cannot simply be made from the architecture side only or the physical design side only. To enable the co-optimization between 3D microarchitectural and physical design, we need a true 3D packing engine which can choose the implementation while performing the packing optimization.

Acknowledgments The authors would like to acknowledge the support from the Gigascale Silicon Research Center, IBM under a DARPA subcontract, the National Science Foundation under CCF-0430077 and CCF-0528583, the National Science Foundation of China under 60606007, 60720106003, 60728205, the Tsinghua Basic Research Fund under JC20070021, and the Tsinghua National Laboratory for Information Science and Technology (TNList) Cross-discipline Foundation under 042003011; this support led to a number of results reported in this chapter.

References

1. Y. C. Chang, Y. W. Chang, G. M. Wu, and S. W. Wu, B*-trees: An new representation for nonslicing floorplans, *Proceedings of ACM/IEEE DAC 2000*, pp. 458–463, 2000.
2. L. Cheng, L. Deng, and M. D. Wong, Floorplanning for 3D VLSI design, *Proceedings of IEEE/ACM ASP-DAC 2005*, pp. 405–411, 2005.
3. J. Cong, J. Wei, and Y. Zhang, A thermal-driven floorplanning algorithm for 3D ICs, *Proceedings of ICCAD 2004*, pp. 306–313, 2004.
4. J. Cong and G. Luo, A multilevel analytical placement for 3D ICs, *Proceedings of the 14th ASP-DAC*, Yokohama, Japan, pp. 361–366, January 2009.
5. B. Goplen and S. Sapatnekar, Efficient thermal placement of standard cells in 3D ICs using a force directed approach, *Proceedings of ICCAD 2003*, pp. 86–89, Nov. 2003.
6. P.-N. Guo, C.-K. Cheng, and T. Yoshimura, An O-tree representation of nonslicing floorplan and its application, *Proceedings of ACM/IEEE DAC 1999*, pp. 268–273, 1999.
7. X. Hong, G. Huang, Y. Cai, J. Gu, S. Dong, C. K. Cheng, and J. Gu, Corner block list: An effective and efficient topological representation of nonslicing floorplan, *Proceedings of IEEE/ACM ICCAD 2000*, pp. 8–12, 2000.
8. S. Kirkpatrick, C. D. Gelatt, and M. P. Vecchi Jr, Optimization by simulated annealing, Science, pp. 671–680, May 1983.
9. M. B. Kleiner, S. A. Kuhn, P. Ramm, and W. Weber, Performance and improvement of the memory hierarchy of risc-systems by application of 3-D technology, *IEEE Transactions on Components, Packaging, and Manufacturing Technology*, 19(4): 709–718, 1996
10. Z. Li, X. Hong, Q. Zhou, Y. Cai, J. Bian, H. Yang, P. Saxena, and V. Pitchumani, A divide-and-conquer 2.5-D floorplanning algorithm based on statistical wirelength estimation, *Proceedings of ISCAS 2005*, pp. 6230–6233, 2005.
11. Z. Li, X. Hong, Q. Zhou, Y. Cai, J. Bian, H. H. Yang, V. Pitchumani, and C.-K. Cheng, Hierarchical 3-D floorplanning algorithm for wirelength optimization, *IEEE Transaction on Circuits and Systems I*, 53(12): 2637–2646, 2007.
12. Y. Liu, Y. Ma, E. Kursun, J. Cong, and G. Reinman, Fine grain 3D integration for microarchitecture design through cube packing exploration, *Proceedings of IEEE ICCD 2007*, pp. 259–266, Oct 2007.
13. J. M. Lin and Y. W. Chang, TCG: A transitive closure graph-based representation for non-slicing floorplans, *Proceedings of ACM/IEEE DAC 2001*, pp. 764–769, 2001.
14. J. M. Lin and Y. W. Chang, TCG-S: Orthogonal coupling of P*-admissible representations for general floorplans, *Proceedings of ACM/IEEE DAC 2002*, pp. 842–847, 2002.

15. Y. Ma, X. Hong, S. Dong, Y. Cai, C. K. Cheng, and J. Gu, Floorplanning with abutment constraints and L-shaped/T-shaped blocks based on corner block list, *Proceedings of DAC 2001*, pp. 770–775, 2001.
16. Y. Ma, X. Hong, S. Dong and C. K.Cheng, 3D CBL: an efficient algorithm for general 3-dimensional packing problems, *Proceedings of the 48th MWS-CAS 2005*, 2, pp. 1079–1082, 2005.
17. F. K. Miyazawa and Y. Wakabayashi, An algorithm for the three-dimensional packing problem with asymptotic performance analysis, *Algorithmica*, 18(1): 122–144, May 1997.
18. H. Murata, K. Fujiyoshi, S. Nakatake, and Y. Kajitani, Rectangle packing based module placement, *Proceedings of IEEE ICCAD 1995*, pp. 472–479, 1995.
19. S. Nakatake, K. Fujiyoshi, H. Murata, and Y. Kajitani, Module placement on BSG-structure and IC laylot applications, *Proceedings of IEEE/ACM ICCAD 1999*, pp. 484–491, 1999.
20. T. Ohtsuki, N. Suzigama, and H. Hawanishi, An optimization technique for integrated circuit layout design, *Proceedings of ICCST 1970*, pp. 67–68, 1970.
21. B. Obermeier and F. Johannes, Temperature aware global placement, *Proceedings of ASPDAC 2004*, pp. 143–148, 2004.
22. R. H. J. M. Otten, Automatic floorplan design, *Proceedings of ACM/IEEE DAC 1982*, pp. 261–267, 1982.
23. R. H. J. M Otten, Efficient floorplan optimization, *Proceedings of IEEE ICCD 1983*, pp. 499–502, 1983.
24. S. Palacharla, N. P. Jouppi, and J. E. Smith, Complexity-effective superscalar processors, *Proceedings of the 24th ISCA*, pp. 206–218, June 1997.
25. K. Puttaswamy and G. Loh, The impact of 3-dimensional integration on the design of arithmetic units, *Proceedings of ISCAS 2006*, pp. 4951–4954, May, 2006.
26. K. Puttaswamy and G. Loh, Dynamic instruction schedulers in a 3-dimensional integration technology, *Proceedings of ACM/IEEE GLS-VLSI 2006*, pp. 153–158, May 1, 2006, USA.
27. G. Reinman and N. Jouppi, Cacti 2.0: An integrated cache timing and power model, *In Technical Report*, 2000.
28. R. Ronnen, A. Mendelson, K. Lai, S. Liu, F. Pollack, and J. Shen, Coming challenges in microarchitecture and architecture, *Proceedings of the IEEE*, 89(3): 325–340, 2001.
29. Z. C. Shen and C. C. N. Chu, Bounds on the number of slicing, mosaic, and general floorplans, *IEEE Transaction on CAD*, 22(10): 1354–1361, 2003.
30. Y. Tsai, Y. Xie, N. Vijaykrishnan, and M. Irwin, Three-dimensional cache design exploration using 3D CACTI, *Proceedings of ICCD 2005*, pp. 519–524, October 2005.
31. M. Tremblay, B. Joy, and K. Shin, A three dimensional register file for superscalar processors, *Proceedings of the 28th HICSS*, pp. 191–201, 1995.
32. X. Tang, R. Tian and D. F. Wong, Fast evaluation of Sequence Pair in block placement by longest common subsequence computation, *Proceedings of DATE 2000*, pp. 106–111, 2000.
33. X. Tang and D. F.Wong, FAST-SP: A fast algorithm for block placement based on Sequence Pair, *Proceedings of ASPDAC 2001*, pp. 521–526, 2001.
34. D. F. Wong and C. L. Liu, A new algorithm for floorplan design, *Proceedings of the 3rd ACM/IEEE DAC*, pp. 101–107,1986.
35. H. Yamazaki, K. Sakanushi, S. Nakatake, and Y. Kajitani, The 3D-packing by meta data structure and packing heuristics, IEICE Transaction on Fundamentals, E82-A(4): 639–645, 2000.
36. T. Yan, Q. Dong, Y. Takashima, and Y. Kajitani, How does partitioning matter for 3D floorplanning?, *Proceedings of the 16th ACM GLS-VLSI*, pp. 73-78, 2006.
37. Y. Yang, Z. P. Gu, C. Zhu, R. P. Dick, and L. Shang, ISAC: Integrated space and time adaptive chip-package thermal analysis, *IEEE Transaction on Computer-Aided Design of Integrated Circuits and Systems*, 26(1): 86–99, January 2007.
38. B. Yao, H. Chen, C. K. Cheng and R. Graham, Floorplan representations: complexity and connections, *ACM Transaction on Design Automation of Electronic Systems*, 8(1): 55–80, 2003.

39. E. F. Y. Young, C. C. N. Chu, and Z. C. Shen, Twin Binary Sequences: A nonredundant representation for general nonslicing floorplan, *IEEE Transaction on CAD*, 22(4): 457–469, 2003.
40. S. Zhou, S. Dong, C.-K. Cheng, and J. Gu, ECBL: An extended Corner Block List with solution space including optimum placement, *Proceedings of ISPD 2001*, pp. 150–155, 2001.
41. P. Zhou, Y. Ma, Z. Li, R. P. Dick, L. Shang, H. Zhou, X. Hong, and Q. Zhou, 3D-STAF: scalable temperature and leakage aware floorplanning for three-dimensional integrated circuits, *Proceedings of ICCAD 2007*, pp. 590–597, 2007.
42. C. Zhuang, Y. Kajitani, K. Sakanushi, and L. Jin, An enhanced Q-Sequence augmented with empty-room-insertion and parenthesis trees, *Proceedings of DATE 2002*, pp. 61–68, 2002.

Chapter 5
Thermal-Aware 3D Placement

Jason Cong and Guojie Luo

Abstract Three-dimensional IC technology enables an additional dimension of freedom for circuit design. Challenges arise for placement tools to handle the through-silicon via (TS via) resource and the thermal problem, in addition to the optimization of device layer assignment of cells for better wirelength. This chapter introduces several 3D global placement techniques to address these issues, including partitioning-based techniques, quadratic uniformity modeling techniques, multilevel placement techniques, and transformation-based techniques. The legalization and detailed placement problems for 3D IC designs are also briefly introduced. The effects of various 3D placement techniques on wirelength, TS via number, and temperature, and the impact of 3D IC technology to wirelength and repeater usage are demonstrated by experimental results.

5.1 Introduction

Placement is an important step in the physical design flow. The performance, power, temperature and routability are significantly affected by the quality of placement results. Three-dimensional IC technology brings even more challenges to the thermal problem: (1) the vertically stacked multiple layers of active devices cause a

---

J. Cong (✉)
UCLA Computer Science Department, California NanoSystems Institute, Los Angeles, CA 90095, USA
e-mail: cong@cs.ucla.edu

This chapter includes portions reprinted with permission from the following publications: (a) J. Cong, G. Luo, J. Wei, and Y. Zhang, Thermal-aware 3D IC placement via transformation, *Proceedings of the 2007 Conference on Asia and South Pacific Design Automation*, Yokohama, Japan, pp. 780–785, 2007, © 2007 IEEE. (b) J. Cong, and G. Luo, A multilevel analytical placement for 3D ICs, *Proceedings of the 2009 Conference on Asia and South Pacific Design Automation*, Yokohama, Japan, pp. 361–366, 2009, © 2009 IEEE. (c) B. Goplen and S. Sapatnekar, Placement of 3D ICs with thermal and interlayer via considerations, *Proceedings of the 44th annual conference on Design automation*, pp. 626–631, 2007, © 2007 IEEE.

Y. Xie et al. (eds.), *Three-Dimensional Integrated Circuit Design*,
Integrated Circuits and Systems, DOI 10.1007/978-1-4419-0784-4_5,

rapid increase in power density; (2) the thermal conductivity of the dielectric layers between the device layers is very low compared to silicon and metal. For instance, the thermal conductivity at room temperature (300 K) for SiO_2 is 1.4 W/mK [28], which is much smaller than the thermal conductivity of silicon (150 W/mK) and copper (401 W/mK). Therefore, the thermal issue needs to be considered during every stage of 3D IC designs, including the placement process. Thus, a thermal-aware 3D placement tool is necessary to fully exploit 3D IC technology. The reader may refer to Section 3.2 for a detailed introduction to thermal issues and methodologies on thermal analysis and optimization.

5.1.1 Problem Formulation

Given a circuit $H = (V, E)$, the device layer number K, and the per-layer placement region $R = [0, a] \times [0, b]$, where V is the set of cell instances (represented by vertices) and E is the set of nets (represented by hyperedges) in the circuit H (represented by a hypergraph), a placement (x_i, y_i, z_i) of the cell $v_i \in V$ satisfies that $(x_i, y_i) \in R$ and $z_i \in \{1, 2, ..., K\}$. The 3D placement problem is to find a placement (x_i, y_i, z_i) for every cell $v_i \in V$, so that the objective function of weighted total wirelength is minimized, subject to constraints such as overlap-free constraints, performance constraints and temperature constraints. In this chapter we focus on temperature constraints, as the performance constraints are similar to that of 2D placement. The reader may refer to [18, 35] for a survey and tutorial of 2D placement.

5.1.1.1 Wirelength Objective Function

The quality of a placement solution can be measured by the performance, power, and routability, but the measurement is not trivial. In order to model these aspects during optimization, the weighted total wirelength is a widely accepted metric for placement qualities [34, 35]. Formally, the objective function is defined as

$$\text{OBJ} = \sum_{e \in E} (1 + r_e) \cdot (\text{WL}(e) + \alpha_{\text{TSV}} \cdot \text{TSV}(e)) \tag{5.1}$$

The objective function depends on the placement $\{(x_i, y_i, z_i)\}$, and it is a weighted sum of the wirelength WL(e) and the number of through-silicon vias (TS vias) TSV(e) over all the nets. The weight $(1 + r_e)$ reflects the criticality of the net e, which is usually related to performance optimization. The unweighted wirelength is represented by setting r_e to 0. This weight is able to model thermal effects by relating it to the thermal resistance, electronic capacitance, and switching activity of net e [27].

The wirelength WL(e) is usually estimated by the half-perimeter wirelength [27, 19]:

$$\mathrm{WL}(e) = \left(\max_{v_i \in e}\{x_i\} - \min_{v_i \in e}\{x_i\}\right) + \left(\max_{v_i \in e}\{y_i\} - \min_{v_i \in e}\{y_i\}\right) \tag{5.2}$$

Similarly, TSV(e) is modeled by the range of $\{z_i\colon v_i \in e\}$ [27, 26, 19]:

$$\mathrm{TSV}(e) = \max_{v_i \in e}\{z_i\} - \min_{v_i \in e}\{z_i\} \tag{5.3}$$

The coefficient a_{TSV} is the weight for TS vias; it models a TS via as a length of wire. For example, 0.18 μm silicon-on-insulator (SOI) technology [22] evaluates that a 3 μm thickness TS via is roughly equivalent to 8–20 μm of metal-2 wire in terms of capacitance, and it is equivalent to about 0.2 μm of metal-2 wire in terms of resistance. Thus a coefficient α_{TSV} between 8 and 20 μm can be used for optimizing power or delay in this case.

5.1.1.2 Overlap-Free Constraints

The ultimate goal of overlap-free constraints can be expressed as the following:

$$\begin{aligned} &\left|x_i - x_j\right| \geq (w_i + w_j)/2 \\ &\qquad \text{or} \qquad\qquad \text{for all cell pairs } v_i,\ v_j \text{ with } z_i = z_j \\ &\left|y_i - y_j\right| \geq (h_i + h_j)/2 \end{aligned} \tag{5.4}$$

where (x_i, y_i, z_i) is the placement of cell i, and w_i and h_i are its width and height, respectively. The same applies to cell j. Such constraints were used directly in some analytical placers early on, such as [5].

However, this formulation leads to a number of $O(n^2)$ either–or constraints, where n is the total number of cells. This amount of constraint is not practical for modern large-scale designs.

To formulate and handle these pairwise overlap-free constraints, modern placers use a more scalable procedure to divide the placement into *coarse legalization* and *detailed legalization*. Coarse legalization relaxes the pairwise non-overlap constraints by using regional density constraints:

$$\sum_{\substack{\text{for all cell}_i \\ \text{with } z_i = k}} \mathrm{overlap}(\mathrm{bin}_{m,n,k}, \mathrm{cell}_i) \leq \mathrm{area}(\mathrm{bin}_{m,n,k}) \quad (\text{for all m, n, k}) \tag{5.5}$$

For a 3D circuit with K device layers, each layer is divided into $L \times M$ bins. If every $\mathrm{bin}_{l,m,k}$ satisfies inequality (5.5), the coarse legalization is finished. Examples of the density constraints on one device layer are given in Fig. 5.1.

After coarse legalization, the detailed legalization is to satisfy pairwise non-overlap constraints, using various discrete methods and heuristics, which will be described in Section 5.6.

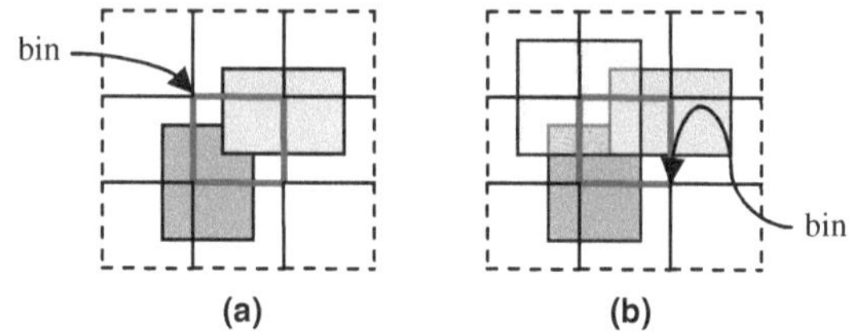

Fig. 5.1 (**a**) Density constraint is satisfied; (**b**) density constraint is not satisfied

5.1.1.3 Thermal Awareness

In existing literature, temperature issues are not directly formulated as constraints. Instead, a thermal penalty is appended to the wirelength objective function to control the temperature. This penalty can either be the weighted temperature penalty that is transformed to thermal-aware net weights [27], or the thermal distribution cost penalty [41], or the distance from the cell location to the heat sink during legalization [19].

In this chapter we will describe the thermal-aware net weights in Section 5.2, thermal distribution cost function in Section 5.3.3, and thermal-aware legalization in Section 5.6.2.2.

5.1.2 Overview of Existing 3D Placement Techniques

The state-of-the-art algorithms for 2D placement could be classified into flat placement techniques, top-down partitioning-based techniques, and multilevel placement techniques [35]. These techniques exhibit scalability for the growing complexity of modern VLSI circuits. In order to handle the scalability issues, these techniques divide the placement problem into three stages of global placement, legalization, and detailed placement. Given an initial solution, the global placement refines the solution until the cell area in every pre-defined region is not greater than the capacity of that region. These regions are handled in a top-down fashion from coarsest level to finest level by the partitioning-based techniques and the multilevel placement techniques, and are handled in a flat fashion at the finest level by the flat placement techniques. After the global placement, legalization proceeds to determine the specific location of all cells without overlaps, and the detailed placement performs local refinements to obtain the final solution.

As the modern 2D placement techniques evolve, a number of 3D placement techniques are also developed to address the issues of 3D IC technology. Most of the existing techniques, especially at the global placement stage, could be viewed as extensions of 2D placement techniques. We group the 3D placement techniques into the following categories:

- Partitioning-based techniques [21, 1, 3, 27] insert the partition planes that are parallel to the device layers at some suitable stages in the traditional partition-based process. The cost of partitioning is measured by a weighted sum of the estimated wirelength and the TS via number, where the nets are further weighted by thermal-aware or congestion-aware factors to consider temperature and routability.

- Flat placement techniques are mostly quadratic placements and their variations, including the force-directed techniques, cell-shifting techniques, and the quadratic uniformity modeling techniques. Since the unconstrained quadratic placement will introduce a great amount of cell overlaps, different variations are developed for overlap removal. The minimization of a quadratic function could be transformed to the problem of solving a linear system. The force-directed techniques [26, 33] append a vector, which is called the repulsive force vector, to the right-hand side of the linear system. These repulsive force vectors are equivalent to the electric field force where the charge distribution is the same as the cell area distribution. The forces are updated each iteration until the cell area in every pre-defined region is not greater than the capacity of that region. The cell-shifting techniques [29] are similar to the force-directed techniques, in the sense that they also append a vector to the right-hand side of the linear system. This vector is a result of the net force from pseudo pins, which are added according to the desired cell locations after cell shifting. The quadratic uniformity modeling techniques [41] append a density penalty function to the objective function, and it locally approximates the density penalty function by another quadratic function at each iteration, so that the whole global placement could be solved by minimizing a sequence of quadratic functions.
- The multilevel technique [13] constructs a physical hierarchy from the original netlist, and solves a sequence of placement problems from the coarsest level to the finest level.
- In addition to these techniques, the 3D placement approach proposed in [19] makes use of existing 2D placement results and constructs a 3D placement by transformation.

In the remainder of this chapter, we shall discuss these techniques in more detail. The legalization and detailed placement techniques specific to 3D placement are also introduced.

5.2 Partitioning-Based Techniques

Partitioning-based techniques [21, 1, 3, 27] can efficiently reduce TS via numbers with their intrinsic min-cut objective. These are constructive methods and can obtain good placement results even when I/O pad connectivity information is missing.

Partitioning-based placement techniques use a recursive two-way partitioning (bisection) approach applied to 3D circuits. At each step of bisection, a *partition* (V_0, R_0) consists of a subset of cells $V_0 \subseteq V$ in the netlist and a certain physical portion R_0 of the placement region R. When a partition is bisected, two new partitions (V_1, R_1) and (V_2, R_2) are created from the bisected list of cells $V_0 = V_1 \cup V_2$ and the bisected physical regions $R_0 = R_1 \cup R_2$, where the section plane is usually orthogonal to the axis of x, y, or z. A balanced bisection of the cell list V_0 into $V_1 \cup V_2$ is usually preferred, which satisfies a balance criterion on the area

$W_i = \sum_{v \in V_i} \text{area}(v)$ for $i = 1, 2$ such that $|W_1 - W_2| \leq \tau (W_1 + W_2)$ with tolerance τ. The area ratio between R_1 and R_2 relates to the cell area ratio between V_1 and V_2. After a certain amount of bisection steps, the regional density constraints defined in Section 5.1.1.2 are automatically satisfied due to the nature of the bisection process.

The placement solution of partitioning-based techniques is determined by the objective function of bisection and the choice of bisection direction, which are described below.

The idea of min-cut-based placement is to minimize the cut size between partitions, so that the cells with high connectives tend to stay in the same partition and stay close to each other for shorter wirelength.

For a bisection of (V_0, R_0) into $(V_1, R_1) \cup (V_2, R_2)$, a net is cut if it has both cells in R_1 and R_2. The total weighted cut size is $\sum_{e \text{ is cut}} (1 + r_e)$. The objective during bisection is to minimize the total weighted cut size, which can be solved using Fiduccia–Mattheyses (FM) heuristics [24] with a multilevel scheme hMetis [32].

Terminal propagation [23] is a successful technique for considering the external connections to the partition. A cell outside a partition is modeled by a fixed terminal on the boundary of this partition, where the location of the terminal is calculated as the closest location to the net center.

However, the cut size function does not directly reflect the wirelength objective function of the 3D placement problem defined in Section 5.1.1.1, where the cut size is unaware of the weights α_{TSV}. When the cut plane is orthogonal to the x-axis or the y-axis, the minimization of cut size only has an implicit effect on the 2D wirelength $\sum_{e \in E} (1 + r_e)\text{WL}(e)$; when the cut plane is orthogonal to the z axis, the cut size is equal to $\sum_{e \in E} (1 + r_e)\alpha_{\text{TSV}}\text{TSV}(e)$. *The only way to trade-off these two objectives is to control the order of bisection directions.* The studies in [21] note that the trade-off between total wirelength and TS via number could be achieved by varying the order of when the circuit is partitioned into device layers. Intuitively, partitioning in z dimension first will minimize the TSV number, while partitioning in x and y dimensions will minimize the total wirelength. References [21, 27] use the weighting factor α_{TSV} to determine the order of bisection direction. Assume the physical region is R, the cut direction for each bisection is selected as orthogonal to the largest of the width $|x_U - x_L|$, height $|y_U - y_L|$, or weighted depth $\alpha_{TSV} |z_U - z_L|$ of the region. By doing this, the min-cut objective minimizes the number of connections in the most costly direction at the expense of allowing higher connectivity in the less costly orthogonal directions.

Equation (5.6) shows a thermal awareness term [27] appended to the unweighted wirelength objective function. We will show that this function could be replaced by a weighted total wirelength.

$$\sum_{e \in E} (\text{WL}(e) + \alpha_{\text{TSV}}\text{TSV}(e)) + \alpha_{\text{TEMP}} \sum_{v_i \in V} T \qquad (5.6)$$

where T_j is the temperature of $cell_j$, and the temperature awareness $\alpha_{TEMP} \sum_{v_i \in V} T_i$ is considered during partitioning. However, using the temperature term directly in the objective function can result in expensive recalculations for each individual cell movement. Therefore, simplification needs to be made for enhanced efficiency. The total thermal resistance from cell v_i to ambient can be calculated as

$$R_i = \left(R_{\text{left},i}^{-1} + R_{\text{right},i}^{-1} + R_{\text{front},i}^{-1} + R_{\text{rear},i}^{-1} + R_{\text{bottom},i}^{-1} + R_{\text{top},i}^{-1}\right)^{-1} \tag{5.7}$$

where $R_{\text{left},\,i}, R_{\text{right},\,i}, R_{\text{front},\,i}, R_{\text{rear},\,i}, R_{\text{bottom},\,i}, R_{\text{top},\,i}$ are the approximated thermal resistances analyzed by finite difference method (FDM, Section 3.2.2.1) considering only heat conduction in that direction. For example, $R_{\text{left},\,i}$ is computed as the thermal resistance from the cell location (x_i, y_i, z_i) to the left boundary $(x = 0)$ of the 3D chip with cross-sectional area equal to cell width times cell thickness.

Thus the objective used in practice is

$$\begin{aligned} &\sum_{e \in E} (\text{WL}(e) + \alpha_{\text{TSV}}\text{TSV}(e)) + \alpha_{\text{TEMP}} \sum_{v_i \in V} \Delta T_i \\ &= \sum_{e \in E} (\text{WL}(e) + \alpha_{\text{TSV}}\text{TSV}(e)) + \alpha_{\text{TEMP}} \sum_{v_i \in V} R_i P_i \end{aligned} \tag{5.8}$$

where ΔT_j is the temperature contribution of v_i and it is a dominant term of T_j; R_i is the thermal resistance from v_i to ambient; P_i is the power dissipation of v_i. In order to achieve thermal awareness, the optimizations of P_i and R_i are performed.

The dynamic power associated with net e is

$$P_e = 0.5 a_e f V_{\text{DD}}^2 \left(C_{\text{per WL}}\text{WL}(e) + C_{\text{per TSV}}\text{TSV}(e) + C_{\text{per pin}} n_e^{\text{input pins}}\right) \tag{5.9}$$

where a_e is the activity factor, f is the clock frequency, V_{DD} is the supply voltage, $C_{\text{per WL}}$ is the capacitance per unit wirelength, $C_{\text{per TSV}}$ is the capacitance per TS via, $C_{\text{per pin}}$ is the capacitance per input pin, and $n_e^{\text{input pins}}$ is the number of cell input pins that net e drives. Because the inherent resistance of a cell is usually much larger than the wire resistance [27], the power P_e dissipates at the driver cell i and contributes to P_i. The sum of these power contributions is the total power dissipation of cell v_i:

$$\begin{aligned} P_i &= \sum_{\substack{\text{net } e \\ \text{driven by } v_i}} P_e \\ &= \sum_{\substack{\text{net } e \\ \text{driven by } v_i}} 0.5 a_e f V_{\text{DD}}^2 \left(C_{\text{per WL}}\text{WL}(e) + C_{\text{per TSV}}\text{TSV}(e) + C_{\text{per pin}} n_e^{\text{input pins}}\right) \end{aligned} \tag{5.10}$$

If dropping the terms $C_{\text{perpin}} n_e^{\text{input pins}}$, which are constant during optimization, and replacing $C_{\text{per TSV}}$ by $C_{\text{per WL}}\alpha_{\text{TSV}}$, where α_{TSV} is as defined in Section 5.1.1.1, Equation (5.8) can be expressed as

$$\begin{aligned}
&\sum_{e \in E} (\mathrm{WL}(e) + \alpha_{\mathrm{TSV}}\mathrm{TSV}(e)) + \alpha_{\mathrm{TEMP}} \sum_{v_i \in V} R_i P_i \\
&= \sum_{e \in E} (\mathrm{WL}(e) + \alpha_{\mathrm{TSV}}\mathrm{TSV}(e)) + \alpha_{\mathrm{TEMP}} \sum_{v_i \in V} R_i \\
&\quad \cdot \sum_{\substack{\text{net } e \\ \text{driven by } v_i}} 0.5 a_e f V_{\mathrm{DD}}^2 C_{\mathrm{per\,WL}} (\mathrm{WL}(e) + \alpha_{\mathrm{TSV}}\mathrm{TSV}(e)) \\
&= \sum_{e \in E} (\mathrm{WL}(e) + \alpha_{\mathrm{TSV}}\mathrm{TSV}(e)) + \alpha_{\mathrm{TEMP}} \sum_{e \in E} \\
&\quad \sum_{\substack{\text{cell } v_i \\ \text{driving net } e}} R_i \cdot 0.5 a_e f V_{\mathrm{DD}}^2 C_{\mathrm{per\,WL}} (\mathrm{WL}(e) + \alpha_{\mathrm{TSV}}\mathrm{TSV}(e)) \\
&= \sum_{e \in E} \left(1 + \alpha_{\mathrm{TEMP}} \sum_{\substack{\text{cell } v_i \\ \text{driving net } e}} R_i \cdot 0.5 a_e f V_{\mathrm{DD}}^2 C_{\mathrm{per\,WL}} \right) (\mathrm{WL}(e) + \alpha_{\mathrm{TSV}}\mathrm{TSV}(e))
\end{aligned} \tag{5.11}$$

Compared to the general weighted wirelength defined in Equation (5.1), these thermal-aware net weights can be implemented by setting

$$r_e = \alpha_{\mathrm{TEMP}} \sum_{\substack{\text{cell } v_i \\ \text{driving net } e}} R_i \cdot 0.5 a_e f V_{\mathrm{DD}}^2 C_{\mathrm{per\,WL}} \tag{5.12}$$

The thermal-aware net weight r_e is not a constant during the partitioning process. Instead, the thermal resistance R_i is determined by the distance between the cell v_i and the chip boundaries. A simple calculation [27] can be done by assuming that the heat flows in straight paths from the cell location toward the chip boundaries in all three directions, and the overall thermal resistance is calculated from these separated directional thermal resistances. These thermal resistances are evaluated during the partitioning process for the computation of the gain by moving a cell from one partition to another.

In addition to the thermal-aware net-weighting objective function, the temperature is also optimized by pseudo-nets that pull the cells to the heat sink [27].

5.3 Quadratic Uniformity Modeling Techniques

Different from the discrete partitioning-based techniques, the quadratic placement-based techniques are continuous. The idea is to relax the device layer assignment of a cell $z \in \{1, ..., K\}$ by a weaker constraint, where $z \in [1, K]$. The 3D placement problem is solved by minimizing a quadratic cost function, or finding the solution to a derived linear system. The regional density constraints are handled by appending a force vector to the linear system (force-directed techniques [26, 33] and cell-shifting techniques [29]) or appending a quadratic penalty to the quadratic cost function (quadratic uniformity modeling techniques [41]). The 3D global placement is solved by minimizing a sequence of quadratic cost functions. In this section, we will discuss the quadratic uniformity modeling techniques.

The complete placement flow is shown in Fig. 5.2. The flow is divided into global placement and detailed placement, where global placement is solved by the quadratic uniformity modeling technique, and the detailed placement can be solved with simple layer-by-layer 2D detailed placement or other advanced legalization and detailed placement techniques discussed in Section 5.6.

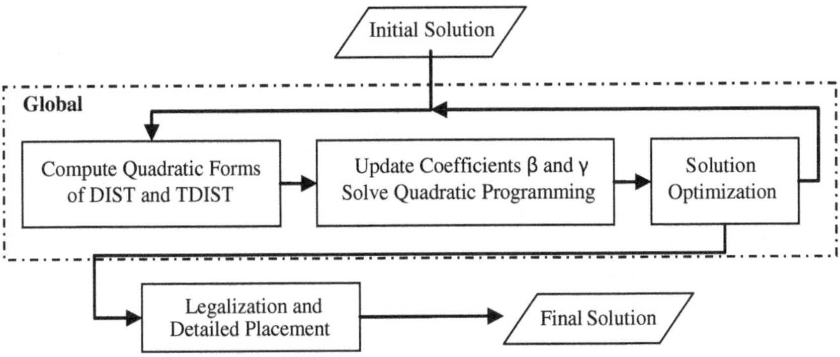

Fig. 5.2 Quadratic placement flow

The unified quadratic cost function is defined as

$$\mathrm{OBJ}^{+} = \mathrm{OBJ} + \beta \times \mathrm{DIST} + \gamma \times \mathrm{TDIST} \tag{5.13}$$

where OBJ is the wirelength objective defined in Section 5.1.1.1; DIST is the cell distribution cost; β is the weight of the cell distribution cost; TDIST is the thermal distribution cost; and γ is the thermal distribution cost. Moreover, all these functions OBJ, DIST, and TDIST are expressed in quadratic forms as in Equation (5.14), which will be explained in the following sections.

$$\begin{aligned}
\mathrm{OBJ} &= \sum_{i-1}^{n}\left(\sum_{j-1}^{n} q_{x,ij}x_i x_j + p_{x,i}x_i\right) + \sum_{i=1}^{n}\left(\sum_{j=1}^{n} q_{y,ij}y_i y_j + p_{y,i}y_i\right) \\
&\quad + \sum_{i=1}^{n}\left(\sum_{j=1}^{n} q_{z,ij}z_i z_j + p_{z,i}z_i\right) + r \\
\mathrm{DIST} &\approx \sum_{i=1}^{n}(a_{x,i}x_i^2 + b_{x,i}x_i) + \sum_{i=1}^{n}(a_{y,i}y_i^2 + b_{y,i}y_i) \\
&\quad + \sum_{i=1}^{n}(a_{z,i}z_i^2 + b_{z,i}z_i) + C \\
\mathrm{TDIST} &\approx \sum_{i=1}^{n}(a_{x,i}^{(T)}x_i^2 + b_{x,i}^{(T)}x_i) + \sum_{i=1}^{n}(a_{y,i}^{(T)}y_i^2 + b_{y,i}^{(T)}y_i) \\
&\quad + \sum_{i=1}^{n}(a_{z,i}^{(T)}z_i^2 + b_{z,i}^{(T)}z_i) + C^{(T)}
\end{aligned} \tag{5.14}$$

5.3.1 Wirelength Objective Function

In order to construct a quadratic wirelength function to approximate the wirelength objective defined in Section 5.1.1.1, the multiple-pin nets are decomposed to two-pin nets by either the star model or clique model. In the resulting graph, the quadratic wirelength is defined as

$$\mathrm{OBJ} = \sum_{\substack{e \in E \\ v_i, v_j \in e}} (1 + r_e)\left(\left(s_{e,x}(x_i - x_j)^2 + s_{e,y}(y_i - y_j)^2\right) + \alpha_{TSV} s_{e,z}(z_i - z_j)^2\right) \tag{5.15}$$

where $(1+r_e)$ is the net weight, and α_{TSV} is the TS via coefficient defined in Section 5.1.1.1; net e is the decomposed two-pin net connecting v_i at (x_i, y_i, z_i) and v_j at (x_j, y_j, z_j). The coefficients $s_{e,x}, s_{e,y}, s_{e,z}$ could linearize the quadratic wirelength to approximate the HPWL wirelength and the TS via number defined in Equations (5.2) and (5.3) [38].

It is obvious that this quadratic function OBJ can be rewritten in the matrix form:

$$\begin{aligned} \mathrm{OBJ} = &\sum_{i=1}^{n}\left(\sum_{j=1}^{n} q_{x,ij}x_i x_j + p_{x,i}x_i\right) + \sum_{i=1}^{n}\left(\sum_{j=1}^{n} q_{y,ij}y_i y_j + p_{y,i}y_i\right) \\ &+ \sum_{i=1}^{n}\left(\sum_{j=1}^{n} q_{z,ij}z_i z_j + p_{z,i}z_i\right) + r \end{aligned} \tag{5.16}$$

where x_i, y_i, z_i are the problem variables and the coefficients $q_{x,ij}, p_{x,i}, q_{y,ij}, p_{y,i}, q_{z,ij}, p_{z,i}$, and r can be directly computed from Equation (5.15). The coefficients $p_{x,i}, p_{y,i}, p_{z,i}$ and r are related to the locations of I/O pins and fixed cells in the circuit.

5.3.2 Cell Distribution Cost Function

The original idea of using discrete cosine transformation (DCT) to evaluate cell distribution and help spread cells is from [42] in 2D placement. This idea is extended and applied to 3D placement.

Similar to the bin density defined in Section 5.1.1.2, another bin density for the relaxed problem with continuous variables (z_i) is defined as

$$d_{m,n,l} = \frac{\sum_{\text{for all cell } i} \mathrm{intersection}(\mathrm{bin}_{m,n,l}, \mathrm{cell}_i)}{\mathrm{volume}(\mathrm{bin}_{m,n,l})} \tag{5.17}$$

Assuming a 3D circuit has K device layers, with die width W and die height H, the relaxed placement region $[0, W] \times [0, H] \times [0, K]$ is divided into $M \times N \times L$ bins, where cell_i at (x_i, y_i, z_i) is mapped to the region $[x_i - \frac{w_i}{2}, x_i + \frac{w_i}{2}] \times [y_i - \frac{h_i}{2}, y_i + \frac{h_i}{2}] \times [z_i, z_i + 1]$.

The 3D DCT transformation of $\{f_{p,q,v}\} = \text{DCT}\left(\{d_{m,n,l}\}\right)$ is defined as

$$f_{p,q,v} = \sqrt{\frac{8}{MNL}} C(p)C(q)C(v) \sum_{m=0}^{M-1} \sum_{n=0}^{N-1} \sum_{l=0}^{L-1} d_{m,n,l} \cos\left(\frac{(2m+1)p\pi}{2M}\right) \cos\left(\frac{(2n+1)q\pi}{2N}\right) \cos\left(\frac{(2l+1)v\pi}{2L}\right) \tag{5.18}$$

where m, n, l are the coordinates in the spatial domain, and p, q, v are coordinates in the frequency domain. The coefficients are $C(v) = \begin{cases} 1/\sqrt{2} & t = 0 \\ 1 & \text{otherwise} \end{cases}$

The cell distribution cost is defined as

$$\text{DIST} = \sum_{p,q,t} u_{p,q,t} f_{p,q,t}^2 \tag{5.19}$$

where $u_{p,q,t} = 1/(p+q+t+1)$ is set heuristically.

Note that (5.19) is not a quadratic function with respect to the placement variables (x_i, y_i, z_i). In order to construct a quadratic form, approximation is made as follows:

$$\text{DIST} \approx \sum_{i=1}^{n} (a_{x,i}x_i^2 + b_{x,i}x_i) + \sum_{i=1}^{n} (a_{y,i}y_i^2 + b_{y,i}y_i) + \sum_{i=1}^{n} (a_{z,i}z_i^2 + b_{z,i}z_i) + C \tag{5.20}$$

Although the coefficients $a_{x,i}, b_{x,i}, a_{y,i}, b_{y,i}, a_{z,i}, b_{z,i}$ depend on the intermediate placement, they are assumed to be constant in this quadratic function. These coefficients are updated when the intermediate placement changes. Since the variables are decoupled well in this approximation, the coefficients can be computed one by one. To compute $a_{x,i}, b_{x,i}$ all the variables except x_i can be fixed, thus the cost function is a quadratic function of x_i:

$$\text{DIST}(x_i) \approx a_{x,i}x_i^2 + b_{x,i}x_i + C'_{i,x} \tag{5.21}$$

The three coefficients $a_{x,i}, b_{x,i}$, and $C'_{i,x}$ are computed from the three costs $\text{DIST}(x_i)$, $\text{DIST}(x_i - \delta)$, and $\text{DIST}(x_i + \delta)$. Through the computation, we can see that the first-order and second-order derivatives of the quadratic approximation satisfy

$$\begin{aligned} 2a_{x,i}x_i + b_{x,i} &= \frac{\text{DIST}(x_i+\delta) - \text{DIST}(x_i-\delta)}{2\delta} \approx \frac{\partial \text{DIST}(x_i)}{\partial x_i} \\ 2a_{x,i} &= \frac{\text{DIST}(x_i+\delta) - 2\text{DIST}(x_i) + \text{DIST}(x_i-\delta)}{\delta^2} \approx \frac{\partial^2 \text{DIST}(x_i)}{\partial x_i^2} \end{aligned} \tag{5.22}$$

so that the first-order and second-order derivatives of this quadratic function locally approximates the first-order and second-order derivatives of the area distribution cost function DIST, respectively.

The computation of multiple DIST functions avoids 3D DCT transformation by pre-computations [42]. It spends $O(M^2 N^2 L^2)$ space for $O(n)$ runtime during the computation of matrix coefficients in Equation (5.20).

5.3.3 Thermal Distribution Cost Function

The thermal cost is treated like cell distribution cost, by replacing the cell densities $\{d_{m,n,l}\}$ with thermal densities $\{t_{m,n,l}\}$. The thermal density is defined as

$$t_{m,n,l} = T_{m,n,l} / T_{avg} \tag{5.23}$$

where $T_{m,n,l}$ is the average temperature in $\text{bin}_{m,n,l}$, and T_{avg} is the average temperature of the whole chip.

As the cell distribution cost, the thermal distribution is transformed by 3D DCT, and the distribution cost function is approximated by a quadratic form.

Besides the computation of matrix coefficients in the quadratic approximation of thermal distribution function TDIST, another significant cost of runtime is the computation of thermal densities $\{t_{m,n,l}\}$, because an accurate computation requires thermal analysis. To save runtime from the thermal analysis during computing $\text{TDIST}(x_i)$, $\text{TDIST}(x_i - \delta)$, $\text{TDIST}(x_i + \delta)$, etc., approximation is made in computing a new $\{t_{m,n,l}\}$. The work in [41] uses two methods of approximation, both of which may be lack of accuracy but are fast to be integrated in the distribution cost computation.

The first approximation makes use of *thermal contribution* of cells. Let $P_{\text{bin}}(i)$ and $T_{\text{bin}}(i)$ be the power and average temperature in $\text{bin}_{m(i),n(i),l(i)}$, the thermal contribution of a cell in this bin is defined as

$$T_{\text{cell}} = \frac{P_{\text{cell}}}{P_{\text{bin}}(i)} \cdot T_{\text{bin}}(i) \tag{5.24}$$

When the cell is moved from $\text{bin}_{m(i),n(i),l(i)}$ to $\text{bin}_{m(j),n(j),l(j)}$, the temperature of bins are updated as

$$\begin{aligned} T_{\text{bin}}(i) &\leftarrow T_{\text{bin}}(i) - \beta \cdot T_{\text{cell}} \\ T_{\text{bin}}(j) &\leftarrow T_{\text{bin}}(j) - \beta \cdot T_{\text{cell}} \end{aligned} \tag{5.25}$$

where $\beta = l(j)/l(i)$ is the influence of the cell on bin temperature.

The second approximation updates the bin temperature in the same ratio as the power density updates as

$$T'_{\text{bin}}(i) = \frac{P'_{\text{bin}}(i)}{P_{\text{bin}}(i)} \cdot T_{\text{bin}}(i) \tag{5.26}$$

5.4 Multilevel Placement Technique

Multilevel heuristics [15] have proved to be effective in large-scale designs. The application of multilevel heuristics to the partitioning problem [32] also shows that it could also improve the solution quality; this is also implicated by the partitioning-based techniques discussed in Section 5.2. Moreover, the solvers of the quadratic placement-based problem usually apply the multigrid method, which is the origin of multilevel heuristics.

In this section, we will introduce an analytical 3D placement engine that explicitly makes use of multilevel heuristics.

5.4.1 3D Placement Flow

The overall placement flow is shown in Fig. 5.3. The global placement starts from scratch or takes in the given initial placement. The global placement incorporates the analytical placement engine (Section 5.4.2) into the multilevel framework that is used in [15]. The global placement is then processed layer-by-layer with the 2D detailed placer [16] to obtain the final placement.

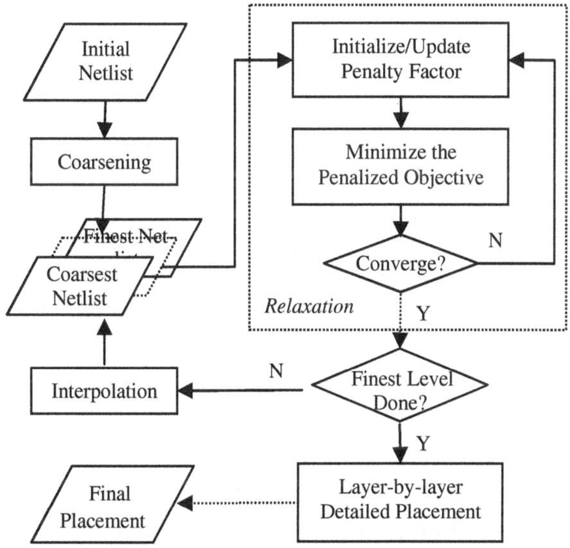

Fig. 5.3 Multilevel analytical 3D placement flow

5.4.2 Analytical Placement Engine

Analytical placement is not a unique engine for multilevel heuristics. In fact, any flat 3D placement technique like the one introduced in Section 5.3 can also be used.

In this section, we focus on the analytical engine [13] which was the first work to apply multilevel heuristics for 3D placement.

The analytical placement engine solves 3D global placement problem by transforming the non-overlap constraints to density penalties.

$$\begin{aligned} &\text{minimize} \sum_{e \in E} (\mathrm{WL}(e) + \alpha_{\mathrm{TSV}} \cdot \mathrm{TSV}(e)) \\ &\text{subject to Penalty } (\vec{x}, \vec{y}, \vec{z}) = 0 \end{aligned} \tag{5.27}$$

The wirelength WL(e) (Section 5.4.2.2), the TS via number TSV(e) (Section 5.4.2.3), and the density penalty function *Penalty* $(\vec{x}, \vec{y}, \vec{z})$ (Section 5.4.2.4) will be described in the following sections in detail.

In order to solve this constrained problem, penalty methods [37] are usually applied:

$$\mathrm{OBJ}(\vec{x}, \vec{y}, \vec{z}) = \sum_{e \in E} (\mathrm{WL}(e) + \alpha_{\mathrm{TSV}} \cdot \mathrm{TSV}(e)) + \mu \cdot \mathrm{Penalty}\ (\vec{x}, \vec{y}, \vec{z}) \tag{5.28}$$

This penalized objective function is minimized by each iteration, with a gradually increasing penalty factor μ to reduce the density violations. It can be shown that the minimizer of Equation (5.28) is equivalent to problem (5.27) when $\mu \to \infty$ if the penalty function is non-negative.

5.4.2.1 Relaxation of Discrete Variables

As mentioned in Section 5.1.1, the placement variables are represented by triples (x_i, y_i, z_i), where z_i is a discrete variable in $\{1, 2, ..., K\}$. The range of z_i is relaxed from the set $\{1, 2, \ldots, K\}$ to a continuous interval $[1, K]$. After relaxation, a nonlinear analytical solver can be used in our placement engine. The relaxed solution is mapped back to the discrete values before the detailed placement phase.

5.4.2.2 Log-Sum-Exp Wirelength

The half-perimeter wirelength WL(e) defined in Equation (5.2) is replaced by a differentiable approximation with the log-sum-exp function [4], which is introduced to placement by [36]

$$\begin{aligned} \mathrm{WL}(e) \approx \eta(&\log \sum_{v_i \in e} \exp(x_i/\eta) + \log \sum_{v_i \in e} \exp(-x_i/\eta) \\ &+ \log \sum_{v_i \in e} \exp(y_i/\eta) + \log \sum_{v_i \in e} \exp(-y_i/\eta)) \end{aligned} \tag{5.29}$$

For numerical stability, the placement region R is scaled into $[0, 1] \times [0, 1]$, thus variables of (x_i, y_i) are in a range between 0 and 1, and the parameter η is set to 0.01 in implementation as [6].

5.4.2.3 TS Via Number

The TS via number $TSV(e)$ estimation defined in Equation (5.3) is also replaced by the log-sum-exp approximation:

$$\mathrm{TSV}(e) \approx \eta(\log \sum_{v_i \in e} \exp(z_i/\eta) + \log \sum_{v_i \in e} \exp(-z_i/\eta)) \tag{5.30}$$

5.4.2.4 Density Penalty Function

The density penalty function is for overlap removal in both the (x, y)-direction and the z-direction. The minimization of the density penalty function should lead to a non-overlap placement in theory.

Assume that every cell v_i has a legal device layer assignment (i.e., $z_i \in \{1, 2, \ldots, K\}$), then we can define K density functions for these K device layers. Intuitively, the density function $D_k(u, v)$ indicates the number of cells that cover the point (u, v) on the k-th device layer. This is defined as

$$D_k(u, v) = \sum_{i:z_i = k} d_i(u, v) \tag{5.31}$$

which is the sum of the density contribution $d_i(u, v)$ of cell v_i assigned to this device layer at point (u, v). The density contribution $d_i(u, v)$ is 1 inside the area occupied by v_i, and is 0 outside this area. An example is given in Fig. 5.4 showing the density function with two overlapping cells.

During global placement, it is possible that cell v_i stays between two device layers, so that the variable $z_i \in [1, K]$ is not aligned to any of the two device layers. We borrow the idea from the bell-shaped function in [31] to define the density function for this case:

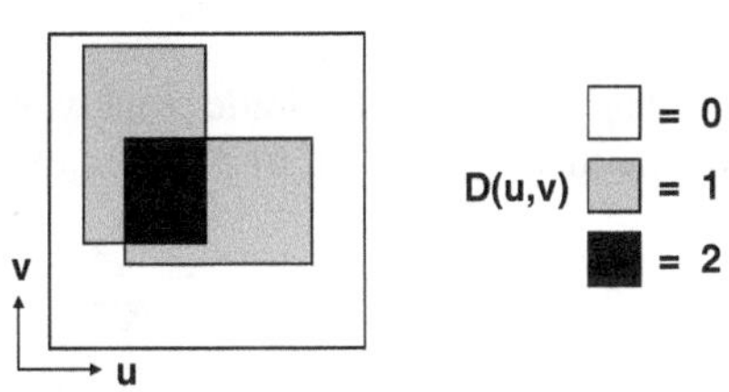

Fig. 5.4 An example of the density function

$$D_k(u,v) = \sum_i \eta(k, z_i) d_i(u,v), \text{ for } 1 \le k \le K \tag{5.32}$$

where

$$\eta(k,z) = \begin{cases} 1 - 2(z-k)^2 & |z-k| \le 1/2 \\ 2(|z-k| - 1)^2 & 1/2 < |z-k| \le 1 \\ 0 & \text{otherwise} \end{cases} \tag{5.33}$$

We call (5.33) the *bell-shaped density projection function*, which extends the density function (5.31) from integral layer assignments to the definition (5.32) for relaxed layer assignments. It is obvious that (5.32) is consistent with (5.31) when the layer assignments $\{z_i\}$ are integers.

An example of how this extension works for a four-layer 3D placement is given in Fig. 5.5. The x-axis is the relaxed layer assignment in z-direction, while the y-axis indicates the amount of area to be projected in the actual device layers. The four curves, the dash-dotted curve, the dotted curve, the solid curve and the dashed curve, represent the functions $\eta(1, z), \eta(2, z), \eta(3, z)$ and $\eta(4, z)$ for device layers 1, 2, 3, and 4, respectively. In this example, a cell is temporarily placed at $z = 2.316$ (the triangle on the x-axis) between layer 2 and layer 3. The bell-shaped density projection functions project 80% of its area to layer 2 (the upper triangle on the y-axis) and 20% of its area to layer 3 (the lower triangle on the y-axis). In this way, we establish a mapping from a relaxed 3D placement to the area distributions in discrete layers.

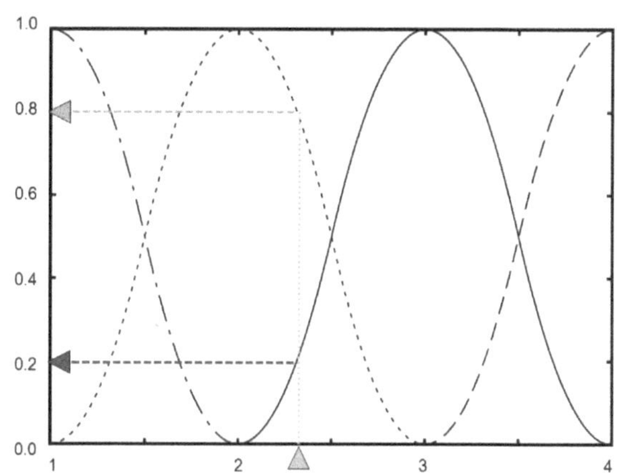

Fig. 5.5 An example of the *bell-shaped* density projections

Inspired by the quadratic penalty terms in 2D placement methods [6, 31, 9], we define this *density penalty function* to measure the amount of overlaps:

$$P(\vec{x}, \vec{y}, \vec{z}) = \sum_{k=1}^{K} \int_0^1 \int_0^1 (D_k(u,v) - 1)^2 \, du dv \tag{5.34}$$

Lemma 1 *Assume the total area of cells equals the placement area (i.e., $\sum_i \text{area}(v_i) = K$, no empty space), every legal placement $(\vec{x}^*, \vec{y}^*, \vec{z}^*)$, which satisfies $D_k(u,v) = 1$ for every k and (u,v) without any non-integer z_i^*, is a minimizer of $P(\vec{x}, \vec{y}, \vec{z})$.*

The proof of Lemma 1 is trivial and thus is omitted. Therefore, minimizing $P(\vec{x}, \vec{y}, \vec{z})$ provides a necessary condition for a legal placement. However, there exist minimizers that cannot form a legal placement. An example is shown in Fig. 5.6 where placement (b) also minimizes the density penalty function but it is not legal.

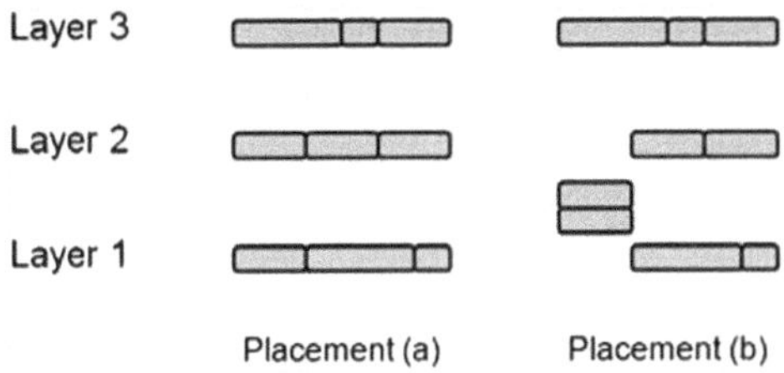

Fig. 5.6 Two placements with the same density penalties

To avoid reaching such minimizers, we introduce the *interlayer density function*:

$$E_k(u,v) = \sum_i \eta(k + 0.5, z_i) d_i(u,v), \text{for } 1 \leq k \leq K - 1 \tag{5.35}$$

and also the interlayer density penalty function:

$$Q(\vec{x}, \vec{y}, \vec{z}) = \sum_{k=1}^{K-1} \int_0^1 \int_0^1 (E_k(u,v) - 1)^2 \, \mathrm{d}u \mathrm{d}v \tag{5.36}$$

Similar to the density penalty function $P(\vec{x}, \vec{y}, \vec{z})$, the following *Lemma 2* is also true.

Lemma 2 *Assume the total area of cells equals the placement area, every legal placement is a minimizer of $Q(\vec{x}, \vec{y}, \vec{z})$.*

Combining the density penalty functions $P(\vec{x}, \vec{y}, \vec{z})$ and $Q(\vec{x}, \vec{y}, \vec{z})$, we define the following density penalty function:

$$\text{Penalty}(\vec{x}, \vec{y}, \vec{z}) = P(\vec{x}, \vec{y}, \vec{z}) + Q(\vec{x}, \vec{y}, \vec{z}) \tag{5.37}$$

Theorem 1 *Assume the total area of cells equals the placement area, every legal placement $(\vec{x}^*, \vec{y}^*, \vec{z}^*)$ is a minimizer of* Penalty $(\vec{x}, \vec{y}, \vec{z})$, *and vice versa.*

Proof It is obvious that every legal placement is a minimizer of Penalty $(\vec{x}, \vec{y}, \vec{z})$ by combining *Lemma 1* and *Lemma 2*. We shall prove that every minimizer

$(\vec{x}^*, \vec{y}^*, \vec{z}^*)$ of Penalty $(\vec{x}, \vec{y}, \vec{z})$ is a legal placement. From the proof of *Lemma 1* and *Lemma 2*, we know the minimum value of Penalty $(\vec{x}, \vec{y}, \vec{z})$ is achieved if and only if $D_k(u, v) = 1$ and $E_k(u, v) = 1$ for every k and (u, v). First, if all the components of $\vec{z}^*$ are integers, it is easy to see that the placement is legal, because all the cells are assigned to a certain device layer, and for any point (u, v) on any device layer k there is only one cell covering this point (no overlaps).

Next, we show that there does not exist a z_i^* with a non-integer value (proof by contradiction). If a cell v_i has a non-integer z_i^*, we know that there are K cells covering (x_i^*, y_i^*) because $\sum_{k=1}^{K} D_k(x_i^*, y_i^*) = K$. According to the pigeon-hole principal, among these K cells there are at least two cells v_{i1}, v_{i2} with the z-direction distance $\left|z_{i1}^* - z_{i2}^*\right| < 1$, since all the variables $\{z_i^*\}$ are in the range of $[1, K]$. Without loss of generality we may assume $z_{i1}^* \leq z_{i2}^*$, therefore there exists an integer $k \in \{1, 2, \ldots K\}$ such that either $z_{i1}^* \in (k, k+0.5]$ and $z_{i2}^* \in (k, k+1.5)$, or $z_{i1}^* \in (k-0.5, k]$ and $z_{i2}^* \in (k-0.5, k+1)$. It is easy to verify that in the former case $\left|z_{i1}^* - (k+0.5)\right| + \left|z_{i2}^* - (k+0.5)\right| < 1$ and $E_k(x_i^*, y_i^*) \geq \eta(k+0.5, z_{i1}^*) + \eta(k+0.5, z_{i2}^*) > 1$; in the latter case $\left|z_{i1}^* - k\right| + \left|z_{i2}^* - k\right| < 1$ and $D_k(x_i^*, y_i^*) \geq \eta(k, z_{i1}^*) + \eta(k, z_{i2}^*) > 1$. Both cases lead to either $E_k(x_i^*, y_i^*) > 1$ or $D_k(x_i^*, y_i^*) > 1$, which conflict with the assumption that $(\vec{x}^*, \vec{y}^*, \vec{z}^*)$ is a minimizer of Penalty $(\vec{x}, \vec{y}, \vec{z})$.

Therefore there does not exist a non-integer z_i^*, and every minimizer of Penalty $(\vec{x}, \vec{y}, \vec{z})$ is a legal placement in the z-dimension. ■

In the analytical placement engine, the densities $D_k(u, v)$ and $E_k(u, v)$ are replaced by smoothed densities $\widehat{D}_k(u, v)$ and $\widehat{E}_k(u, v)$ for differentiability. As in [6], the densities are smoothed by solving Helmholtz equations:

$$
\begin{aligned}
\widehat{D}_k(u, v) &= -\left(\frac{\partial^2}{\partial u^2} + \frac{\partial^2}{\partial v^2} - \varepsilon\right)^{-1} D_k(u, v) \\
\widehat{E}_k(u, v) &= -\left(\frac{\partial^2}{\partial u^2} + \frac{\partial^2}{\partial v^2} - \varepsilon\right)^{-1} E_k(u, v)
\end{aligned}
\tag{5.38}
$$

and the smoothed density penalty function

$$
\begin{aligned}
\text{Penalty}(\vec{x}, \vec{y}, \vec{z}) = &\sum_{k=1}^{K} \int_0^1 \int_0^1 \left(\widehat{D}_k(u, v) - 1\right)^2 \mathrm{d}u\,\mathrm{d}v \\
&+ \sum_{k=1}^{K-1} \int_0^1 \int_0^1 \left(\widehat{E}_k(u, v) - 1\right)^2 \mathrm{d}u\,\mathrm{d}v
\end{aligned}
\tag{5.39}
$$

is used in our implementation, whose gradient is computed efficiently with the method in [12].

5.4.3 Multilevel Framework

The optimization problem below summarizes our analytical placement engine:

$$\begin{cases} \text{minimize} & \sum\limits_{e \in E} (\mathrm{WL}(e) + \alpha_{\mathrm{TSV}} \mathrm{TSV}(e)) \\ & + \mu \left(\sum\limits_{k=1}^{K} \int_0^1 \int_0^1 \left(\widehat{D}_k(u,v) - 1 \right)^2 \mathrm{d}u\mathrm{d}v \right. \\ & \left. + \sum\limits_{k=1}^{K-1} \int_0^1 \int_0^1 \left(\widehat{E}_k(u,v) - 1 \right)^2 \mathrm{d}u\mathrm{d}v \right) \\ \text{increase } \mu \text{ until the density penalty is small enough} & \end{cases} \tag{5.40}$$

This analytical engine is incorporated into the multilevel framework in [15], which consists of coarsening, relaxation, and interpolation.

The purpose of coarsening is to build a hierarchy for the multilevel diagram, where we use the best-choice hypergraph clustering [2].

After the hierarchy is set up, multiple placement problems are solved from the coarsest level to the finest level. In a coarser level, clusters are modeled as cells and the connections between clusters are modeled as nets, so that there is one placement problem for each level. The placement problem at each level is solved (relaxed) by the analytical engine (5.40).

These placement problems are solved in the order from the coarsest level to the finest level, where the solution at a coarser level is interpolated to obtain an initial solution of the next finer level. The cell with highest degree in a cluster is placed in the center of this cluster (C-points), while the other cells are placed at the weighted average locations of their neighboring C-points, where the weights are proportional to the connectivity to those clusters.

5.5 Transformation-Based Techniques

The basic idea of transformation-based approaches [19] is to generate 3D thermal-aware placement from existing 3D placement results in a two-step procedure: 3D transformation and refinement through layer reassignment. In this section we will introduce 3D transformation, including local stacking transformation, folding-based transformation, and the window-based stacking/folding transformation. The refinement through layer reassignment is general to all techniques and will be introduced in Section 5.6.3.

The framework of transformation-based 3D placement techniques is shown in Fig. 5.7. The components with a dashed boundary are the existing 2D placement tools that the transformation-based approaches make use of. A 2D wirelength-driven and/or thermal-driven placer is first used to generate a 2D placement for the target design, in a placement region with area equal to the total 3D placement areas. The quality of the final 3D placement highly depends on this initial placement. The 2D placement is then transformed into a legalized 3D placement according to the given

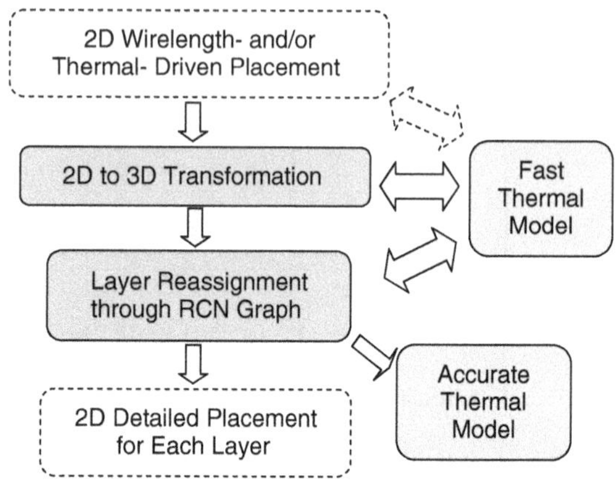

Fig. 5.7 Framework of transformation-based techniques

3D technology. During the transformation wirelength, TS via number, and temperature are considered. A refinement process through layer reassignment will be carried out after 3D transformation to further reduce the TS via number and bring down the maximum on-chip temperature. Finally, a 2D detailed placer will further refine the placement result for each device layer.

The transformation-based techniques start with a 2D placement with the placement area K times larger than one device layer of the 3D chip, where K is the number of device layers. Given a 2D placement solution with optimized wirelength, we may perform *local stacking transformation* to achieve even shorter wirelength for the same circuit under 3D IC technology. We may also apply folding-based transformation schemes, *folding-2* or *folding-4*, which can generate 3D placement with a very low TS via number. Moreover, TS via number and wirelength trade-offs can be achieved by the window-based stacking/folding. All these transformation methods can guarantee wirelength reduction over the initial 2D placements.

5.5.1 Local Stacking Transformation Scheme

Local stacking transformation (LST) consists of two steps, stacking and legalization, as shown in Fig. 5.8. The stacking step shrinks the chip uniformly but does not shrink cell areas so that cells are stacked in a region K times smaller and remain in the original relative locations. The legalization step minimizes maximum on-chip temperature and TS via number through the position assignment of cells. The result of LST is a legalized 3D placement.

For K device layer designs, if the original 2D placement is of size S, then the 3D cell area of each layer is S/K. During the stacking step, the width and length of the original placement are shrunk by ratio of $\sqrt{K}$, so that the chip region can maintain the original chip aspect ratio. Cell locations (x_i, y_i) for cell i are also transformed to new locations (x_i', y_i'), where $x_i' = x_i/\sqrt{K}$ and $y_i' = y_i/\sqrt{K}$.

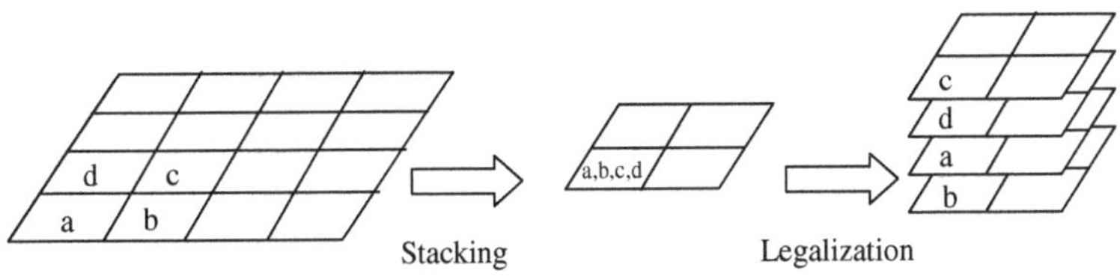

Fig. 5.8 Local stacking transformation

After such a transformation, the initial 2D placement is turned into a 2D placement of size S/K with an average cell density of K, which later will be distributed to K device layers in the legalization step. The Tetris-style legalization (Section 5.6.2.2) could be applied to determine the layer assignment, which may also optimize the TS via number and temperature. As shown in Fig. 5.8, a group of neighboring cells stacking on each other are distributed to different device layers after the transformation process.

5.5.2 Folding Transformation Schemes

LST achieves short wirelength by stacking the neighboring cells together. However, a great number of TS vias will be generated when the cells of local nets are put on top of one another. If the target 3D IC technology allows only limited TS via density, transformations that generate fewer TS vias are required.

Folding-based transformation folds the original 2D placement like a piece of paper without cutting off any parts of the placement. The distance between any two cells will not increase and the total wirelength is *guaranteed* to decrease. TS vias are only introduced to the nets crossing the folding lines (shown as the dashed lines in Fig. 5.9). With an initial 2D placement of minimized wirelength, the number of such long nets should be fairly small, which implies that the connections between the folded regions should be limited, resulting in much fewer TS vias (compared to that of the LST transformation, where many dense local connections cross different device layers). Figure 5.9a shows one way of folding, named folding-2, by folding once at both x- and y-directions. Figure 5.9 b shows another way of folding,

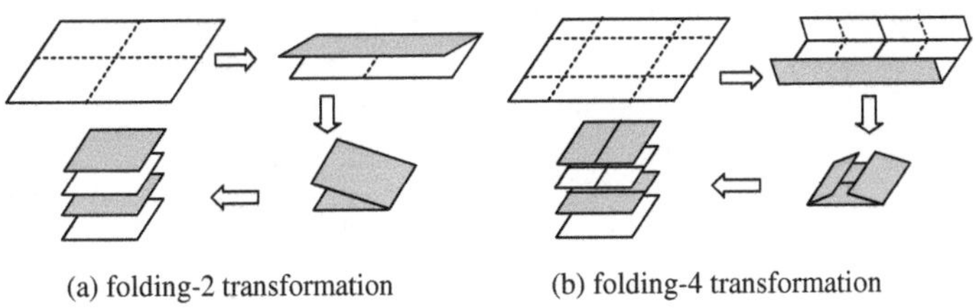

Fig. 5.9 Two folding-based transformation schemes

named folding-4, by folding twice at both x and y-directions. The folding results are legalized 3D placements, so no legalization step is necessary.

After folding-based transformations, only the lengths of the global nets that go across the folding lines (dotted lines in Fig. 5.9) get reduced. Therefore, folding-based transformations cannot achieve as much wirelength reduction as LST. Furthermore, if we want to maintain the original aspect ratio of the chip, folding-based transformations are limited to even numbers of device layers.

5.5.3 Window-Based Stacking/Folding Transformation Scheme

As stated above, LST achieves the greatest wirelength reduction at the expense of a large amount of TS vias, while folding results in a much smaller TS via number but longer wirelength and possibly high via density along the folding lines.

An ideal 3D placement should have short wirelength with TS via density satisfying what the vertical interconnection technology can support. Moreover, we prefer even TS via density for routability reason. Therefore, we propose a window-based stacking/folding method for better TS via density control.

In this method, 2D placement is first divided into $N \times N$ windows. Then the stacking or folding transformation is applied in every window. Each window can use different stacking/folding orders. Figure 5.10 shows the cases for $N = 2$. The circuit is divided into 2×2 windows (shown with solid lines). Each window is again divided into four *squares*(shown with dotted lines). The number in each square indicates the layer number of that square after stacking/folding. The four-layer placements of each window are packed to form the final 3D placement.

Wirelength reduction is due to the following reasons: the wirelength of the nets inside the same square is preserved; the wirelength of nets inside the same window is most likely reduced due to the effect of stacking/folding; and the wirelength of nets that cross the different windows is reduced. Therefore the overall wirelength quality is improved.

Meanwhile, the TS vias are distributed evenly among different windows and can be reduced by choosing proper layer assignments. TS vias are introduced by the nets that cross the boundaries between neighboring squares with different layer numbers, and we call this boundary between two neighboring squares a *transition*. Fewer transitions result in fewer TS vias. Intra-window transitions cannot be reduced because we need to distribute intra-window squares to different layers, so we focus on reducing inter-window transitions. Since the sequential layer assignment in Fig. 5.10a

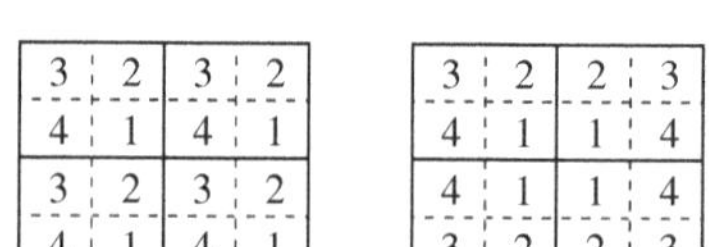

(a) sequential (b) symmetric

Fig. 5.10 2×2 windows with different layer assignments

creates lots of transitions, we use another layer assignment as in Fig. 5.10b, called symmetric assignment, to reduce the amount of inter-window transitions to zero. So this layer assignment generates the smallest TS via number, while the wirelength is similar.

The wirelength versus TS via number trade-offs can be controlled by the number of windows.

5.6 Legalization and Detailed Placement Techniques

A final location for any cell is not desired in the global placement stage. The legalization is in charge of removing the remaining overlaps between cells, and the detailed placement performs further refinement for the placement quality.

Coarse legalization (Section 5.6.1) bridges the gap between global placement and detailed placement. Even for the discrete partitioning-based techniques discussed in Section 5.2, overlap exists after recursive bisection if the device layer number K is not a power of two. The other continuous techniques discussed in Sections 5.3 and 5.4 usually stop before the regional density constraints are strictly satisfied for purpose of runtime reduction. The coarse legalization distributes cells more evenly, so that the latter detailed legalization stage (Section 5.6.2) can assume that local displacement of cells is enough to obtain a legal placement. Another legalization technique called Tetris-style legalization will also be described in Section 5.6.2.2.

The detailed placement performs local swapping of cells to further refine the objective function. If the swapping is inside a device layer, it is not different from that of the 2D detail placement. The swapping between device layers is new in the context of 3D placement. A swapping technique that uses Relaxed Conflict Net (RCN) graph to reduce the TS via number will be introduced in Section 5.6.3.

5.6.1 Coarse Legalization

Placements produced after coarse legalization still contains overlaps, but the cells are evenly distributed over the placement area so that the computational intensive localized calculations used in detailed legalization are prevented from acting over excessively large areas. Coarse legalization [27] utilizes a spreading heuristic called *cell shifting* to prepare a placement for detailed legalization and refinement.

To utilize the cell-shifting heuristic, the placement region $[0, W] \times [0, H] \times [0, K]$ is divided into $M \times N \times L$ bins, where $cell_i$ at (x_i, y_i, z_i) is mapped to the region $[x_i - \frac{w_i}{2}, x_i + \frac{w_i}{2}] \times [y_i - \frac{h_i}{2}, y_i + \frac{h_i}{2}] \times [z_i - 1, z_i]$. During cell shifting, the cells are shifted in one direction at a time, and are shifted three times in three directions.

A demonstration of cell shifting in the x-direction is shown in Fig. 5.11. In this example, the boundaries of the bins in the row with gray color are shifted according

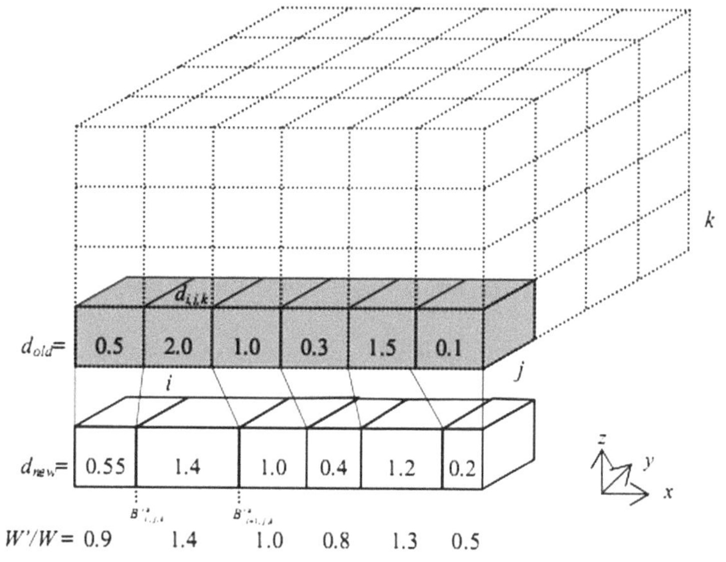

Fig. 5.11 Cell shifting in x-direction [27]

to the bin densities. The numbers labeled inside the bins are the bin densities, where the d_{old} and d_{new} are the densities before and after cell shifting, respectively. The ratios W_b'/W_b between the new bin width W_b' and the old bin width W_b are approximately 0.9, 1.4, 1.0, 0.8, 1.3, and 0.5, respectively for the bins from left to right in this row. Thus the cells inside these bins are also shifted in the x-direction and the bin densities are adjusted to meet the density constraints. The ratio W_b'/W_b is related to the bin density d, which is visualized in Fig. 5.12. In this figure, the x-axis is for the bin density d, and the y-axis is for the ratio W_b'/W_b. The coefficients a_U, a_L, and b are the same for each row (like the one in gray color), but may be different for different rows, which are adjusted to keep the total bin widths in a row be constant.

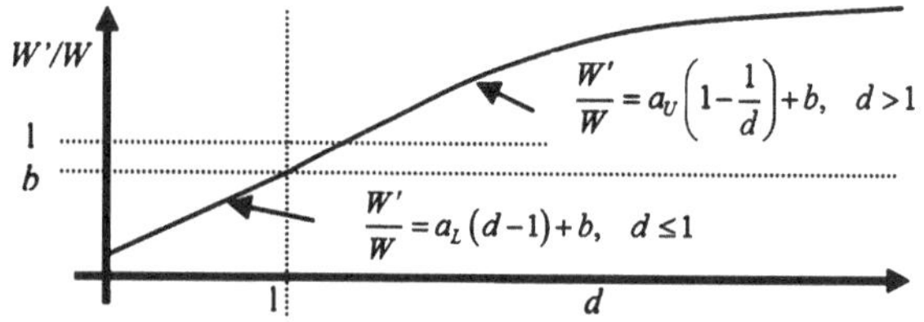

Fig. 5.12 Cell-shifting bin width versus density [27]

After cell shifting, the cell density in every bin is guaranteed not to exceed its volume. But this heuristic does not consider the objective function that should be optimized. Therefore, cell-moving and cell-swapping operations are done after cell shifting; this optimizes the objective function (5.8) and maintains the density underflow properties inside every bin.

5.6.2 Detailed Legalization

Detailed legalization puts cells into the nearest available space that produces the least degradation in the objective function. We describe two detailed legalization techniques that perform this task. The DAG-based legalization assumes that the cell distribution has already been evened with coarse legalization and tries to move cells only locally. The Tetris-style legalization only assumes that the cell distribution is even in the projection on the (x, y) plane, and is able to determine the layer assignments if they are not given, or to minimize the displacement if initial layer assignments are given.

5.6.2.1 DAG-Based Legalization

This detailed legalization process creates a much finer density mesh than what was used with coarse legalization and consists of bins similar in size to the average cell. Bin densities are calculated in a more fine-grained fashion by dividing the precise amount of cell width (rather than area) in the bin by the bin width. To ensure that densities are precisely balanced between different halves of the placement, the amount of space available or the amount of space lacked is calculated for each side of the dividing planes formed by the bin boundaries. A directed acyclic graph (DAG) is constructed in which directed edges are created from bins having an excess amount of cell area to adjacent bins that can accept additional cell area. From this DAG, the dependencies on the processing order of bins can be derived, and cells are placed into their final position in this order. In addition, an estimate of the objective function's sensitivity to cell movement is also used in determining the cell processing order. Using this processing order, the algorithm looks for the best available position for each cell within a target region around its original position. The objective function is used to determine which available position in the target region produces the best results. If an available position is not found, the target region is gradually expanded until enough free space is found within the row segments that it contains. If already processed cells need to be moved apart to legally place the cell, the effect of their movement on the objective function is included in the cost for placing the cell in that position.

5.6.2.2 Tetris-Style Legalization

The Tetris-style legalization technique [19] is applicable to 3D global placements where the projection of cell areas on the (x, y) plain is well distributed. To prepare for the legalization, all the cells are sorted by their x-coordinates in the increasing order. Starting from the leftmost cell, the location of cells is determined one by one in a way similar to the method used in 2D placement legalization [30]. Each time, the leftmost legal position of every row at every layer is considered. We pick one position by minimizing the relocation cost R:

$$R = \alpha \cdot d + \beta \cdot v + \gamma \cdot t \tag{5.41}$$

where d is the cell displacement from the global placement result, v is the TS via number, and t is the thermal cost. Coefficients α, β, γ are predetermined weights. The cost d is related to the (x, y) locations of the cells, and the costs v and t are related to the layer assignment of the cells.

In this legalization procedure, temperature optimization is considered through the layer assignment of the cells. Under the current 3D IC technologies [40], the heat sink(s) are usually attached at the bottom (and/or top) side(s) of the 3D IC stack, with other boundaries being adiabatic. So the main dominant heat flow within the 3D IC stack is vertical toward the heat sink. The study in [17] shows that the z location of a cell will have a larger influence on the final temperature than the (x, y) location of the cell. However, the lateral heat flow can be considered if the initial 2D placement is thermal-aware, so that hot cells will be evenly distributed to avoid hot spots.

The full resistive thermal model is used for the final temperature verification. During the inner loops of the optimization process, a much simpler and faster thermal model [17] is used for the temperature optimization to speedup the placement process. Each tile stack is viewed as an independent thermal-resistive chain. The maximum temperature of such a tile stack then can be written as follows:

$$T = \sum_{i=1}^{k} \left(R_i \sum_{j=i}^{k} P_j \right) + R_b \sum_{i=1}^{k} P_i = \sum_{i=1}^{k} P_i \left(\sum_{j=1}^{i} R_j + R_b \right) \tag{5.42}$$

Besides the fast speed, such a simple close-form equation can also provide a direct guide to thermal-aware cell layer assignment. Equation (5.42) tells us that the maximum temperature of a tile stack is the weighted sum of the power number at each layer, while the weight of each layer is the sum of the resistances below that layer. Device layers that are closer to the heat sink will have smaller weights.

The thermal cost $t_{i,j}$ of assigning cell j to layer i in Equation (5.41) can be written as

$$t_{i,j} = P_j \left(\sum_{k=1}^{i} R_k + R_b \right) \tag{5.43}$$

This thermal cost of layer assignment is also used both in Equation (5.41) and during placement refinement, which will be presented in Section 5.6.3.

5.6.3 Layer Reassignment Through RCN Graph

During the 3D transformations proposed in Section 5.5, layer assignment of cells is based on simple heuristics. To further reduce the TS via number and the temperature, a novel layer assignment algorithm to reassign the cell layers is proposed in [19].

5.6.3.1 Conflict-Net Graph

The metal wire layer assignment algorithm proposed in [8] is extended for cell layer assignment in 3D placement. For a given legalized 3D placement, a conflict-net (CN) graph is created, as shown in Fig. 5.13, where both the cells and the vias are nodes. One via node is assigned for each net. There are two types of edges: net edges and conflicting edges. Within each net, all cells are connected to the via node by net edges in a star mode. A conflict edge is created between cells that overlap with each other if they are placed in the same layer.

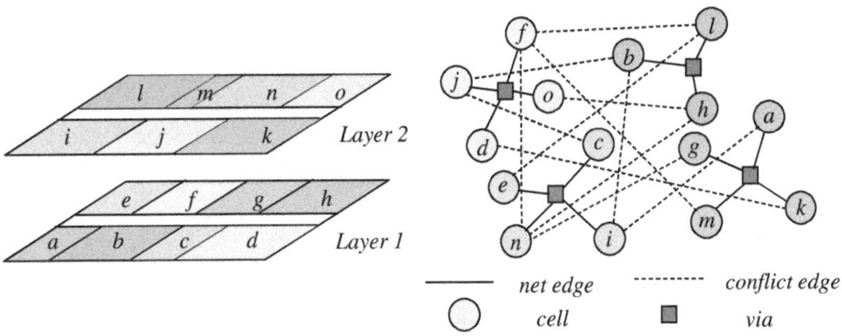

Fig. 5.13 Relaxed conflict-net graph

A layer assignment of each cell node in the graph is preferred, with which the total costs, including edge costs and node costs, are minimized. Cost 0 is assigned for all net edges. If two cells connected by a conflicting edge are assigned to the same layer, the cost of the conflicting edge is set to $+\infty$; otherwise, the cost is set to 0. The cost of a via node is the height of that via, which represents the total TS via number in that net. The heights of the vias are determined by the layers of the cells connecting them. The cost of a cell node v_j is the thermal cost $t_{i,j}$ of assigning v_j to layer i. The cost of a path is the sum of the edge costs and the node costs along that path.

The resulting graph is a directional acyclic graph. A dynamic programming optimization method can be used to find the optimal solution for each induced sub-tree of the graph in linear time. An algorithm that constructs a sequence of maximal induced sub-trees from the CN graph is then used to cover a large portion of the original graph. It turns out that average node number of the induced sub-trees can be as many as 40–50% of the total nodes in the graph. After the iterative optimization of the sub-trees, we can achieve a globally optimized solution. Please refer to [8] for the detailed algorithm for solving the layer assignment problem with the CN graph.

5.6.3.2 Relaxed Non-overlap Constraint

To further reduce the TS via number and the maximum on-chip temperature, the non-overlap constraints can be relaxed so that a small amount of overlap r is allowed in exchange for more freedom in layer reassignment of the cells.

The relaxed non-overlap is defined as follows:

$$\text{overlap}(i,j) = \begin{cases} \text{false}, & if \dfrac{o(i,j)}{s(i)+s(j)} \leq r \\ \text{true}, & if \dfrac{o(i,j)}{s(i)+s(j)} > r \end{cases} \tag{5.44}$$

where $o(i,j)$ is the area of the overlapped region of cell v_i and cell v_j and $s(i)$ is the area of cell i. The relaxation r is a positive real number between 0 and 0.5. This is illustrated in Fig. 5.14. However, with the relaxed non-overlap constraint, the layer assignment result is no longer a legalized 3D placement and another round of legalization is needed to eliminate the overlap.

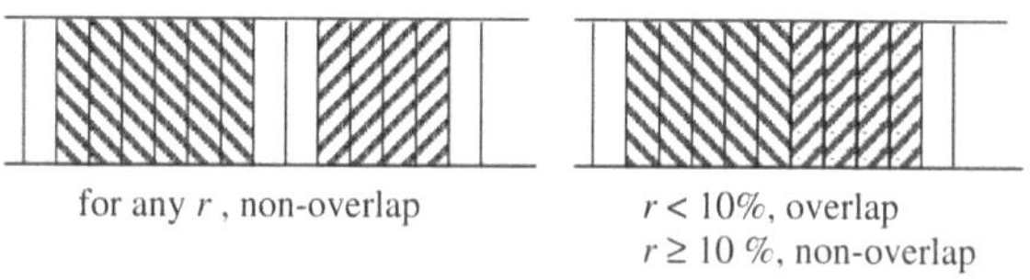

Fig. 5.14 Relaxation of non-overlap constraint

5.7 3D Placement Flow

The 3D placement is divided into stages of global placement, coarse legalization, and detailed legalization, where we focus on global placement techniques in most of previous sections.

We may use partitioning-based techniques, quadratic uniformity modeling techniques, analytical techniques (introduced as an example engine of multilevel techniques), or transformation-based techniques discussed in Sections 5.2–5.6 for global placement. To speedup the runtime and to achieve better qualify, multilevel techniques may be applied, where either of the above global placement techniques can be used as the placement engine.

The coarse legalization is not always necessary, and its application depends on the requirements of detailed legalization. The DAG-based detailed legalization requires a roughly even distribution of density in the given bins, so that coarse legalization is necessary if the global placement results cannot meet the area distribution requirements. The Tetris-style legalization works for any given placement, but still prefer an evenly optimized global placement for better legalized placement quality.

After detailed legalization, RCN-based layer assignment refinement may be applied, as well as the layer-by-layer 2D detailed placement. Legalization may need to be performed if overlaps (e.g., 10%) are allowed during the RCN-based refinement. Several iterations of RCN refinement and legalization can be performed if the placement quality keeps improving. The entire 3D placement flow terminates when a legalized 3D placement is reached.

5.8 Effects of Various 3D Placement Techniques

In this section we shall summarize the experimental results for the various 3D placement techniques.

Section 5.8.1 includes the experimental results on the wirelength and TS via optimization. The ability to trade-off between wirelength and TS via number in the transformation-based techniques and the multilevel analytical placement techniques is demonstrated and they are compared to each other. The results of partitioning-based techniques are also extracted from [27] and are converted for comparisons, and readers may refer to [41] for the results of uniformity quadratic modeling placement techniques. During detailed placement, RCN graph-based refinement also has effects on the trade-offs between wirelength and TS via number, and thus the results are also shown.

Section 5.8.2 focuses on the thermal optimization during 3D placement. The experimental results for the thermal net weights and the thermal-aware Tetris-style legalization are presented in this section.

5.8.1 Trade-Offs Between Wirelength and TS Via Number

Table 5.1 lists the statistics of the 18 circuits in the benchmark [43], which is used for testing 3D placers [26, 27, 42, 19, 13]. We shall use this benchmark to compare the 3D placement results without thermal awareness. The geometric averages are computed to measure and compare the overall results.

We first compare the results of various transformation-based placement techniques (Section 5.5) without thermal awareness, as shown in Table 5.2. The results are generated from different transformation schemes, including local-stacking transformation "LST," window-based transformation "LST (8 × 8 win)," and a folding transformation "Folding-2." The LST and Folding-2 are the same as described in Sections 5.5.1 and 5.5.2, and the LST (8 × 8 win) is the window-based transformation arrived at by dividing the placement region into 8 × 8 windows and running LST in each window. Compared to Folding-2, LST can reduce the wirelength by 44% with the cost of a 17X increase in TS via number; LST (8 × 8 win) can reduce the wirelength by 20% with the cost of a 5X increase in TS via number. These results show the capability of transformation-based methods in the trade-offs between wirelength and TS via number, which can be achieved by varying the number of windows in the hybrid window-based transformation. The selection

Table 5.1 Benchmark characteristics and 2D placement results by mPL6 [7]

Circuit	#cell	#net	2D WL ($\times 10^7$)
ibm01	12282	11507	0.47
ibm02	19321	18429	1.35
ibm03	22207	21621	1.23
ibm04	26633	26163	1.50
ibm05	29347	28446	3.50
ibm06	32185	33354	1.84
ibm07	45135	44394	2.87
ibm08	50977	47944	3.14
ibm09	51746	50393	2.61
ibm10	67692	64227	5.50
ibm11	68525	67016	3.88
ibm12	69663	67739	6.62
ibm13	81508	83806	4.92
ibm14	146009	143202	11.15
ibm15	158244	161196	12.51
ibm16	182137	181188	16.22
ibm17	183102	180684	25.76
ibm18	210323	200565	18.35
Geo-mean	–	–	4.15

Table 5.2 Three-dimensional placement results by transformation-based techniques

	LST		LST (8 × 8 win)		Folding-2	
Circuit	WL ($\times 10^7$)	#TSV ($\times 10^3$)	WL ($\times 10^7$)	#TSV ($\times 10^3$)	WL ($\times 10^7$)	#TSV ($\times 10^3$)
ibm01	0.24	21.03	0.34	6.69	0.43	1.57
ibm02	0.66	33.31	0.85	14.60	1.25	3.09
ibm03	0.61	36.38	0.79	12.73	1.03	3.38
ibm04	0.76	44.95	1.01	15.63	1.35	3.63
ibm05	2.36	50.67	2.63	25.85	3.99	8.17
ibm06	0.94	57.91	1.28	18.61	1.69	3.26
ibm07	1.46	77.60	2.03	25.16	2.65	5.81
ibm08	1.59	83.50	2.21	26.00	2.91	5.03
ibm09	1.34	87.44	2.03	22.92	2.41	4.33
ibm10	2.80	116.92	4.22	32.52	5.08	5.67
ibm11	2.00	117.03	3.05	29.25	3.55	6.01
ibm12	3.36	124.61	4.78	39.67	6.19	6.49
ibm13	2.53	144.73	3.83	34.26	4.55	6.61
ibm14	5.70	247.46	8.93	56.67	10.34	9.45
ibm15	6.40	284.74	9.91	59.55	11.29	12.07
ibm16	8.30	326.99	13.38	73.66	15.04	12.53
ibm17	13.16	332.80	19.51	92.66	24.21	14.53
ibm18	9.37	359.07	14.81	75.27	16.43	14.19
Geo-mean	2.14	101.44	3.07	29.65	3.84	5.95

of transformation schemes depends on the importance of total wirelength and the manufacturing cost of TS vias.

Table 5.3 presents the results for the multilevel analytical placement techniques (Section 5.4) with the TS via weight $\alpha_{TSV} = 10$. Three sets of results are collected: for one-level placement, two-level placement, and three-level placement, and the detailed placement uses the layer-by-layer 2D placements. The one-level placement is used to run the analytical placement engine directly without any clustering, while the two-level or three-level placements construct a two-level or three-level hierarchy by clustering. In these results, we see that with the same weight for TS via number, one-level placement achieves the shortest wirelength, while the three-level placement achieves the fewest TS via number. We compare the multilevel analytical placement techniques and the transformation-based placement techniques by comparing the one-level placement with the LST (r=10%) (the best wirelength case), comparing the two-level placement with the LST (8 × 8 win), and comparing the three-level placement with the Folding-2 method (the best TS via case). From the data shown in Tables 5.2 and 5.3, it is clear that the one-level placement can achieve an average 29% fewer TS vias with only 5% wirelength degradation than LST (r=10%); the three-level placement can also achieve an average 12% shorter wirelength with 24% fewer TS vias number than Folding-2.

Table 5.4 presents the results for the partitioning-based techniques (Section 5.2) with different weights for TS via. These data are converted from the results in [25],

Table 5.3 Three-dimensional placement results by multilevel placement techniques

	1-Level		2-Level		3-Level	
Circuit	WL ($\times 10^7$)	#TSV ($\times 10^3$)	WL ($\times 10^7$)	#TSV ($\times 10^3$)	WL ($\times 10^7$)	#TSV ($\times 10^3$)
ibm01	0.28	8.12	0.37	1.28	0.37	1.09
ibm02	0.73	15.82	1.13	2.26	1.04	3.08
ibm03	0.67	16.67	0.79	3.51	0.89	2.21
ibm04	0.82	28.79	1.12	5.04	1.22	2.17
ibm05	1.88	31.77	2.15	13.20	2.50	9.04
ibm06	1.01	38.17	1.31	6.89	1.54	2.86
ibm07	1.56	54.21	2.05	9.03	2.40	3.33
ibm08	1.69	53.71	2.07	11.64	2.39	4.32
ibm09	1.44	61.65	1.84	10.73	2.24	2.73
ibm10	2.90	88.62	3.90	18.16	4.60	3.79
ibm11	2.12	88.46	2.70	16.16	3.19	4.07
ibm12	3.59	95.89	4.82	19.51	5.80	4.59
ibm13	2.68	110.56	3.35	21.00	4.05	4.12
ibm14	5.95	219.65	6.76	73.71	9.39	9.95
ibm15	6.67	260.17	7.56	84.33	10.36	9.74
ibm16	8.42	300.69	9.58	106.46	13.89	9.89
ibm17	13.28	310.52	15.49	120.77	20.59	12.29
ibm18	9.52	333.75	10.89	107.49	14.60	12.58
Geo-mean	2.24	70.78	2.80	16.12	3.36	4.55

Table 5.4 Three-dimensional placement results by partitioning-based placement techniques

	8.00E-07		2.00E-04		1.30E-02	
Circuit	WL ($\times 10^7$)	#TSV ($\times 10^3$)	WL ($\times 10^7$)	#TSV ($\times 10^3$)	WL ($\times 10^7$)	#TSV ($\times 10^3$)
ibm01	0.30	20.50	0.39	5.39	0.52	0.49
ibm02	0.85	32.44	0.97	11.98	1.51	0.86
ibm03	0.81	34.87	0.95	9.97	1.23	1.95
ibm04	1.02	42.43	1.13	14.24	1.61	2.05
ibm05	2.18	49.78	2.29	20.29	3.07	5.92
ibm06	1.34	55.35	1.47	20.29	2.09	2.47
ibm07	1.91	74.51	2.15	24.77	3.23	2.85
ibm08	2.06	80.86	2.32	26.39	3.35	2.59
ibm09	1.78	83.96	2.10	24.97	2.94	1.79
ibm10	3.33	115.48	3.82	35.25	5.80	2.39
ibm11	2.60	112.90	3.01	33.59	4.31	2.69
ibm12	4.44	121.39	4.89	44.50	7.52	3.97
ibm13	3.26	139.26	3.78	41.85	5.63	2.63
ibm14	7.16	238.96	7.82	80.71	12.23	4.16
ibm15	8.29	275.91	9.10	91.86	13.20	6.40
ibm16	10.43	319.76	11.52	105.99	18.44	5.58
ibm17	15.20	327.27	16.37	125.42	26.13	7.59
ibm18	11.21	350.36	12.41	110.94	19.98	4.58
Geo-mean	2.70	98.27	3.04	32.32	4.48	2.80

which are based on a modified version of the benchmark [43]. In [25], the row spacing is set to 25% of the row height, while the row spacing equals the row height in the original benchmark. To obtain comparable data with Tables 5.2 and 5.3, we assume that the wirelength in [25] has equal amount of x-direction wires and y-direction wires and use the factor $50\% + 50\% \cdot 2/(1 + 25\%) = 1.3$ to scale the wirelength. The three columns in Table 5.4 are with increasing weights for TS vias, where also show the trade-offs between wirelength and TS via. The rightmost column with best TS via number shows 40% reduction in TS via number with 33% wirelength degradation compared with the three-level placement in Table 5.3. But the leftmost column with best wirelength costs 20% longer wirelength and 39% more TS vias compared with one-level placement in Table 5.3. The middle column also does not work as well as two-level placement in Table 5.3. These data indicate that partitioning-based techniques are good at TS via reduction due to the partitioning nature, but they may not as suitable as the multilevel techniques for the cases where more TS vias are manufacturable to achieve shorter wirelength.

As mentioned in Section 5.6.3, the RCN graph-based layer assignment process [19] is used to further optimize the TS via number of the 3D circuits. Tables 5.5 and 5.6 show the effects of the RCN graph-based layer assignment algorithm on the placement by local stacking transformation (Section 5.5.1) and flat analytical technique (Section 5.4.2), respectively. The results of RCN refinement with overlaps $r = 0$ and 10% allowed are reported, where $r = 0\%$ is a strict non-overlap constraint, and $r = 10\%$ allows 10% overlap between neighboring cells during

Table 5.5 Local stacking results and RCN refinement with $r = 0$ and 10%

	LST		After RCN with $r = 0\%$		After RCN with $r = 10\%$	
Circuit	WL ($\times 10^7$)	#TSV ($\times 10^3$)	WL ($\times 10^7$)	#TSV ($\times 10^3$)	WL ($\times 10^7$)	#TSV ($\times 10^3$)
ibm01	0.24	21.03	0.24	20.73	0.24	18.63
ibm02	0.66	33.31	0.66	32.75	0.66	28.87
ibm03	0.61	36.38	0.62	35.38	0.62	30.49
ibm04	0.76	44.95	0.76	43.44	0.77	38.07
ibm05	2.36	50.67	2.36	48.82	2.36	44.37
ibm06	0.94	57.91	0.94	57.29	0.95	50.26
ibm07	1.46	77.60	1.46	74.35	1.47	64.85
ibm08	1.59	83.50	1.59	78.42	1.59	70.46
ibm09	1.34	87.44	1.33	82.79	1.35	73.13
ibm10	2.80	116.92	2.80	112.62	2.81	99.59
ibm11	2.00	117.03	2.00	112.29	2.02	98.77
ibm12	3.36	124.61	3.37	121.31	3.38	107.89
ibm13	2.53	144.73	2.53	138.41	2.54	122.95
ibm14	5.70	247.46	5.70	234.24	5.73	210.08
ibm15	6.40	284.74	6.40	267.28	6.41	248.06
ibm16	8.30	326.99	8.30	311.33	8.34	283.10
ibm17	13.16	332.80	13.16	320.34	13.15	286.26
ibm18	9.37	359.07	9.39	337.12	9.40	300.87
Geo-mean	2.14	101.44	2.14	97.46	2.15	86.73

Table 5.6 Flat analytical results and RCN refinement with $r = 0$ and 10%

	1-Level		After RCN with r= 0%		After RCN with r= 10%	
Circuit	WL ($\times 10^7$)	#TSV ($\times 10^3$)	WL ($\times 10^7$)	#TSV ($\times 10^3$)	WL ($\times 10^7$)	#TSV ($\times 10^3$)
ibm01	0.28	8.12	0.28	8.03	0.29	7.87
ibm02	0.73	15.82	0.73	15.69	0.76	15.59
ibm03	0.67	16.67	0.67	16.45	0.69	16.10
ibm04	0.82	28.79	0.82	27.99	0.84	26.56
ibm05	1.88	31.77	1.88	30.94	1.89	30.20
ibm06	1.01	38.17	1.01	37.24	1.04	35.58
ibm07	1.56	54.21	1.56	52.82	1.59	49.57
ibm08	1.69	53.71	1.69	52.66	1.71	50.97
ibm09	1.44	61.65	1.44	59.88	1.47	56.37
ibm10	2.90	88.62	2.90	86.26	2.97	81.19
ibm11	2.12	88.46	2.12	85.39	2.15	79.63
ibm12	3.59	95.89	3.59	93.51	3.64	87.73
ibm13	2.68	110.56	2.68	106.74	2.71	99.67
ibm14	5.95	219.65	5.95	209.11	5.92	188.71
ibm15	6.67	260.17	6.67	246.45	6.62	224.01
ibm16	8.42	300.69	8.42	288.13	8.35	261.84
ibm17	13.28	310.52	13.28	297.61	13.18	267.90
ibm18	9.52	333.75	9.52	318.80	9.45	286.02
Geo-mean	2.24	70.78	2.24	68.67	2.27	64.61

refinement. In Table 5.5, the average TS via reduction is 4% without any wirelength degradation when $r = 0\%$, and the average TS via reduction is 15% with rare wirelength degradation when $r = 10\%$. In Table 5.6, the average TS via reduction is 3% without wirelength degradation when $r = 0\%$, and the average TS via reduction is 9% with 1% wirelength degradation when $r = 10\%$. From these results, we see that the placement by local stacking transformation has more room to be improved than the flat analytical placement, which also imply that analytical placement approaches produce better solutions than transformation-based placement.

5.8.2 Effects of Thermal Optimization

5.8.2.1 Effects of the Thermal-Aware Net Weights on Temperature

The thermal-aware term defined in Equation (5.6) is for control of temperature during wirelength optimization. A large thermal coefficient α_{TEMP} provides more emphasis on the temperature reduction in the cost of longer wirelength and larger TS via number. The thermal-aware net weights defined in Equation (5.12) are an equivalent way to implement thermal awareness, which is proportional to the thermal coefficient α_{TEMP}.

The thermal-aware net weights are implemented in the partitioning-based 3D placer [27], whose effect on the temperature reduction and the impact on the wirelength and TS via number are shown in Fig. 5.15. Experiments are also performed on the benchmark [43] with minor modification. With the TS via coefficient α_{TSV} set to 10(μm), the impact of the thermal coefficient α_{TEMP} to the TS via number (inter-layer via count), wirelength, total power, average temperature, and maximum temperature are computed, and the percentage of change in these aspects is the average percentage change for ibm01 to ibm18 in the benchmark when compared to the unweighted results. When the average temperatures are reduced by 19%, wirelength is increased by only 1% and the TS via number is increased by 10%.

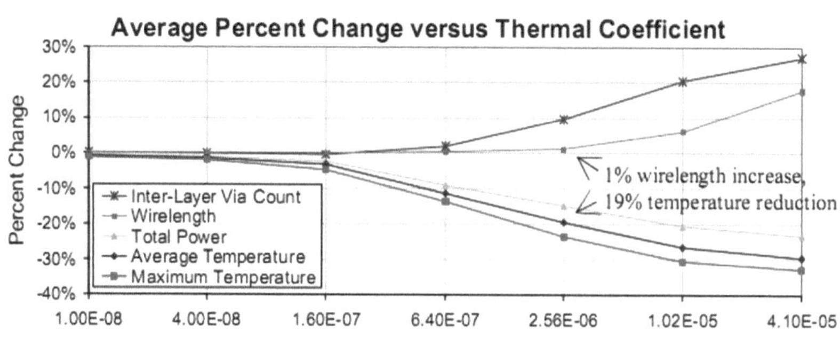

Fig. 5.15 Average percent change as the thermal coefficients are varied [27]

5.8.2.2 Effects of the Legalization on Temperature

Here we compare the two Tetris-style legalization processes, one without thermal awareness and the other with thermal awareness. Cell power dissipation is generated randomly by assigning cell power densities ranging from 10^5 to 10^6 W/m^2 [39]. The temperature evaluation adopts the thermal-resistive network model and the thermal resistance values in [40]. The initial placement is generated by applying the local stacking (LST) scheme of the transformation-based techniques (Section 5.5). The results are shown in Table 5.7, and the temperatures reported are the difference between the maximum on-chip temperature and the heat sink temperature. Compared to the legalization without thermal awareness, the thermal-aware legalization can reduce the maximum on-chip temperature by 39% on average with 8% longer wirelength but 5% fewer TS vias.

Table 5.7 Thermal-aware results of Tetris-style legalization

	Tetris-style legalization without thermal awareness			Tetris-style legalization with thermal awareness		
Circuit	WL ($\times 10^7$)	#TSV ($\times 10^3$)	Temp. (°C)	WL ($\times 10^7$)	#TSV ($\times 10^3$)	Temp. (°C)
ibm01	0.24	21.03	279.002	0.29	19.67	150.422
ibm02	0.66	33.31	207.802	0.72	31.83	117.516
ibm03	0.61	36.38	205.766	0.67	34.13	120.487
ibm04	0.76	44.95	163.279	0.85	42.05	94.648
ibm05	2.36	50.67	138.501	2.44	48.59	78.607
ibm06	0.94	57.91	165.881	1.05	52.12	101.269
ibm07	1.46	77.60	108.015	1.57	72.93	68.382
ibm08	1.59	83.50	101.04	1.68	78.86	61.897
ibm09	1.34	87.44	96.899	1.47	83.35	59.7815
ibm10	2.80	116.92	58.335	3.01	112.95	36.3501
ibm11	2.00	117.03	283.705	2.18	108.96	172.396
ibm12	3.36	124.61	206.811	3.65	120.89	122.211
ibm13	2.53	144.73	254.684	2.76	134.61	157.983
ibm14	5.70	247.46	128.623	6.07	235.17	83.4365
ibm15	6.40	284.74	137.455	6.76	274.44	87.672
ibm16	8.30	326.99	98.5005	8.74	318.43	62.428
ibm17	13.16	332.80	84.73	13.62	324.44	52.954
ibm18	9.37	359.07	89.203	9.76	348.26	57.089
Geo-mean	2.14	101.44	141.88	2.32	96.30	86.11

5.9 Impact of 3D Placement on Wirelength and Repeater Usage

In this section we present the quantitative studies [20] of the impact of 3D IC technology on the wirelength and repeater usage. The wirelength is reported in half-perimeter wirelength, and the repeater usage is estimated by the interconnect optimizer IPEM [14] in the post-placement/pre-routing stage, where the 2D and 3D placement are generated by state-of-the-art 2D placer mPL6 [7] and a multilevel analytical 3D placer [13]. Experiments on a placement benchmark suite [43] show that the total number of repeaters can be reduced by 22 and 50% on average with three-layer and four-layer 3D circuits, respectively, compared to 2D circuits.

5.9.1 2D/3D Placers and Repeater Estimation

mPL6 [7] is a large-scale mixed-size placement package which combines a multi-level analytical placer and a robust legalizer and detailed placer. It is designed for wirelength-driven placement and is density sensitive. The results in the ISPD 2006 placement contest [34] show that mPL6 achieves the best wirelength among all the participating placers.

To explore the advantage of the 3D technology, we use the multilevel analytical 3D placer (Section 5.4). It is a 3D placer providing trade-offs between wirelength and TS via number, and shows better trade-off abilities than transformation- and partitioning-based techniques. Please refer Section 5.8.1 for more experimental results.

IPEM [14] is developed to provide a set of procedures that estimate interconnect performance under various performance optimization algorithms for deep submicron technology. These optimization algorithms include OWS (optimal wire sizing), SDWS (simultaneous driver and wire sizing), BIWS (buffer insertion and wire sizing), and BISWS (buffer insertion, sizing, and wire Sizing). While there are extensive interconnect layout optimization tools such as Trio [11], IPEM is targeted at providing fast and accurate estimation of the optimized interconnect delay and area to enable the design convergence as early as possible through using simple closed-form computational procedures. Experimental results [14] show that IPEM has an accuracy of 90% on average with a running speed of 1000× faster than Trio.

5.9.2 Experimental Setup and Results

The experiments are performed on the IBM-PLACE benchmarks [43]. Since these benchmarks do not have source/sink pin information, to obtain relatively more accurate information of the net wirelength, we use the length of the minimum-wirelength-tree of a net to estimate the optimal number of repeaters required in this net.

The rectilinear Steiner minimal tree has been widely used in early design stages such as physical synthesis, floorplanning, interconnect planning, and placement to estimate wirelength, routing congestion, and interconnect delay. It uses the minimum wirelength edges to connect nodes in a given net. A rectilinear Steiner tree construction package FLUTE [10] is used to calculate the Steiner wirelength tree in order to estimate the repeater insertion without performing the detailed routing. FLUTE is based on a pre-computed lookup table to make the Steiner minimum tree construction fast and accurate for low-degree nets. For high-degree nets, the net is divided into several low-degree nets until the table can be used.

To accurately estimate the delay and area of the TS via resistance and capacitance, the approach in [22] is used to model the TS via as a length of wire. Because of its large size, the TS via has a great self-capacitance. By simulations on each via and the lengths of metal-2 wires in each layer, the authors in [22] approximate the capacitance of a TS via with 3 μm thickness as roughly 8–20 μm of wire. The resistance is less significant because of the large cross-sectional area of each TS via (about 0.1 Ω per TS via), which is equivalent to about 0.2 μm of a metal-2 wire. We use 3D IC technology developed by MIT Lincoln lab and the minimum distance between adjacent layers is 2–3.45 μm. Thus, we can approximately transform all the TS vias between adjacent layers as 14 μm wires (an average value of 8–20 μm). This value is doubled when the TS via is going through two layers.

Since FLUTE can only generate a 2D minimum wirelength tree, in order to transform it to a 3D tree for our 3D designs, the following assumptions are made: (1) assume that all the tree wires are placed in a middle layer of the 3D stack layers, (2) the pins in other layers use TS vias to connect to the tree on the middle layer. This assumption minimizes the total traditional wires in a net but overestimates the total number of TS vias. However, it can provide us with more accurate information concerning the total net wirelength compared to the 3D via and wirelength estimation method used in [19], where the number of vias is simply set as the number of the layers the net spans.

The experiments are performed under 32 nm technology. The technology parameters we used to configure IPEM are listed in Table 5.8. We run FLUTE and IPEM for each net in each benchmark.

Table 5.9 shows the comparison of results between 2D designs, 3D designs with three-device layers, and 3D designs with four-device layers for the IBM-PLACE benchmarks. The wirelength (WL, in μm) and repeater number (#repeater) of each circuit are presented in this table, and the overall geometric mean and the normalized geometric mean are also presented. As can be seen, by applying a 3D design with three-device layers, the total wirelength can be reduced by 17%, and the number of repeaters used in interconnection can be reduced by 22% on average compared to

Table 5.8 Technology parameters

Technology	32 nm
Clock frequency	2 GHz
Supply voltage (V_{DD})	0.9 V
Minimum sized repeater's transistor size (w_{min})	70 nm
Transistor output resistance (r_g)	5 KOhm
Transistor output capacitance (c_p)	0.0165 fF
Transistor input capacitance (c_g)	0.105 fF
Metal wire resistance per unit length (r)	1.2 Ohm/μm
Metal wire area capacitance (c_a)	0.148 fF/μm^2
Metal wire effective-fringing capacitance (c_f)	0.08 fF/μm

Table 5.9 Results of the wirelength/repeaters for IBM-PLACE benchmarks

	2D design		3D design with 3 device layers		3D design with 4 device layers	
Circuit	WL ($\times 10^7$)	#Repeater ($\times 10^3$)	WL ($\times 10^7$)	#Repeater ($\times 10^3$)	WL ($\times 10^7$)	#Repeater ($\times 10^3$)
ibm01	0.54	5.26	0.52	4.80	0.37	2.85
ibm02	1.58	18.36	1.62	18.81	0.96	9.49
ibm03	1.40	15.65	1.11	11.52	0.85	7.75
ibm04	1.65	17.69	1.40	14.04	1.02	8.83
ibm05	4.08	51.81	3.09	37.80	2.35	27.21
ibm06	2.16	23.72	1.89	19.72	1.33	12.13
ibm07	3.18	35.61	2.72	29.01	1.94	17.88
ibm08	3.71	42.95	3.22	35.54	2.23	21.80
ibm09	2.94	31.54	2.58	26.07	1.84	15.85
ibm10	6.09	72.10	5.27	60.09	3.52	35.48
ibm11	4.22	45.33	3.83	39.36	2.58	22.09
ibm12	7.42	89.33	6.29	73.05	4.37	46.05
ibm13	5.50	60.63	4.26	42.51	3.34	29.97
ibm14	12.22	141.59	9.36	101.05	7.04	68.48
ibm15	13.88	162.04	10.27	110.37	8.03	80.01
ibm16	18.25	219.26	13.26	147.95	10.21	105.23
ibm17	28.26	358.37	21.31	258.89	15.32	173.60
ibm18	20.75	248.70	14.73	162.79	11.62	120.13
Geo-mean	4.67	53.43	3.87	41.76	2.79	26.74

the case of 2D design. Furthermore, when four layers are used in the 3D design, the wirelength can be further reduced by 40%, and the number of repeaters can be reduced by 50%.

As shown in Table 5.9, the reduction in the number of repeaters through 3D IC compared to that of the 2D cases is always more than the reduction of the total wirelength. This is because increasing the number of layers will efficiently decrease the length of the nets with a large minimum wirelength tree, and nets with a very small minimum wirelength tree always do not need repeaters. As can be seen in the IPEM results, wires less than 500 μm usually result in zero repeaters. Therefore, by reducing the nets with a large length of the minimum wirelength tree, we can significantly reduce the number repeaters and the area/power of the on-chip interconnection.

5.10 Summary and Conclusion

Three-dimensional IC technology enables an additional dimension of freedom for circuit design. It enhances device-packing density and shortens the length of global interconnects, thus benefiting functionality, performance, and power of 3D circuits.

However, this technology also challenges placement tools. The manufacturing of TS vias is not trivial, thus the placement tools should be aware of TS via cost and perform trade-offs to avoid eliminating the benefits from the shortened wirelength. The thermal issues are also key challenges of 3D circuits due to the stacking of heat sources and the long thermal dissipating path.

In this chapter we give a formulation of the thermal-aware 3D placement problem, and an overview of 3D placement techniques existing in the literatures. We especially describe the details of several representative 3D placement techniques, including the partitioning-based techniques, uniformity quadratic modeling techniques, multilevel placement techniques, and transformation-based techniques. The legalization and detailed placement techniques specific to 3D placement are also introduced.

The partitioning-based techniques are presented in Section 5.2. These partitioning-based techniques insert the partition planes that are parallel to the device layers at some suitable stages in the traditional partition-based process. The cost of partitioning is measured by a weighted sum of the estimated wirelength and the TS via number, where the nets are further weighted by thermal-aware or congestion-aware factors to consider temperature and routability.

The uniformity quadratic modeling techniques belong to the category of quadratic placements techniques, of which the flat placement techniques consist. Since the unconstrained quadratic placement will introduce a great amount of cell overlaps, different variations are developed for overlap removal. The quadratic uniformity modeling techniques [41] append a density penalty function to the objective function, and it approximates the density penalty function by another quadratic function at each iteration, so that the whole global placement could be solved by minimizing a sequence of quadratic functions.

The multilevel technique [13] presented in Section 5.4 constructs a physical hierarchy from the original netlist, and solves a sequence of placement problems from the coarsest level to the finest level.

Besides these techniques above, the transformation-based techniques presented in Section 5.5 make use of existing 2D placement results and construct a 3D placement by transformation.

In addition to various 3D global placement techniques, the legalization and detailed placement techniques that are specific in the 3D placement context are discussed in Section 5.10.

Finally, experimental data are presented to demonstrate the effectiveness of various 3D placement techniques on wirelength, TS via number and temperature, and the impact of 3D IC technology on wirelength and repeater usage. These experimental data indicate that partitioning-based 3D placement techniques are good at TS via minimization, but are not as effective as the multilevel analytical techniques for wirelength optimization for the cases where more TS vias are manufacturable. For the multilevel analytical placement technique, going through more levels for placement optimization leads to fewer TS vias at a cost of increase of wirelength. Finally, the RCN graph-based layer assignment process is effective for both TS via and thermal optimization.

Acknowledgment This study was partially supported by the Gigascale Silicon Research Center, by IBM under a DARPA subcontract, and by the National Science Foundation under CCF-0430077 and CCF-0528583.

References

1. C. Ababei, H. Mogal, and K. Bazargan, Three-dimensional place and route for FPGAs, *Proceedings of the 2005 Conference on Asia South Pacific Design Automation*, pp. 773–778, 2005.
2. C. Alpert, A. Kahng, G.-J. Nam, S. Reda, and P. Villarrubia, A semi-persistent clustering technique for VLSI circuit placement, *Proceedings of the 2005 International Symposium on Physical Design*, pp. 200–207, 2005.
3. K. Balakrishnan, V. Nanda, S. Easwar, and S. K. Lim, Wire congestion and thermal aware 3D global placement, *Proceedings of the 2005 Conference on Asia South Pacific Design Automation*, pp. 1131–1134, 2005.
4. D. P. Bertsekas, Approximation procedures based on the method of multipliers, *Journal of Optimization Theory and Applications*, 23(4), 487–510, 1977.
5. T. F. Chan, J. Cong, T. Kong, and J. R. Shinnerl, Multilevel optimization for large-scale circuit placement, *Proceedings of the 2000 IEEE/ACM International Conference on Computer-aided Design*, pp. 171–176, 2000.
6. T. F. Chan, J. Cong, and K. Sze, Multilevel generalized force-directed method for circuit placement, *Proceedings of the 2005 International Symposium on Physical Design*, pp. 185–192, 2005.
7. T. F. Chan, J. Cong, J. R. Shinnerl, K. Sze, and M. Xie, mPL6: enhancement multilevel mixed-size placement with congestion control, in *Modern Circuit Placement*, G.-J. Nam and J. Cong, Eds., Springer, New York, NY, 2007.
8. C.-C. Chang and J. Cong, An efficient approach to multilayer layer assignment with an application to via minimization, *IEEE Transactions on Computer-Aided Design of Integrated Circuits and Systems*, 18(5): 608–620, 1999.
9. T.-C. Chen, Z.-W. Jiang, T.-C. Hsu, H.-C. Chen, and Y.-W. Chang, A high-quality mixed-size analytical placer considering preplaced blocks and density constraints, *Proceedings of the 2006 IEEE/ACM International Conference on Computer-Aided Design*, pp. 187–192, 2006.
10. C. Chu and Y. Wong, FLUTE: Fast lookup table based rectilinear steiner minimal tree algorithm for VLSI design, *IEEE Transactions on Computer-Aided Design of Integrated Circuits and Systems*, 27(1): 70–83, 2008.
11. J. Cong and L. He, Theory and algorithm of local refinement based optimization with application to device and interconnect sizing, *IEEE Transactions on Computer-Aided Design*, pp. 1–14, 1999.
12. J. Cong and G. Luo, Highly efficient gradient computation for density-constrained analytical placement methods, *Proceedings of the 2008 International Symposium on Physical Design*, pp. 39–46, 2008.
13. J. Cong and G. Luo, A multilevel analytical placement for 3D ICs, *Proceedings of the 2009 Conference on Asia and South Pacific Design Automation*, Yokohama, Japan, pp. 361–366, 2009.
14. J. Cong and D. Z. Pan, Interconnect estimation and planning for deep submicron designs, *Proceedings of the 26th ACM/IEEE Design Automation Conference*, New Orleans, LA, pp. 507–510,\, 1999.
15. J. Cong and J. Shinnerl, *Multilevel Optimization in VLSICAD*, Kluwer Academic Publishers, Boston, MA, 2003.
16. J. Cong and M. Xie, A robust mixed-size legalization and detailed placement algorithm, *IEEE Transactions on Computer-Aided Design of Integrated Circuits and Systems*, 27(8): 1349–1362, 2008.

17. J. Cong and Y. Zhang, Thermal via planning for 3-D ICs, *Proceedings of the 2005 IEEE/ACM International Conference on Computer-Aided Design*, pp. 745–752, 2005.
18. J. Cong, J. R. Shinnerl, M. Xie, T. Kong, and X. Yuan, Large-scale circuit placement, *ACM Transactions on Design Automation Electronic Systems*, 10(2): 389–430, 2005.
19. J. Cong, G. Luo, J. Wei, and Y. Zhang, Thermal-aware 3D IC placement via Transformation, *Proceedings of the 2007 Conference on Asia and South Pacific Design Automation*, pp. 780–785, 2007.
20. J. Cong, C. Liu, and G. Luo, Quantitative studies of impact of 3D IC design on repeater usage, *Proceedings of the International VLSI/ULSI Multilevel Interconnection Conference*, 2008.
21. S. Das, *Design Automation and Analysis of Three-Dimensional Integrated Circuits*, PhD Dissertation, Massachusetts Institute of Technology, Cambridge, MA, 2004.
22. W. R. Davis, J. Wilson, S. Mick, J. Xu, H. Hua, C. Mineo, A. M. Sule, M. Steer, and P. D. Franzon, Demystifying 3D ICs: The pros and cons of going vertical, *IEEE Design & Test of Computers*, 22(6): 498–510,\, 2005.
23. A. E. Dunlop and B. W. Kernighan, A procedure for placement of standard-cell VLSI circuits, *IEEE Transactions on Computer-Aided Design of Integrated Circuits and Systems*, 4(1): 92–98, 1985.
24. C. M. Fiduccia and R. M. Mattheyses, A linear-time heuristic for improving network partitions, *Proceedings of the 19th ACM/IEEE Conference on Design Automation*, pp. 175–181, 1982.
25. B. Goplen, *Advanced Placement Techniques for Future VLSI Circuits*, PhD Dissertation, University of Minnesota, Minneapolis, MN, 2006.
26. B. Goplen and S. Sapatnekar, Efficient thermal placement of standard cells in 3D ICs using a force directed approach, *Proceedings of the 2003 IEEE/ACM International Conference on Computer-Aided design*, p. 86, 2003.
27. B. Goplen and S. Sapatnekar, Placement of 3D ICs with thermal and interlayer via considerations, *Proceedings of the 44th Annual Conference on Design Automation*, pp. 626–631, 2007.
28. A. S. Grove, *Physics and Technology of Semiconductor Devices*, John Wiley & Sons, Inc., Hoboken, NJ, 1967.
29. R. Hentschke, G. Flach, F. Pinto, and R. Reis, 3D-vias aware quadratic placement for 3D VLSI circuits, *IEEE Computer Society Annual Symposium on VLSI*, pp. 67–72, 2007.
30. D. Hill, Method and system for high speed detailed placement of cells within an integrated circuit design, US Patent 6370673, 2001.
31. A. B. Kahng, S. Reda, and Q. Wang, Architecture and details of a high quality, large-scale analytical placer, *Proceedings of the 2005 IEEE/ACM International Conference on Computer-Aided Design*, pp. 891–898, 2005.
32. G. Karypis and V. Kumar, Multilevel k-way hypergraph partitioning, *Proceedings of the 36th ACM/IEEE Conference on Design Automation*, pp. 343–348, 1999.
33. I. Kaya, S. Salewski, M. Olbrich, and E. Barke, Wirelength reduction using 3-D physical design, *Proceedings of the 14th International Workshop on Power and Timing Optimization and Simulation*, pp. 453–462, 2004.
34. G.-J. Nam, ISPD 2006 placement contest: benchmark suite and results, *Proceedings of the 2006 International Symposium on Physical Design*, pp. 167–167, 2006.
35. G.-J. Nam and J. Cong (Eds.), *Modern Circuit Placement: Best Practices and Results*, Springer, New York, NY, 2007.
36. W. C. Naylor, R. Donelly, and L. Sha, Non-linear optimization system and method for wire length and delay optimization for an automatic electric circuit placer, US Patent 6301693, 2001.
37. J. Nocedal and S. J. Wright, *Numerical Optimization*, 2nd ed., Springer, New York, NY, 2006.
38. P. Spindler and F. M. Johannes, Fast and robust quadratic placement combined with an exact linear net model, *Proceedings of the 2006 IEEE/ACM International Conference on Computer-Aided Design*, pp. 179–186, 2006.

39. C.-H. Tsai and S.-M. Kang, Cell-level placement for improving substrate thermal distribution, *IEEE Transactions on Computer-Aided Design of Integrated Circuits and Systems*, 19(2): 253–266, 2000.
40. P. Wilkerson, A. Raman, and M. Turowski, Fast, automated thermal simulation of three-dimensional integrated circuits, *Proceedings of the 9th Intersociety Conference on Thermal and Thermomechanical Phenomena in Electronic Systems*, Las Vegas, Nevada, 2004.
41. H. Yan, Q. Zhou, and X. Hong, Thermal aware placement in 3D ICs using quadratic uniformity modeling approach, *Integration, the VLSI Journal*, 42(2): 175–180,\, 2009.
42. B. Yao, H. Chen, C.-K. Cheng, N.-C. Chou, L.-T. Liu, and P. Suaris, Unified quadratic programming approach for mixed mode placement, *Proceedings of the 2005 International Symposium on Physical Design*, pp. 193–199, 2005.
43. http://er.cs.ucla.edu/benchmarks/ibm-place/

Chapter 6
Thermal Via Insertion and Thermally Aware Routing in 3D ICs

Sachin S. Sapatnekar

Abstract Thermal challenges in 3D chips motivate the need for on-chip thermal conduction networks to deliver the heat to the heat sink. The most prominent example is a passive network of thermal vias, which serves the function of heat conduction without necessarily serving any electrical function. This chapter begins with an overview of techniques for thermal via insertion. Next, it addresses the problem of 3D routing, overcoming challenges as conventional 2D routing is stretched to a third dimension and as electrical routes must vie with thermal vias for scarce on-chip routing resources, particularly intertier vias.

6.1 Introduction

Three-dimensional integration technologies pack together multiple tiers of active devices, permitting increased levels of integration for a given footprint. The advantages of 3D are numerous and include the potential for reduced interconnect lengths and/or latencies, as well as enhancements in system performance, power, reliability, and portability. However, 3D designs also bring forth notable challenges in areas such as architectural design, thermal management, power delivery, and physical design.

To enable the design of 3D systems, it is essential to develop CAD infrastructure that moves from current-day 2D systems to 3D topologies. One aspect of this is topological, with the addition of a third dimension in which wires can be routed (or can create blockages that prevent the routing of other wires). Strictly speaking, 3D technologies do not allow complete freedom in the third dimension, since the allowable coordinates are restricted to a small number of possibilities, corresponding to the number of 3D tiers. As a result, physical design in this domain is often said to

---

S.S. Sapatnekar (✉)
Department of Electrical and Computer Engineering, University of Minnesota, Minneapolis, MN 55455, USA
e-mail: sachin@umn.edu

Y. Xie et al. (eds.), *Three-Dimensional Integrated Circuit Design*,
Integrated Circuits and Systems, DOI 10.1007/978-1-4419-0784-4_6,

correspond to a 2.5D problem. A second aspect is related to performance issues in general and thermal issues in particular.

Both of these aspects necessitate a 3D design/CAD flow that has significant differences from a corresponding flow for 2D IC design. While Chapters 4 and 5 of this book discuss the floorplanning and placement problems in 3D, this chapter particularly focuses on issues in this flow that are related to thermal mitigation and routing through the use of interconnects. The process of adding thermal vias can be considered a post-placement step prior to routing or integrated within a routing framework. This chapter begins by addressing the problem of inserting thermal vias into a placed 3D circuit to mitigate the temperature profile within a 3D system. Next, methods for simultaneous routing and thermal via allocation, to better manage the blockages posed by thermal vias, are discussed.

6.2 Thermal Vias

The potential for high temperatures in 3D circuits has two roots: first, from the increased power dissipation, as more active devices are stacked per unit footprint, and second from inadequate thermal conduction paths from the devices to the package, and thence the ambient.

While the first can be addressed through the use of low-power design techniques that have been extensively researched, the second requires improvements in the effective thermal conductivity from the devices to the package. Silicon is a good thermal conductor, with half or more of the conductivity of typical metals, but many of the materials used in 3D technologies are strong insulators. These materials include epoxy bonding materials used to attach 3D tiers, or field oxide, or the insulator in an SOI technology. Such a thermal environment places severe restrictions on the amount of heat that can be removed, even under the best placement solution that optimally redistributes the heat sources to control the on-chip temperature. Therefore, the use of deliberate metal connections that serve as heat removing channels, called "thermal vias," is an important ingredient of the total thermal solution. In the absence of thermal vias, simulations have shown that the peak on-chip temperature on a 3D chip can be about 150°C; this can be relieved through the judicious insertion of thermal vias as a post-processing step after placement. In realistic 3D technologies, the dimensions of these intertier thermal vias are of the order of microns on a side; an example of such a via is illustrated in Fig. 6.1.

The idea of using thermal vias to alleviate thermal problems has long been utilized in the design of packaging and printed circuit boards (PCBs). The specific addition of thermal vias was generally unnecessary for 2D chips, since bulk silicon is a very good thermal conductor. Thermal vias have become attractive in the 3D domain, since the concentration of heat flux is large, and adjacent tiers are separated by thermally insulating materials in many processes. In such a scenario, on-chip thermal vias can play a significant role in directing the heat toward the package and the heat sink, and reducing on-chip temperatures.

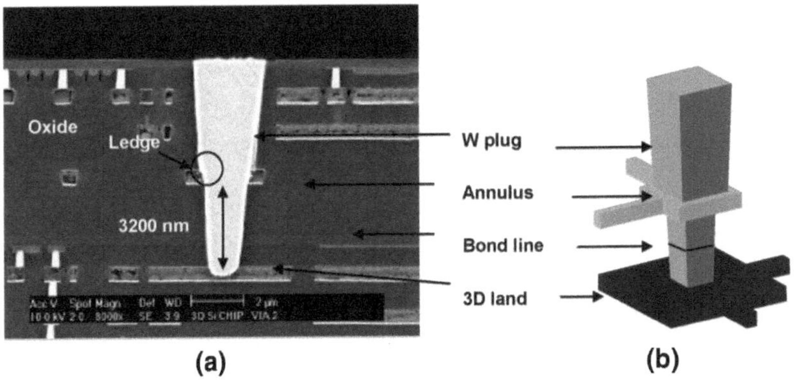

Figure 6.1 (**a**) Cross-sectional SEM and (**b**) isometric drawing of a 3D intertier via [1]. ©2006 IEEE

Historically, in the multichip module (MCM) context, Lee et al. [2] studied arrangements of thermal vias and found that as the size of thermal via islands increased, more heat removal was achieved, but less space was available for routing. The relationships between design parameters and thermal resistance of thermal via clusters in PCBs and packaging were studied in [3]. These relationships were determined by simplifying the via cluster into parallel networks using the observation that heat transfer is much more efficient vertically through the thickness than laterally from heat spreading. Pinjala et al. performed further thermal characterizations of thermal vias in packaging [4]. Although these papers have limited application for the placement of thermal vias inside chips, they demonstrate the basic use and properties of thermal vias. It is important to realize that there is a tradeoff between routing space and heat removal, indicating that thermal vias should be used sparingly. Simplified thermal calculations can be used for thermal vias, and the direction of heat conduction is primarily in the orientation of the thermal via. Chiang et al. suggested that dummy thermal vias can be added to the chip substrate as additional electrically isolated vias to reduce effective thermal resistances and potential thermal problems [5]. Several other early papers addressed the potential of integrating thermal vias directly inside chips to reduce thermal problems internally, e.g., [6, 7]. Because of the insulating effects of numerous dielectric layers, thermal problems are greater and thermal vias can have a larger impact on 3D ICs than 2D ICs. In addition, interconnect structures can create efficient thermal conduits and greatly reduce chip temperatures.

6.3 Inserting Thermal Vias into a Placed Design

While thermal vias can play a major role in moving heat toward the sink and the ambient, it is also important to consider that these vias introduce restrictions on the design. These may be summed up as follows:

- First, the landing pad of each via is considerable, of the order of microns for advanced technologies. These restrictions on the pitch arise from the need to facilitate reliable connections after wafer alignment.
- Second, a through-silicon via creates mechanical stress in its neighborhood, implying that there is a keep-out region for circuit structures in the vicinity of the via.
- Third, a thermal via may act as a routing blockage and may introduce congestion bottlenecks.

In order to manage these constraints, it is necessary to impose a design discipline whereby certain areas of the chip are reserved for placing thermal vias, thereby providing some predictability on the location of the vias, and hence the obstacles and keep-out regions. The work in [8] uses the notion of thermal via regions, illustrated in Fig. 6.2, that lie between rows of cells: any inserted vias must lie within these regions, though it is not necessary for all of these vias to be utilized. The density of these routing obstacles is limited in any particular area so that the design does not become unroutable.

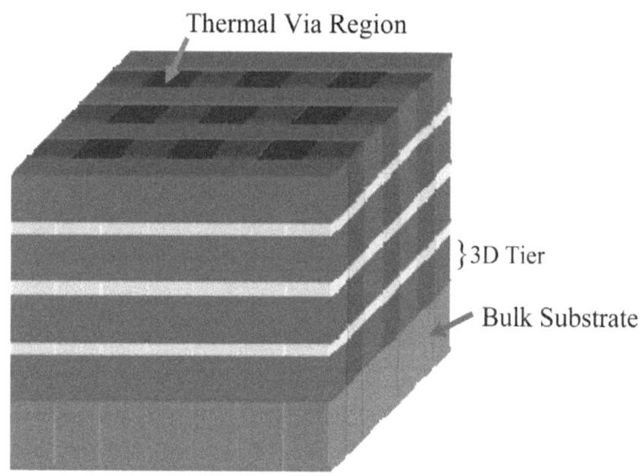

Figure 6.2 A thermal mesh for a 3D IC with thermal via regions

The value of the thermal conductivity, K, in any particular direction corresponds to the density of thermal vias that are arranged in that direction. For all practical purposes, the addition of vertical thermal vias in a 3D IC helps with conduction in only the z direction, toward the heat sink, and lateral conduction due to thermal vias is negligible.

Any thermal optimization must necessarily be linked to thermal analysis. In this chapter, we will draw upon the techniques described in Chapter 3 and highlight specifics that are germane to our discussion here.

In principle, the problem of placing thermal vias can be viewed as one of determining one of two conductivities (corresponding to the presence or absence of metal) at every candidate point where a thermal via may be placed in the chip. However, in practice, it is easy to see that such an approach could lead to an extremely large search space that is exponential in the number of possible positions. Moreover, from a practical standpoint, it is unreasonable to perform full-chip thermal analysis, particularly in the inner loop of an optimizer, at the granularity of individual thermal vias. At this level of detail, individual elements would have to correspond to the size of a thermal via, and the size of the finite element analysis (FEA) stiffness matrix would become extremely large.

Fortunately, there are reasonable ways to overcome these issues. To control the FEA stiffness matrix size, one could work with a two-level scheme with relatively large elements, where the average thermal conductivity of each region is a design variable. Once this average conductivity is chosen, it could be translated back into a precise distribution of thermal vias within the element that achieves that average conductivity.

The procedure in [8] uses an iterative approach for thermal via insertion to control temperatures in a 3D IC. The procedure uses a finite element-based thermal analysis method to compute on-chip temperatures and adds thermal vias to alter thermal conductivities through the chip, thereby lowering temperatures in the 3D stack. Starting from an initial configuration, the thermal conductivities in the z direction are iteratively updated.

During each iteration, the thermal conductivities of thermal via regions are modified, and these thermal conductivities reflect the density of thermal vias needed to be utilized within the region. The new thermal conductivities are derived from the element FEA equations. In each iteration, a small perturbation is made to the thermal conductivities of specific elements by adding thermal vias. For each element, this method assumes that the power flux passing through it remains unchanged under this perturbation, i.e.,

$$K_c^{\text{old}}\mathbf{T}_{\text{old}} = K_c^{\text{new}}\mathbf{T}_{\text{new}} \tag{6.1}$$

where T_q and K_c^q, $q \in \{\text{old,new}\}$, correspond to the element stiffness stamp and the temperature at the corners of the element. Based on a mathematical analysis of the element stamps for an 8-node rectangular prism, it can be shown that

$$k_i^{\text{old}}\,\Delta T_i^{\text{old}} = k_i^{\text{new}}\,\Delta T_i^{\text{new}} \tag{6.2}$$

where $i \in \{x, y, z\}$, and along the given direction i, k_i^q, $q \in \{\text{old,new}\}$, is the effective thermal conductivity of the vias in the given element and ΔT_i^{new} is the change in temperature in the corresponding direction. Defining the thermal gradient as

$$g_i^q = \frac{\Delta T_i^q}{d_i}, i \in \{x,y,z\}, q \in \{\text{old,new}\} \tag{6.3}$$

where d_i is the dimension of the element in direction i; it can be demonstrated through simple algebra that

$$k_i^{\text{new}} = \frac{k_i^{\text{old}} \Delta T_i^{\text{old}}}{\Delta T_i^{\text{new}}} = \frac{k_i^{\text{old}} g_i^{\text{old}}}{g_i^{\text{new}}}, i \in \{x,y,z\} \tag{6.4}$$

A key observation in this approach is that the gradient of the temperature is the most important metric in controlling the temperature. Intuitively, the idea is that if a region locally has a high thermal gradient, then adding thermal vias will help even out the thermal profile. In fact, upper layers are typically hotter than lower layers, but adding thermal vias to reduce the temperature of a lower layer, closer to the heat sink, can help reduce the temperature elsewhere in the layout. Given a target thermal gradient, g_{ideal}, the gradient in the previous iteration, g_i^{old}, can be updated to that in the new iteration using the following calculation:

$$g_i^{\text{new}} = g_{\text{ideal}} \left(\frac{|g_i^{\text{old}}|}{g_{\text{ideal}}} \right)^{\alpha}, i \in \{x,y,z\} \tag{6.5}$$

where $\alpha \in (0,1)$ is a user-defined parameter. Combining (6.4) and (6.5) yields

$$k_i^{\text{new}} = k_i^{\text{old}} \left(\frac{|g_i^{\text{old}}|}{g_{\text{ideal}}} \right)^{1-\alpha} \tag{6.6}$$

This decreases the value of k when the thermal gradient is above g_{ideal} and increases it when it is below that value. The approach in [8] defines prescriptions for the choice of g_{ideal} for various objective functions such as maximum thermal gradient, average thermal gradient, maximum temperature, average temperature, maximum thermal via density, and average thermal via density.

Once the thermal conductivities have been determined using the above approach, the next step is to translate these into thermal via densities for each thermal via region. The percentage of thermal vias or metallization, m, also called thermal via density, in a thermal via region is given by the following equation:

$$m = \frac{n A_{\text{via}}}{wh} \tag{6.7}$$

where n is the number of individual thermal vias in the region (clearly this is upper-bounded by the capacity of the region), A_{via} is the cross-sectional area of each thermal via, w is the width of the region, and h is the height of the region. The relationship between the percentage of thermal vias and the effective vertical thermal conductivity is given by

$$K_z^{\text{eff}} = m K_{\text{via}} + (1 - m) K_z^{\text{layer}} \tag{6.8}$$

where K_{via} is the thermal conductivity of the via material and K_z^{layer} is the thermal conductivity of the region without any thermal vias. Using this equation, the percentage of thermal vias can be found for any K_z^{new}, provided that $K_z^{\text{layer}} \leq K_z^{\text{new}} \leq K_{\text{via}}$:

$$m = \frac{K_z^{\text{new}} - K_z^{\text{layer}}}{K_{\text{via}} - K_z^{\text{layer}}} \tag{6.9}$$

During each iteration, the new vertical thermal conductivity is used to calculate the thermal via density, m, and the lateral thermal conductivities for each thermal via region. The effective lateral thermal conductivities, K_x^{new} and K_y^{new}, can be computed as

$$K_{x[y]}^{\text{new}} = \left(1 - \sqrt{m}\right) K_{x[y]}^{\text{layer}} + \frac{\sqrt{m}}{\frac{1-\sqrt{m}}{K_{x[y]}^{\text{layer}}} + \frac{\sqrt{m}}{K_{\text{via}}}} \tag{6.10}$$

The overall pseudocode of the approach is shown in Algorithm 1.

Algorithm 1: Thermal_via_placement(*objective*)

```
input : Objective for thermal via placement in a thermal via region
output: Thermal conductivities for a thermal via region
Set g_ideal to g_ave
Set Ks to their minimum values
Calculate the temperature profile
while not converged do
    foreach thermal via region do
        K_z = K_z (|g_z| / g_ideal)^(1-α)
        Update K_x, K_y
    Calculate temperature profile
    Update g_ideal using objective
Compute thermal via densities
```

The technique in [8] has been applied to a range of benchmark circuits, with over 158,000 cells, and the insertion of thermal vias shows a reduction in the average temperature of about 30%, with runtimes of a couple of minutes. Therefore, thermal via addition has a more dramatic effect on temperature reduction than thermal placement.

Figure 6.3 shows the 3D layout of the benchmark `struct` before and after the addition of thermal vias. The dark and light regions in the thermal map represent hot and cool regions, respectively. The greatest concentration of thermal vias is *not* in the hottest regions, as one might expect at first. The intuition behind this is as follows: if we consider the center of the uppermost tier, it is hot principally because the tier below it is at an elevated temperature. Adding thermal vias to remove heat from the second tier, therefore, effectively also significantly reduces the temperature of the top tier. For this reason, the regions where the insertion of thermal vias is most effective are those that have high thermal gradients. For detailed experiments, for various thermal objectives, the reader is referred to [10].

The work in [11] presents an approach for thermal via insertion based on transient analysis. The method exploits the electrical–thermal duality and the relationship between the power grid problem and the thermal problem. Like [12], which uses

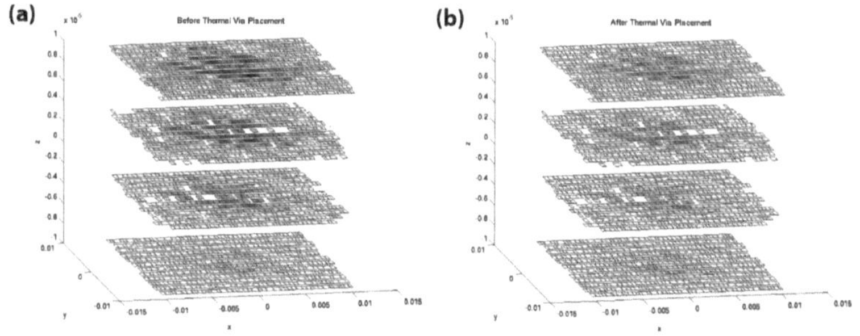

Figure 6.3 Thermal profile of struct before and after thermal via insertion [9]. ©2006 IEEE

the total noise violation metric, taken as the integral over time of the amount by which the waveform exceeds the noise threshold, this method uses the integral of the total thermal violation based on a specified temperature threshold. The layout is tessellated into grids, and constraints for the optimization are the amount of space available in each grid tile for thermal via insertion and the total amount of thermal via area, although it is not clearly explained why both constraints are necessary. A model order reduction technique is used as the simulation engine, and the optimization problem is solved using sequential quadratic programming.

Subsequent work in [13] uses power grids to conduct heat and optimizes the power grid by determining where to insert TSVs to ensure that voltage drop constraints and temperature constraints are met. As in the work described in the previous paragraph, the layout is tessellated into tiles, and the via density in each tile is computed.

6.4 Routing Algorithms

Once the cells have been placed and the locations of the thermal vias determined, the routing stage finds the optimal interconnections between the wires. As in 2D routing, it is important to optimize the wire length, the delay, and the congestion. In addition, several 3D-specific issues come into play. First, the delay of a wire increases with its temperature, so that more critical wires should avoid the hottest regions, as far as possible. Second, intertier vias are a valuable resource that must be optimally allocated among the nets. Third, congestion management and blockage avoidance are more complex with the addition of a third dimension. For instance, a signal via or thermal via that spans two or more tiers constitutes a blockage that wires must navigate around.

Each of the above issues can be managed through exploiting the flexibilities available in determining the precise route within the bounding box of a net, or perhaps even considering detours outside the bounding box, when an increase in the wire length may improve the delay or congestion or may provide further flexibility for intertier via assignment.

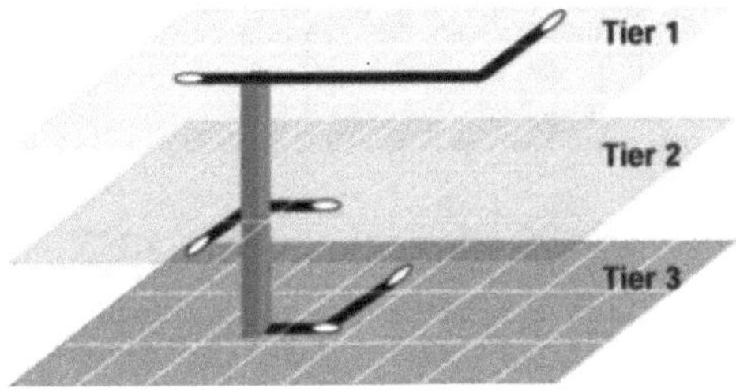

Figure 6.4 An example route for a net in a three-tier 3D technology [14]. ©2005 IEEE

Consider the problem of routing in a three-tier technology, as illustrated in Fig. 6.4. The layout is gridded into rectangular tiles, each with a horizontal and vertical capacity that determines the number of wires that can traverse the tile and an intertier via capacity that determines the number of free vias available in that tile. These capacities account for the resources allocated for nonsignal wires (e.g., power and clock wires) as well as the resources used by thermal vias. For a single net, as shown in the figure, the degrees of freedom that are available are in choosing the locations of the intertier vias and selecting the precise routes within each tier. The locations of intertier vias will depend on the resource contention for vias within each grid. Moreover, critical wires should avoid the high-temperature tiles as far as possible.

The basic grid graph on which routing is performed is similar to the standard 2D routing grid, extended to three dimensions. Each tier is tessellated into a 2D grid, with vertices corresponding to grids, and edges between adjacent grids, with weights corresponding to the capacity of the grid boundary. Connections between vertices in adjacent tiers correspond to the presence of available intertier vias at these locations.

6.4.1 A Multilevel Approach

The work in [15] presented an initial approach to 3D routing with thermal via insertion, and this was subsequently refined in an improved method in [16]. Both methods lie on top of a multilevel routing framework, similar to [17]. The stages in this multilevel framework include recursive coarsening, initial solution generation, and level-to-level refinement and are illustrated in Fig. 6.5. The allocation of TSVs is first performed at the coarsest level and then progressively at finer levels.

The work in [15] uses a compact thermal resistive model from [18], which is essentially the resistive model presented in Chapter 3. The idea of this approach is to iterate between two steps that determine the number of thermal vias and the number of signal vias. The thermal via distribution between two tiers in a given grid uses a simple heuristic, choosing the number to be proportional to the difference between the temperatures in those two grids. Signal via insertion at each level of multilevel routing is performed using a network flow formulation described below.

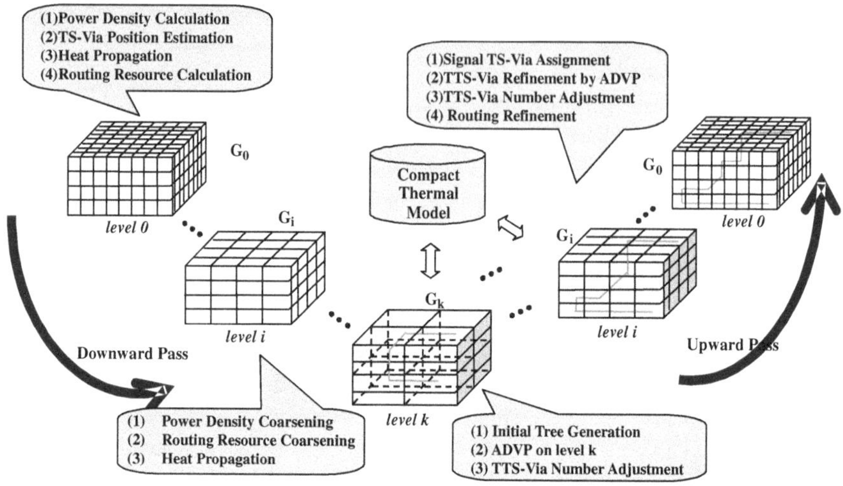

Figure 6.5 A multilevel routing framework that includes TSV planning [16]. ©2005 IEEE

At each level of the multilevel scheme, the intertier via planning problem assigns vias in a given region at level k–1 of the multilevel hierarchy to grid tiles at level k. The problem is formulated as a mincost flow problem, which has the form of a transportation problem. The flow graph, illustrated in Fig. 6.6, is constructed as follows:

Figure 6.6 A network flow formulation for signal intertier via planning [15]. ©2005 IEEE

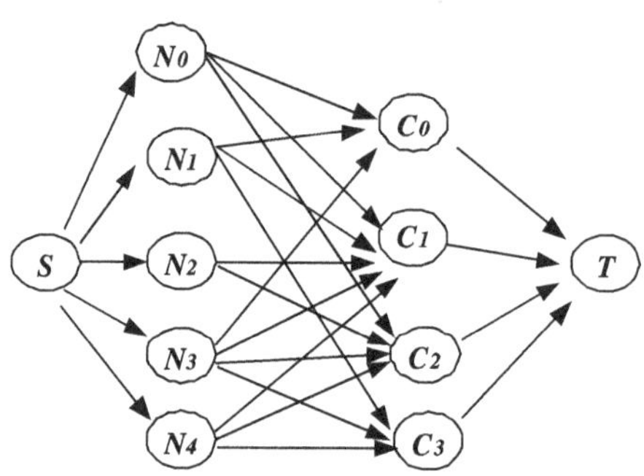

- The source node of the flow graph is connected through directed edges to a set of nodes N_i representing candidate vias; the edges have unit capacity and zero cost.
- Directed edges connect a second set of nodes, C_j, from each candidate grid tile to the sink node, with capacity equaling the number of vias that the tile can contain, and cost zero. The capacity is computed using a heuristic approach that takes into account the temperature difference between the tile and the one directly in the tier below it (under the assumption that heat flows downward toward the sink).

- The source [sink] has supply [demand] m, which equals the number of intertier vias in the entire region.
- Finally, a node N_i is connected to a tile C_j through an arc with infinite capacity and cost equaling the estimated wirelength of assigning an intertier via N_i to tile C_j.

An extension of this work in [16] is again based on a multilevel routing framework illustrated in Fig. 6.5. In this approach, the via planning method was improved using an approach referred to as the alternating direction TSV planning (ADVP) method. This method also assumes that the primary direction for heat flow is in the vertical dimension. A nonlinear programming formulation for TSV insertion is presented, but is determined to be too expensive, and is only used for comparison purposes in the work. The chief engine that is proposed is an iterative two-step relaxation. First, the (x, y) locations of the TSVs are fixed and their distribution in the z direction is determined. An Elmore delay-like thermal estimated formulation [19] is developed for this vertical direction, and the distribution of the TSVs is based on a theoretical result. However, this result assumes that the number of TSVs is unconstrained, which is typically not true in practice. Next, these vias are moved horizontally within each tier, according to the vertical heat flow in each tile. These two steps are iterated until a solution is found.

6.4.2 A Two-Phase Approach Using Linear Programming

The work in [20] presents an approach for thermally aware routing that simultaneously creates a thermal conduction network while managing congestion constraints. The approach effectively reduces on-chip temperatures by appropriate insertion of thermal vias and thermal wires to generate a routing solution free of thermal and routing capacity violations. As defined earlier, thermal vias correspond to vertical intertier vias that do not have any electrical function but are explicitly added as thermal conduits. Thermal wires as defined in [20] perform a similar function but conduct heat laterally within the same tier, which is especially useful for lateral heat spread (e.g., when adjacent tiers are separated by insulating layers, and the availability of thermal vias is limited). Thermal vias perform the bulk of the conduction to the heat sink, while thermal wires help distribute the heat paths over multiple thermal vias.

Figure 6.7 shows how intertier vias can reduce the routing capacity of a neighboring lateral routing edge. If $v_i \times v_i$ intertier vias pass through grid cell i and $v_j \times v_j$ intertier vias pass through the adjacent grid cell j, the signal routing capacity of boundary e_{ij} will be reduced from the original capacity, C_e, and the signal wire usage W_e is required to satisfy

$$W_e \leq \min\left(C_e - v_i \cdot w, C_e - v_j \cdot w\right) \tag{6.11}$$

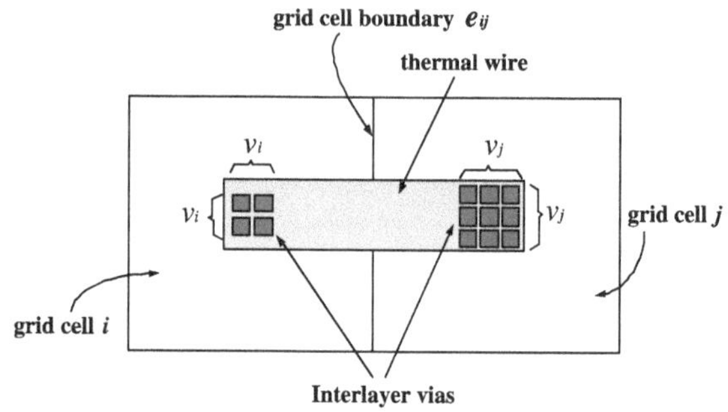

Figure 6.7 Reduction of lateral routing capacity due to the intertier vias in neighboring grid; thermal wires are lumped, and together with thermal vias form a thermal dissipation network [20]. ©2006 IEEE

where w is the geometrical width of an intertier via. Here, the smaller of the two reduced routing widths is defined as the reduced edge capacity, so that there can be a feasible translation from the global routing result to a detailed routing solution. On the other hand, given the actual signal wire usage W_e of a routing edge, Eq. (6.11) can also be used to determine how many intertier vias can go through the neighboring grid cells so that there is no overflow at the routing edge. Since temperature reduction requires insertion of a large number of thermal vias, careful planning is necessary to meet both the temperature and routability requirements.

A simple way of improving lateral thermal conduction could be to identify routing edges where signal wires may not utilize all of the routing tracks. The remaining tracks here may be employed to connect the thermal vias in adjoining grid cells with *thermal wires*. These are connected directly to thermal vias to form an efficient heat dissipation network as shown in Fig. 6.7. Thermal wires enable the conduction of heat in the lateral direction and can thus help vertical thermal vias to reduce hot spot temperature efficiently: for those hot spots where only a restricted number of thermal vias can be added, thermal wires can be used to conduct heat laterally and then remove heat through thermal vias in adjoining grids. Thermal wires also help in providing more uniform metallization, which has advantages from the point of CMP polishing [21].

However, the deferment of the thermal wire or thermal via additions to a post-routing post-processing step, using the resources that are left unused after routing, is clearly suboptimal: ideally, these should be allocated *during* routing. In other words, since thermal vias and wires contend for routing resources with signal wires and vias, they should be well planned to satisfy the temperature and routability requirements. The approach in [20] provides a framework to achieve this.

The global routing approach in [20] proceeds in two phases, and its overall flow is shown in Fig. 6.8. The input to the algorithm is a tessellated 3D circuit with given power distribution. The algorithm proceeds in two phases, where Phase I corresponds to the first three boxes of the flow, and Phase II is represented by the iterative loop. Practically, it is seen that this loop converges to the optimized solution in a small number of iterations.

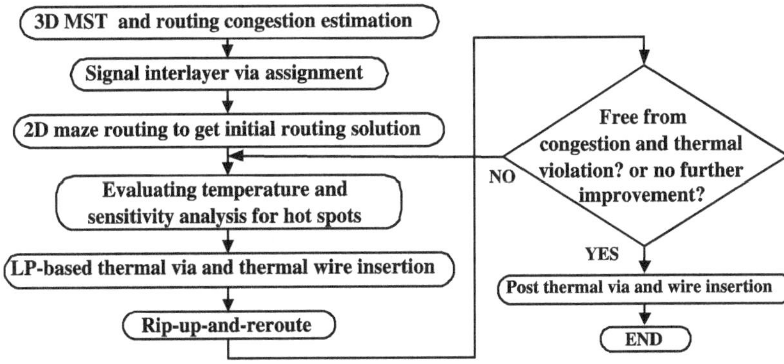

Figure 6.8 Overall flow for the temperature-aware 3D global routing algorithm [20]. © 2006 IEEE

Phase I begins with a minimum spanning tree (MST) generation and routing congestion estimation step. Next, signal intertier vias are assigned using a recursive network flow-based formulation. Once these intertier vias are assigned, the problem reduces to a 2D problem in each tier, and a thermally driven 2D maze router, which augments the standard maze routing cost function with an additional temperature term, is used to solve this problem separately in each tier.

Next, Phase II performs iterative routing involving rip-up-and-reroute and LP-based thermal via/wire insertion. For each of n temperature-violating hot spots with temperature $T_i > T_{\text{target}}$, $i = 1,2,\ldots,n$, fast adjoint sensitivity analysis is performed to find the sensitivity of T_i with respect to the number of thermal vias at each thermal via location. If the sensitivity value exceeds a threshold in a location, i.e., $s_{v,ij} \geq s_{\text{th}}$, then this location j is a candidate for thermal via insertion. Similarly, candidate thermal wire locations can be defined, and their sensitivities, $s_{w,ik}$, obtained.

A linear program is formulated to insert thermal vias and wires using a linearized model based on this sensitivity to achieve a small improvement in the temperature (consistent with the range of the sensitivity-based model). A very small violation in the routing capacities is permitted in this stage, on the understanding that it can probably be rectified using the rip-up-and-reroute step.

Let $N_{v,j}$ be the number of inserted thermal vias at candidate location j, and $N_{w,k}$ be the number of inserted thermal wires at candidate thermal wire location k, so that the temperature at hot spot i can be reduced by ΔT_i. The LP is formulated as:

$$\text{minimize} \sum_{j=1}^{p} N_{v,j} + \sum_{k=1}^{q} N_{w,k} + \Gamma \sum_{i=1}^{n} \delta_i \tag{6.12}$$

$$\begin{gathered}\text{subject to: } \sum_{j=1}^{p} -S_{v,ij}N_{v,j} + \sum_{k=1}^{q} -S_{w,ik}N_{w,k} + \delta_i \geq \Delta T_i, \\ i = 1,2,\ldots,n,\ \Delta T_i = T_i - T_{\text{target}}\end{gathered} \tag{6.13}$$

$$N_{v,j} \leq \min\left((1+\beta)R_{v,j},\, U_j - V_j\right), j = 1,2,...,p \tag{6.14}$$

$$N_{w,k} \leq (1+\beta)R_{w,k},\, k = 1,2,...,q \tag{6.15}$$

$$\delta_i \geq 0,\, i = 1,2,...,n;\, N_{v,j} \geq 0, j = 1,2,...,p; \\ N_{w,k} \geq 0,\, k = 1,2,...,q \tag{6.16}$$

The objective function that minimizes the total usage of thermal vias and wires is consistent with the goal of routing congestion reduction. To guarantee that the problem is feasible, the relaxation variables δ_i, $i = 1, 2...,n$, are introduced. The constant Γ remains the same over all iterations and is chosen to be a value that is large enough to suppress the value of δ_i to be 0 when the thermal via and thermal wire resources in constraints (6.14) and (6.15) are enough to reduce temperature as desired.

Constraint (6.13) requires that the temperature reduction at hot spot i, plus a relaxation variable δ_i (introduced to ensure that the problem is feasible), should be at least ΔT_i during the current iteration, where ΔT_i is the difference between current temperature T_i and target temperature T_{target}.

Constraints (6.14) and (6.15) are related to capacity constraints on the thermal vias and thermal wires, respectively, based on lateral boundary capacity overflows in the same tier, and intertier via capacity overflows across tiers. Constraint (6.14) sets the upper limit for the number of thermal via insertions $N_{v,\,j}$, with two limiting factors. $R_{v,\,j}$ is the maximum number of additional thermal vias that can be inserted at location j without incurring lateral routing overflow on a neighboring edge, and it is calculated as $R_{v,j} = v_j - v_{\text{cur.},j}$ in which $v_{\text{cur.},j}$ is the current intertier via usage at location j and v_j is the maximum number of intertier vias that can be inserted at location j without incurring lateral overflow. Adding more intertier vias in the most sensitive locations can be very influential in temperature reduction, and therefore, the constraint is intentionally amplified by a factor β to temporarily permit a violation of this constraint but which will allow better temperature reduction. This can potentially result in lateral routing overflow after the thermal via assignment, but this overflow can be resolved in the iterative rip-up-and-rerouting phase.

A second limiting factor for $N_{v,\,j}$ is that the total intertier via usage cannot exceed U_j, which is the intertier via capacity at position j, and the constraint formulation takes the minimum of the two limiting factors. Similarly, constraint (6.15) sets a limit on the number of thermal wire insertions with the consideration of lateral routing overflow. $R_{w,\,k}$ is the maximum number of additional thermal wires that can be inserted at location k without incurring lateral routing overflow, and it is calculated as $R_{w,k} = m_k - m_{\text{cur.},k}$, where $m_{\text{cur.},k}$ is the current thermal wire usage at location k, and m_k is the maximum number of thermal wires at location k without incurring lateral overflow. In the same spirit of encouraging temperature reduction, $R_{w,\,k}$ is relaxed by a factor of β, and any potential overflow will be resolved in the rip-up-and-rerouting phase.

Details of the experimental results for this approach are described in [20]. Four sets of results are generated: (i) temperature-aware routing (TA) for the temperature-aware routing algorithm described above; (ii) post-insertion routing (P) for a scheme that uses Phase I of the above algorithm but then inserts thermal vias and thermal wires into all the available space; (iii) thermal vias only (V), which uses the above approach but only uses thermal vias and no thermal wires; and (iv) uniform via insertion (U), which uses the same number of thermal vias and wires as TA but distributes these uniformly over the entire layout area. Experimental results, illustrated in Fig. 6.9, show that in comparison with TA, the V, P, and U schemes all have significantly higher peak temperatures. While the U case appears, at first glance, to have similar thermal profiles as the TA case, this situation results in massive routing overflows and is not a legal solution. The wire length overhead of TA is found to be just slightly higher than that of P which can be considered to be the thermally unaware routing case.

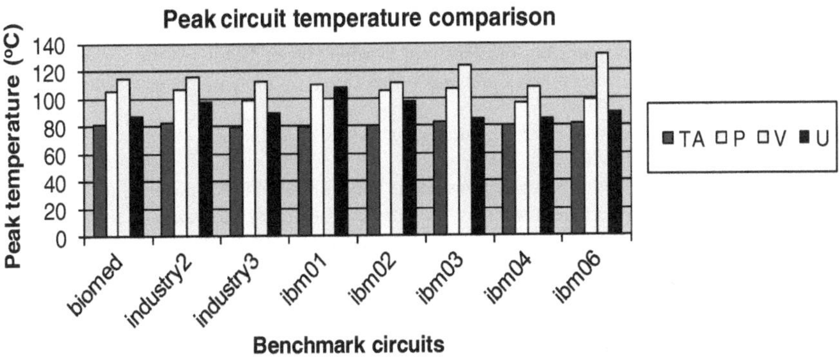

Figure 6.9 Overall flow for the temperature-aware 3D global routing algorithm

6.5 Conclusion

This chapter has presented an overview of approaches for routing and thermal via insertion in 3D ICs. The two problems are related, since they compete for the same finite set of on-chip interconnect resources, and judicious management of these resources can be seen to provide significant improvements in the thermal profile while maintaining routability.

Acknowledgments Thanks to Brent Goplen and Tianpei Zhang, and the UCLA group led by Jason Cong, whose work has contributed significantly to the contents of this chapter.

References

1. J. A. Burns, B. F. Aull, C. K. Chen, C. L. Keast, J. M. Knecht, V. Suntharalingam, K. Warner, P. W. Wyatt, and D. Yost. A wafer-scale 3-D circuit integration technology. *IEEE Transactions on Electron Devices*, 53(10):2507–2516, October 2006.
2. S. Lee, T.F. Lemczyk, and M. M. Yovanovich. Analysis of thermal vias in high density interconnect technology. In *Proceedings of the IEEE Annual Semiconductor Thermal Measurement and Management Symposium (Semi-Therm)*, pp. 55–61, 1992.

3. R.S. Li. Optimization of thermal via design parameters based on an analytical thermal resistance model. In *Proceedings of Thermal and Thermomechanical Phenomena in Electronic Systems*, pp. 475–480, 1998.
4. D. Pinjala, M.K. Iyer, Chow Seng Guan, and I.J. Rasiah. Thermal characterization of vias using compact models. In *Proceedings of the Electronics Packaging Technology Conference*, pp. 144–147, 2000.
5. T-Y Chiang, K. Banerjee, and K. C. Saraswat. Effect of via separation and low-k dielectric materials on the thermal characteristics of cu interconnects. In *IEEE International Electronic Devices Meeting*, pp. 261–264, 2000.
6. A. Rahman and R. Reif. Thermal analysis of three-dimensional (3-D) integrated circuits (ICs). In *Proceedings of the Interconnect Technology Conference*, pp. 157–159, 2001.
7. T-Y. Chiang, S.J. Souri, Chi On Chui, and K.C. Saraswat. Thermal analysis of heterogeneous 3D ICs with various integration scenarios. In *IEEE International Electronic Devices Meeting*, pp. 681–684, 2001.
8. B. Goplen and S. S. Sapatnekar. Thermal via placement in 3D ICs. In *Proceedings of the International Symposium on Physical Design*, pp. 167–174, 2005.
9. B. Goplen and S. S. Sapatnekar. Placement of thermal vias in 3-D ICs using various thermal objectives. *IEEE Transactions on Computer-Aided Design of Integrated Circuits and Systems*, 26(4):692–709, April 2006.
10. B. Goplen. *Advanced Placement Techniques for Future VLSI Circuits*. PhD thesis, University of Minnesota, Minneapolis, MN, 2006.
11. H. Yu, Y. Shi, L. He, and T. Karnik. Thermal via allocation for 3D ICs considering temporally and spatially variant thermal power. In *Proceedings of the ACM International Symposium on Low Power Electronics and Design*, pp. 156–161, 2006.
12. H. Su, S. R. Nassif, and S. S. Sapatnekar. An algorithm for optimal decoupling capacitor sizing and placement for standard cell layouts. In *Proceedings of the International Symposium on Physical Design*, pp. 68–73, 2002.
13. H. Yu, J. Ho, and L. He. Simultaneous power and thermal integrity driven via stapling in 3D ICs. In *Proceedings of the IEEE/ACM International Conference on Computer-Aided Design*, pp. 802–808, 2006.
14. C. Ababei, Y. Feng, B. Goplen, H. Mogal, T. Zhang, K. Bazargan, and S. Sapatnekar. Placement and routing in 3D integrated circuits. *IEEE Design & Test*, 22(6):520–531, November–December 2005.
15. J. Cong and Y. Zhang. Thermal-driven multilevel routing for 3-D ICs. In *Proceedings of the Asia-South Pacific Design Automation Conference*, pp. 121–126, 2005.
16. J. Cong and Y. Zhang. Thermal via planning for 3-D ICs. In *Proceedings of the IEEE/ACM International Conference on Computer-Aided Design*, pp. 745–752, 2005.
17. J. Cong, M. Xie, and Y. Zhang. An enhanced multilevel routing system. In *Proceedings of the IEEE/ACM International Conference on Computer-Aided Design*, pp. 51–58, 2002.
18. P. Wilkerson, M. Furmanczyk, and M. Turowski. Compact thermal modeling analysis for 3D integrated circuits. In *Proceedings of the International Conference on Mixed Design of Integrated Circuits and Systems*, pp. 24–26 2004.
19. S. S. Sapatnekar. *Timing*. Springer, Boston, MA, 2004.
20. T. Zhang, Y. Zhan, and S. S. Sapatnekar. Temperature-aware routing in 3D ICs. In *Proceedings of the Asia-South Pacific Design Automation Conference*, pp. 309–314, 2006.
21. A. B. Kahng and K. Samadi. CMP fill synthesis: A survey of recent studies. *IEEE Transactions on Computer-Aided Design of Integrated Circuits and Systems*, 27(1):3–19, January 2008.

Chapter 7
Three-Dimensional Microprocessor Design

Gabriel H. Loh

Abstract Three-dimensional integration provides many new exciting opportunities for computer architects. There are many potential ways to apply 3D technology to the design and implementation of microprocessors. In this chapter, we discuss a range of approaches from simple rearrangements of traditional 2D components all the way down to very fine-grained partitioning of individual processor functional unit blocks across multiple layers. This chapter also discusses different techniques and trade-offs for situations where die-to-die communication resources are constrained and what the computer architect can do to alter a design deal with this. Three dimensional integration provides many ways to reduce or eliminate wires within the microprocessor, and this chapter also discusses high-level design styles for converting the wire reduction into performance or power benefits.

7.1 Introduction

Three-dimensional integration presents new opportunities for the design (or redesign) of microprocessors. While this chapter focuses on high-performance processors, most of the concepts and techniques can be applied to other market segments such as embedded processors. The chapter focuses on general design techniques and patterns for 3D processors. The best way to leverage this technology for a future 3D processor will depend on many factors, including how the fabrication technology develops and scales, how cooling and packaging technologies progress, performance requirements, power constraints, limitations of engineering effort, and many other issues. As 3D processor architectures evolve and mature, a combination of the techniques described in this chapter will likely be employed.

G.H. Loh (✉)
College of Computing, Georgia Institute of Technology, Atlanta, GA, USA
e-mail: loh@cc.gatech.edu

Y. Xie et al. (eds.), *Three-Dimensional Integrated Circuit Design*,
Integrated Circuits and Systems, DOI 10.1007/978-1-4419-0784-4_7,

This chapter is organized in a forward-looking chronological fashion. We start by exploring near-term opportunities for 3D processor designs that stack large macro-modules (e.g., entire cores), thereby requiring minimal changes to conventional 2D architectures. We then consider designs where the processor blocks (e.g., register file, ALU) are reorganized in 3D, which allows for more flexibility and greater optimization of the pipeline. Finally, we study fine-grained 3D organizations where even individual blocks may be partitioned such that their logic and wiring are distributed across multiple layers. Table 7.1 details the benefits and obstacles for the different granularities of 3D stacking.

Table 7.1 Overview of the pros and cons of 3D stacking at different granularities

Stacking granularity	Potential benefits	Redesign effort
Entire cores, caches	Added functionality, more transistors, mixed-process integration	Low: Reuse existing 2D designs
Functional unit blocks	Reduced latency and power of global routes provide simultaneous performance improvement with power reduction	Must re-floorplan and retime paths. Need 3D block-level place-and-route tools. Existing 2D blocks can be reused
Logic gates (block splitting)	Reduced latency/power of global, semi-global, and local routes. Further area reduction due to compact footprints of blocks and resizing opportunities	Need new 3D circuit designs, methodologies, and layout tools. Reuse existing 2D standard cell libraries

The exact technical details of future, mass-market, volume-production 3D technologies are unknown. We already have a good grasp on what is technically *possible*, but economics, market demands, and other nontechnical factors may influence the future development of this technology. Perhaps the most important technology parameter is the size of the die-to-die or through-silicon vias (TSVs). With a very tight TSV pitch, processor blocks may be partitioned at very fine levels of granularity. With coarser TSVs, the 3D options may be limited to block-level or even core-level stacked arrangements. Throughout the rest of this chapter, we will revisit how the size of the TSVs impacts designs, and in many cases how one can potentially design *around* these constraints.

7.2 Stacking Complete Modules

While 3D microprocessors may eventually make use of finely partitioned structures, with functional units, wiring, and gates distributed over multiple silicon layers, near-term 3D solutions will likely be much simpler. The introduction of 3D integration to a mass-production fabrication plant will already incur some significant technology risks, and therefore risks in other areas of the design (i.e., the processor architecture) should be minimized. With this in mind, the simplest applications for 3D stacking are those that involve reusing existing 2D designs. In this section, we explore three general approaches that fall under this category: enhancing the cache hierarchy, using 3D to provide optional functionality, and system-level integration.

7.2.1 Three-Dimensional Stacked Caches

Stacking additional layers of silicon using 3D integration provides the processor architect with more transistors. In this section, we are explicitly avoiding any 3D designs that require the complete reimplementation of complex macro-units (such as an entire processor pipeline). The easiest way to make use of the additional transistors is to either add more cache and/or add more cores.

Even with an idea as straightforward as using 3D to increase cache capacity, there still exists several design options for constructing a 3D-stacked level 2 (L2) cache [1].[1] Figure 7.1a illustrates a conventional dual-core processor featuring a 4 MB L2 cache. Since the L2 cache occupies approximately one half of the die's silicon area, stacking a second layer of silicon with equal area would provide an additional 8 MB of cache, for a total of 12 MB, as shown in Fig. 7.1b. Note that from the center of the bottom layer where the L2 controller resides, the lateral (in-plane) distance to the furthest cells is approximately the same in all directions. When combined with the fact that the TSV latency is very small, this 3D cache organization has nearly no impact on the L2 access latency. Contrast this to building a 12 MB cache in a conventional 2D technology, as shown in Fig. 7.1c, where the worst-case access must be routed a much greater distance, thereby increasing the latency of the cache.

With the wafer-stacking or die-stacking approaches to 3D integration, the individual layers are manufactured separately prior to bonding. As a result, the fabrication processes used for the individual layers need not be the same. An alternative approach for 3D stacking a large cache is to implement the cache in DRAM instead of a conventional logic/CMOS process. Memory implemented as DRAM provides much greater storage density (bits/cm^2) than SRAM, therefore a cache implemented using DRAM could potentially provide much more storage for the same area. Figure 7.2a illustrates the same dual-core processor as before, but the SRAM-based L2 cache has been completely removed, and a 32 MB DRAM-based L2 cache has been stacked on top of the cores.

While the stacked DRAM design point provides substantially more on-chip storage than the SRAM approach, the cost is that the latency of accessing the DRAM structure is much greater than that for an SRAM-based approach. The SRAM cache

[1] The L2 cache is sometimes also referred to as the last-level cache (LLC) to avoid confusion when comparing to architectures that feature level-3 caches.

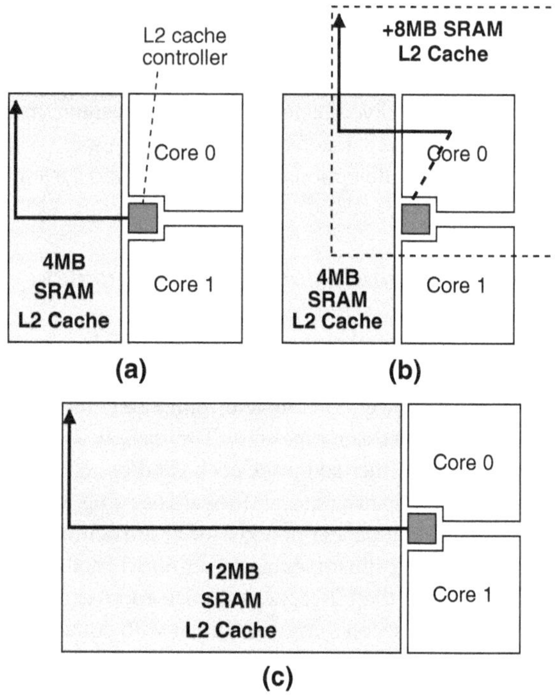

Fig. 7.1 (**a**) Conventional 2D dual-core processor with L2 cache, (**b**) the same processor augmented with MB more of L2 cache in a 3D organization, (**c**) the equivalent 2D layout for supporting 12 MB total of L2 cache

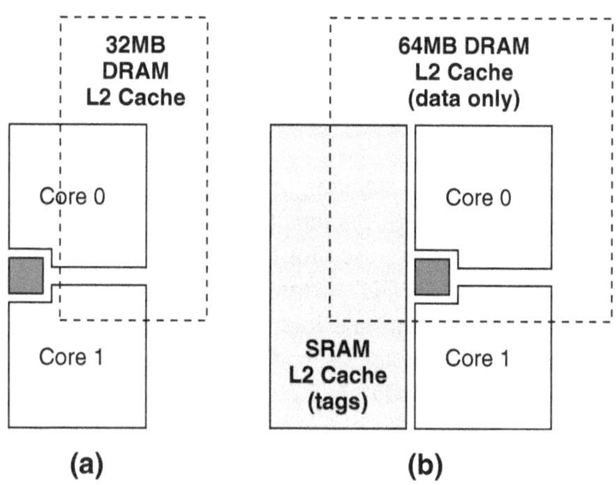

Fig. 7.2 (**a**) Stacking a 32 MB DRAM L2 cache and (**b**) a hybrid organization with SRAM tags and 3D-stacked DRAM for data

has an access latency of 10–20 cycles, whereas the DRAM cache requires 50–150 cycles (depending on row buffer hits, precharge latency, and other memory parameters). Consider the three hypothetical applications shown in Table 7.2. Program A has a relatively small working set that fits in a 4 MB SRAM cache. Program B has a larger working set that does not fit in the 4 MB SRAM cache but does fit

Table 7.2 Access latencies for different cache configurations, the number of hits and misses for three different programs, and the average cycles per memory access (CPMA) assuming a 500-cycle main memory latency. The level-1 cache is ignored for this example

Cache	L2 latency		Program A			Program B			Program C		
organization	Hit	Miss	Hits	Miss	CPMA	Hits	Miss	CPMA	Hits	Miss	CPMA
2D/4 MB (SRAM)	16	16	900	100	6.6	200	800	41.6	100	900	46.6
3D/12 MB (SRAM)	16	16	902	98	6.5	600	400	21.6	100	900	46.6
3D/32 MB (DRAM)	100	100	904	96	14.8	880	120	16.0	100	900	55.0
3D/64 MB (hybrid)	100	16	908	92	13.8	960	40	11.7	100	900	47.4

within the 32 MB DRAM cache. Program C has streaming memory access patterns and has very poor cache hit rates for both cache configurations. For Program A, both cache configurations provide very low miss rates, but the SRAM cache's lower latency yields a lower average number of *cycles* per memory access (CPMA). For Program B's larger working set, the SRAM cache yields a very large number of cache misses, resulting in a high average CPMA metric. While the DRAM cache still has a greater access latency than the SRAM cache, this is still significantly less than the penalty of accessing the off-chip main memory. As a result, the DRAM cache provides a lower CPMA metric for Program B. For Program C, neither configuration can deliver a high cache hit rate, and so the CPMA metric is dominated by the latency required to determine a cache miss. The faster access latency of the SRAM implementation again yields lower average CPMA. The best design point for a 3D-stacked L2 clearly depends on the target application workloads that will be executed on the processor.

The previous example demonstrates that the latency for both cache hits and cache misses may be very important depending on the underlying application's memory access patterns. A third option combines both SRAM and DRAM to build a hybrid cache structure as shown in Fig. 7.2b. The bottom layer uses the SRAM array to store only the *tags* of the L2 cache. The top layer uses DRAM to store the actual individual cache lines (data). On a cache access, the SRAM tags can quickly provide the hit/miss indication. If the access results in a miss, the request can be sent to the memory controller for off-chip access immediately after the fast SRAM lookup. Contrast this to the pure DRAM organization that requires the slower DRAM access regardless of whether there is a hit or a miss. The last row in Table 7.2 shows how this hybrid SRAM-tag/DRAM-data design improves the CPMA metric over the pure DRAM approach for all three programs.

Three-dimensional stacking enables a large last-level cache to be placed directly on top of the processor cores with very short, low-latency interconnects. Furthermore, since this interface does not need full I/O pads which consume a lot of area, but relatively much smaller TSVs, one can build a very wide interface to the stacked cache. The desired width of the interface would likely be the size of a full cache line plus associated address and control bits. For example, a 64-byte linesize would require a 512-bit data bus plus a few dozen extra bits for the block's physical address and command/control information. To make signaling easier, one could even build two separate datapaths with one dedicated for each direction of communication. As transistor sizes continue to decrease, however, the size and pitch of the TSVs may not scale at the same pace. This results in a gradual *increase* in the *relative* size and pitch of the TSVs. To continue to exploit 3D stacking, the interface for the stacked cache will need to constantly adjust to the changing TSV parameters. For example, early designs may use two uni-directional datapaths to communicate to/from the 3D-stacked cache, but as the relative TSV sizes increase, one may need to use a single bi-directional bus. Another orthogonal possibility is to reduce the width of the bus and then pipeline the data transfer over multiple cycles. These are just a few examples to demonstrate that a design can be adapted over a wide range of TSV characteristics.

7.2.2 Optional Functionality

Three-dimensional integration can increase manufacturing costs by increasing the total amount of silicon required for the chip (i.e., the sum of all layers), extra manufacturing steps for bonding, impact on yield rates, and other reasons. Furthermore, not all markets may need the extra functionality provided by this technology. A second approach for leveraging 3D stacking is to use it as a means to optionally *augment* the processor with some additional functionality. For example, a 4 MB L2 cache may be sufficient for many markets, or the added cost and power may not be appropriate for others (e.g., low cost or mobile). In such cases, the original 2D, single-layer microprocessor is more desirable. For markets that do benefit from the additional cache (e.g., servers, workstations), however, 3D can be used to provide this functionality *without requiring a completely new processor design.* That is, with a single design effort, the processor manufacturer can leverage 3D to adapt its product to a wider range of uses.

7.2.2.1 Introspective 3D Processors

Apart from pure performance enhancements, 3D can also be used to provide new capabilities to the microprocessor. In particular, Loi et al. proposed 3D-stacked *Introspective* processors [2]. Some programmers and engineers would greatly benefit from being able to access more dynamic information about the internal state of a microprocessor. Modern *hardware performance monitoring* (HPM) support only allows the user to monitor some basic statistics about the processor, such as the number of cache misses or branch predictor hit rates. There are many richer types of data that would be tremendously useful to software and hardware developers, but adding this functionality into standard processors has some significant costs. Consider Fig. 7.3a that depicts a conceptual processor floorplan. Each dot on

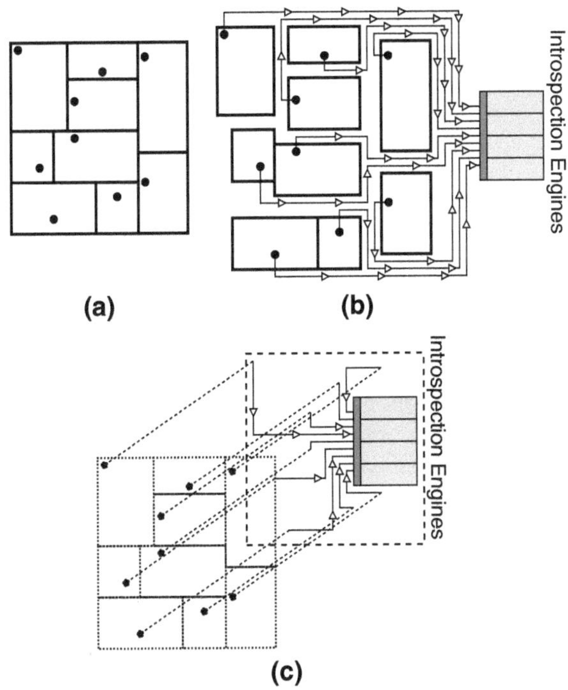

Fig. 7.3 (**a**) Processor floorplan with data observation points marked by dots, (**b**) the same floorplan with extra space for the wiring and repeaters to forward data to the introspection engines, and (**c**) a 3D introspective processor

the floorplan represents a site where we would like to monitor some information (e.g., reorder buffer occupancy statistics, functional unit utilization rates, memory addresses). To expose this information to the user, the information first needs to be collected to some centralized HPM unit. The user can typically configure the HPM unit to select the desired statistics. Additional hardware introspection engines could be included to perform more complicated analysis, such as data profiling, memory profiling, security checks, and so forth. Figure 7.3b shows how the overall processor floorplan could be impacted by required routing. The additional wires, as well as repeaters/buffers, all require the allocation of additional die space. This in turn can increase increase wire distances between adjacent functional unit blocks, which in turn can lead to a decrease in performance. The overall die size may grow which increases the cost of the chip. While this profiling capability is useful for developers, the vast majority of users will not make use of it.

The key ideas behind introspective 3D chips are to first leave the base 2D processor as unmodified as possible to minimize the impact on the high-volume commodity processors, and then to leverage 3D to *optionally* provide the additional profiling support for the relatively small number of hardware designers, software developers, and OEMs (original equipment manufacturers). Figure 7.3c illustrates the two layers of an introspective 3D chip. The optional top layer illustrates a few example profiling engines. It is possible to design multiple types of introspection layers and then stack a different engine or set of engines for different types of developers. The main point is that this approach provides the capability for adding introspection facilities, but the impact on the base processor layer is minimal, as can be seen by comparing the processor floorplans of Fig. 7.3a,c.

7.2.2.2 Reliable 3D Processors

The small size of devices in modern processors already makes them vulnerable to data corruption due to a variety of reasons, such as higher temperatures, power supply noise, interconnect cross-talk, and random impacts from high-energy particles (e.g., alpha particles). While many SRAM structures in current processors already employ error-correcting codes (ECC) to protect against these *soft errors* [3], as device sizes continue to decrease, the vulnerability of future processors will increase.

Given the assumption that a conventional processor may be prone to errors and may yield incorrect results, one approach to safeguard against this is to provide some form of redundancy. With two copies of the processor, the processors can be forced to operate in lock-step. Each result produced by a processor can be checked against the other. If the results ever disagree, one (or both) must have experienced an error. At this point, the system flushes both pipelines and then re-executes the offending instruction. With *triple modular redundancy*, a majority vote can be used among the three pipelines so that re-execution is not needed. The obvious cost is that multiple copies of the pipeline are required, which dramatically increase the cost of the system.

Instead of using multiple pipelines operating in a lock-step fashion, another approach is to organize two pipelines as a leading execution core and a trailing checking core. For each instruction that the leading core executes, the trailing core will re-execute at a later time (*not* lock-step) to detect possible errors. While this sounds very similar to the modular redundancy approach, this organization enables an optimized checker core which reduces costs. For example, instead of implementing an expensive branch predictor, the checker core can simply use the actual branch outcomes computed by the leading core. Except in the very rare occasion of a soft error, the leading core's branch outcome will be correct and therefore the trailing core will benefit from what is effectively perfect branch prediction. Similarly, the leading core acts as a memory prefetcher, so that the trailing checker core almost always hits in the cache hierarchy. There are many other optimizations for reducing the checker core cost that will not be described here [4].

Even with an optimized checker core, the additional pipeline still requires more area than the original unmodified processor pipeline. Similar to the motivation for the introspective 3D processors, not all users may need this level of reliability in their systems, and they certainly do not want to pay more money for features that they do not care about. Three-dimentional stacking can also be used to *optionally* augment a conventional processor with a checker core to provide a high-reliability system [5]. Figure 7.4a shows the organization of a 2D processor with both leading and checking cores. Similar to Fig. 7.3, the extra wiring required to communicate

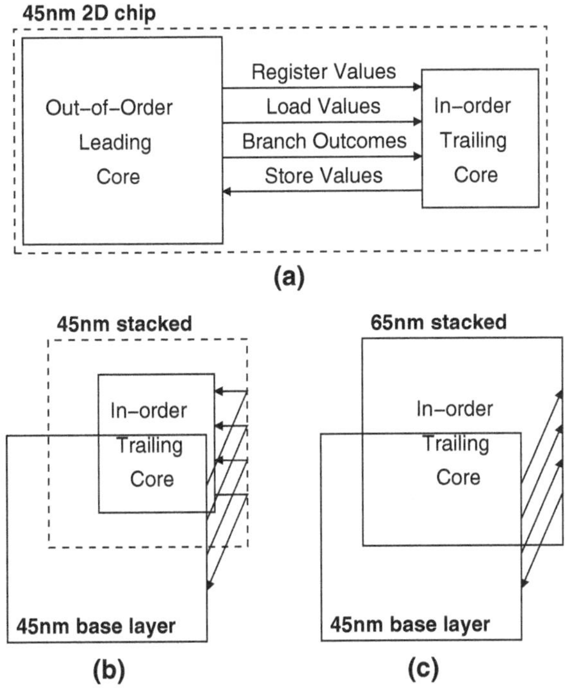

Fig. 7.4 (**a**) A reliable processor architecture with a small verification core, (**b**) 3D-stacked version with the trailing core on top, and (**c**) the same but with the trailing core implemented in an older process technology.

between cores may increase area overhead. The latency of this communication can also impact performance by delaying messages between the cores.

A 3D-stacked organization, illustrated in Fig. 7.4b, greatly alleviates many of the shortcomings of the 2D implementation. First, it allows the checker core to be optional, thereby removing its cost impact on unrelated market segments. Second, the 3D organization minimizes the impact of routing between the leading and the checker cores. This affects the wiring overhead, the disruptions to the floorplanning of the baseline processor core, and the latency of communication between cores.

The checker core requires less area than the original leading core, primarily due to the various optimizations described earlier. This disparity in area profiles may leave a significant amount of silicon area unutilized. One could simply make use of the area to implement additional cache banks. Another interesting approach is to use an older technology generation (e.g., 65 nm instead of 45 nm) for the stacked layer, as shown in Fig. 7.4c. First, chips manufactured in older technologies are cheaper. Second, the feature sizes of transistors in older processes are larger, thereby making them less susceptible to soft errors. Such an approach potentially reduces cost and improves reliability at the same time.

The introspective 3D processor and the 3D-stacked reliability enhancements are only two possible ways to use 3D integration to provide optional functionality to an otherwise conventional processor. There are undoubtedly many other possible applications, such as stacked application-specific accelerators or reconfigurable logic.

7.2.2.3 TSV Requirements

For both the introspective 3D processors and the 3D-stacked reliability checker core, the inter-layer communication requirements are very modest and likely will not be limited by the TSV size and pitch. Current wafer-bonding technologies already provide many thousands (10,000–100,000) of TSVs per cm^2. The total number of signal TSVs required for the introspection layer depends on the number of profiling engines and the amount of data that they need to collect and monitor from the processor layer. For tracking the usage rates or occupancies of various microarchitectural structures, only relatively small counters are needed. To reduce TSV requirements, these can even be partitioned such that the counters' least significant k bits are located on the processor layer. Only once every 2^k events does the counter need to communicate a carry bit to the remainder of the counter on the introspection layer (thus requiring only a single TSV per counter). For security profiling, the introspection layer will likely only need to check memory accesses, which should translate into monitoring a few memory address buses and possibly TLB information totaling no more than a few hundred bits.

The primary communication needs of the 3D-stacked reliability checker core are for communicating data values between the leading and the checker cores. The peak communication rate is effectively limited by the commit rate of the leading core. Typical information communicated between cores includes register results, load values, branch outcomes, and store values. Even assuming 128-bit data values (e.g., multimedia registers), one register result, one load value, one store value, and a branch outcome (including direction and target) requires less than 512 bits. For a four-way superscalar processor, this still only adds up to 2048 bits (or TSVs) to communicate between the leading and the checker cores.

7.2.3 System-Level Integration

The previous applications of 3D integration have all extended the capabilities of a conventional microprocessor in some fashion. Three dimension can also be used to integrate beyond the microprocessor. Example components include system memory (DRAM) [6–9], analog components [10], flash memory, image sensor arrays, as well as other components typically found on a system's motherboard. Since this chapter focuses on 3D *microprocessor* design, we will not explore these system-level opportunities any further. Chapter 9 provides an excellent description and discussion of one possible 3D-integrated server system called PicoServer.

7.3 Stacking Functional Unit Blocks

The previous section described several possible applications of 3D integration that do not require any substantial changes to the underlying microprocessor architecture. For the first few generations of 3D microprocessors, it is very likely that designs will favor such minimally invasive approaches to reduce the risks associated with new technologies. Three-dimensional integration will require many new processes, design automation tools, layout support, verification and validation methodologies, and other infrastructure. The earliest versions of these may not efficiently support complex, finely partitioned 3D structures. As the technology advances, however, the computer architect will be able to reorganize the processor pipeline in new ways.

7.3.1 Removing Wires

Wire delay plays a very significant role in the design of modern processors. While each process technology generation provides faster transistors, the delay of the wires has not kept up at the same pace. As a result, *relative* wire delays have been increasing over time. Whereas logic gates used to be the dominating contributor of a processor's cycle time, wire delay is now a first-class design constraint as well. Figure 7.5a shows the stages in the Intel Pentium III's branch misprediction detection pipeline and Fig. 7.5b illustrates the version used in the Intel Pentium 4 processor [11]. Due to a combination of higher target clock speeds and the longer relative wire delays associated with smaller transistor sizes, the Pentium 4 pipeline requires twice as many stages. Furthermore, there are two pipeline stages (highlighted in the figure) that are simply dedicated for driving signals from one part of the chip to another. The wire delays have become so large that by the time the signal reaches its destination, there is no time left in the clock cycle to perform any useful computations. In a 3D implementation, however, the pipeline stages could be reorganized so that previously distant blocks are now vertically stacked on top of each other. As a result, the pipeline stages consisting of only wire delay can now be completely eliminated, thereby reducing the overall pipeline length.

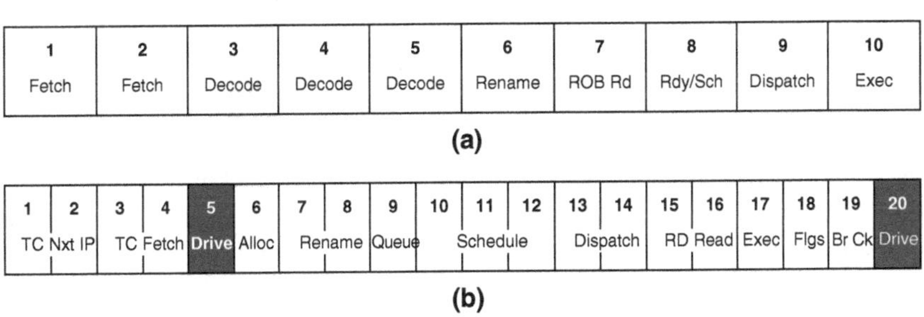

Fig. 7.5 Branch misprediction resolution pipeline for the Intel **(a)** Pentium-III and the **(b)** Pentium 4

Another example of a pipeline designed to cope with increasing wire delays is the Alpha 21264 microprocessor [12]. In particular, a superscalar processor with multiple execution units requires a bypass network to forward results between all of the execution units. This bypass network requires a substantial amount of wiring, and as the number of execution units increases, the lengths of these wires also increase [13]. With a conventional processor organization, the latency of the bypass network would have severely reduced the clock frequency of the Alpha 21264 processor. Instead, the 21264 architects organized the execution units into two groups or *clusters*, as shown in Fig. 7.6a. Each cluster contains its own bypass network for zero-cycle forwarding of results between instructions within the same cluster. If one instruction needs to forward its result to an instruction in the other cluster, then the value must be communicated through a second level of bypassing which incurs an extra cycle of latency. Similar to the extra pipeline stages in the Pentium 4, this extra bypassing is effectively an extra stage consisting of almost nothing but wire delay. In a 3D organization, however, one could conceivably stack the two clusters directly on top of each other, as shown in Fig. 7.6b, to eliminate the long and slow cross-cluster wires, thereby removing the extra clock cycle for forwarding results between clusters. Using 3D to remove extra cycles for forwarding results has also been studied in the context of the data cache to execution unit path and the register file to floating point unit path [1].

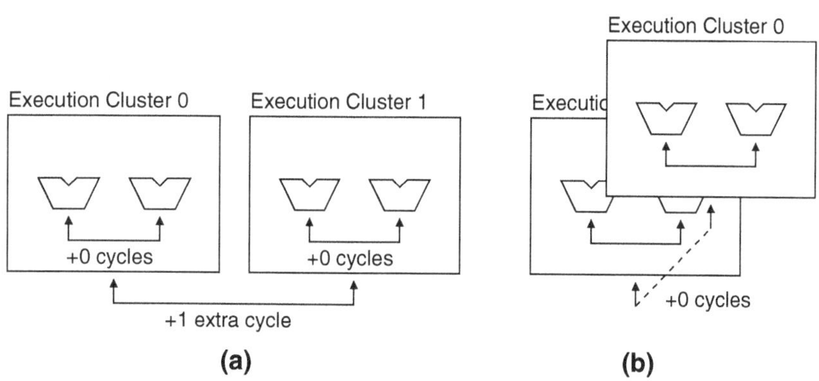

Fig. 7.6 **(a)** Bypass latencies for the Alpha 21264 execution clusters and **(b)** a possible 3D organization

This approach of using 3D integration to stack functional unit blocks provides a much larger level of flexibility in organizing the different pipeline components than compared to the much coarser-grained approach of stacking complete modules discussed in Section 7.2. The benefit is that many more inter-block wires can be shortened or even completely eliminated, which in turn can improve performance and reduce power consumption. Contrast this to traditional microarchitecture techniques for improving performance where increasing performance typically also requires an increase in power, or conversely any attempt to decrease power will often result in a performance penalty. With 3D integration, we are *physically* reducing the amount of wiring in the system. This reduction in the total wire RC directly benefits both latency and energy at the same time.

While stacking functional units provides more opportunities to optimize the processor pipeline, there are some associated costs as well. By removing pipeline stages, the overall pipeline organization may become simpler, but this still requires some nontrivial engineering effort to modify the pipeline and then verify and validate that the new design still works as expected. This represents an additional cost beyond simply reusing a complete 2D processor core. Note that the basic designs of each of the functional unit blocks are still inherently 2D designs. Every block resides on one and one layer only. This allows existing libraries of macros to be reused. In the next section, we will explore the design scenarios enabled when one allows even basic blocks like register files and arithmetic units to be split between layers, but at the cost of even greater design and engineering efforts.

7.3.2 TSV Requirements

The previous technique of stacking complete modules (e.g., cores, cache) on top of each other required relatively few TSVs compared to how many vias 3D stacking can provide. With the stacking of functional unit blocks, however, the number of required TSVs may increase dramatically depending on how the blocks are arranged. For example, the execution cluster stacking of the 21264 discussed in the previous subsection requires bypassing of register results between layers. In particular, each execution cluster can produce up to two 64-bit results per cycle. This requires four results in total (two per direction), which adds up to 256 bits plus additional bits for physical register identifiers. Furthermore, the memory execution cluster can produce two additional load results per cluster which also need to be forwarded to both execution clusters. Assuming the memory execution cluster is located on the bottom cluster, this adds two more 64-bit results for another 128 bits. While in total this still only adds up to a few hundred TSVs, this only accounts for these two or three blocks. If the level-1 data cache is stacked on top of the memory execution cluster, then another two 64-bit data buses and two 64-bit address buses are required for a total of another 256 TSVs. If many pairs of blocks each require a few hundred TSVs, then the total via requirements can very quickly climb to several thousand or even tens of thousands.

In addition to the total TSV requirements, *local* via requirements may also cause problems in the physical layout and subsequent impact of wire lengths. Consider the two blocks in Fig. 7.7a placed side-by-side in 2D with 16 wires connecting them. In this situation, stacking the blocks on top of each other does not cause any problems for the given TSV size, as shown in Fig. 7.7b. Now consider the two blocks in Fig. 7.7c, where there are still 16 wires, but the overall height of the blocks is much shorter. As a result, there is not enough room to fit all of the TSVs;

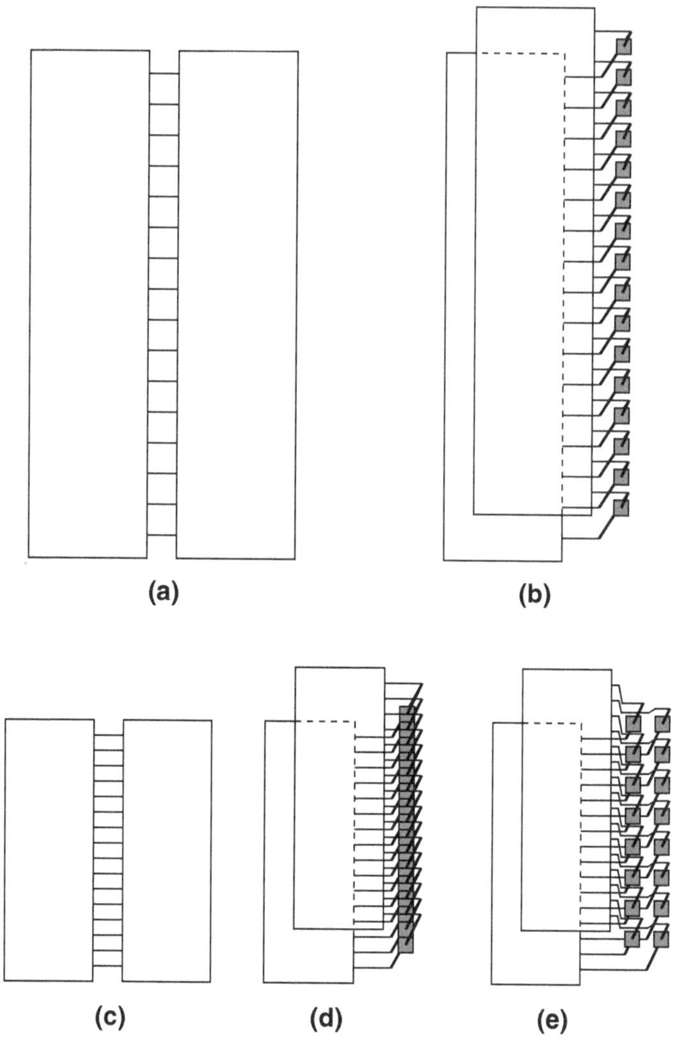

Fig. 7.7 Two communicating blocks with 16 connections: (**a**) original 2D version, (**b**) 3D-stacked version, (**c**) 2D version with tighter pitch, (**d**) nonfunctional 3D version where the TSVs do not fit, and (**e**) an alternate layout.

Fig. 7.7d shows the TSVs all short-circuited together. With different layouts of the TSVs, it may still be possible to re-arrange the connections so that TSV spacing rules are satisfied. Figure 7.7e shows that with some local in-layer routing, one can potentially still make everything fit. Note that the local in-layer routing now re-introduces some wiring overhead, thereby reducing the wire reduction benefit of the 3D layout. In extreme cases, if the TSV requirement is very high and the area is very constrained, the total in-layer routing could completely cancel the original wire reduction benefits of the 3D organization. Such issues need to be considered in early-stage assignments of blocks to layers and the overall floorplanning of the processor data and control paths.

7.3.3 Design Space Issues

Reorganizing the pipeline to stack blocks on blocks may also introduce thermal problems as was discussed in Chapters 4–6. Using the 21264 cluster example again, stacking one execution cluster on top of the other may reduce the lengths of critical wires, but it may simultaneously stack one hot block directly on top of another hot block. The resulting increase in chip temperature can cause the processor's thermal protection mechanism to kick in more frequently. This in turn can cause a lower average voltage and clock speed, resulting in a performance penalty *worse* than that caused by an extra cycle of bypassing. With the greater design flexibility enabled by 3D stacking, we now have more ways to build better products, but we also face a commensurately larger design space that we must carefully navigate while balancing the often conflicting design objectives or high-performance, low-power, low-chip temperatures, low redesign effort, and many other factors.

7.4 Splitting Functional Unit Blocks

Beyond stacking functional unit blocks on top of each other, the next level of granularity that one could apply 3D to is that of actual logic gates. This can enable splitting individual functional units across multiple layers. Some critical blocks in modern high-performance processors have critical paths delays dominated by wire RC. In such cases, reorganizing the functional unit block into a more compact 3D arrangement can help to reduce the lengths of the intra-block wiring and thereby improve the operating frequencies of these blocks. In this section, we study only two example microprocessor blocks, but the general techniques and approaches can be extended or modified to split other blocks as well. The techniques discussed are not meant to be an exhaustive list, but they provide a starting point for thinking about creative ways to organize circuits across multiple layers.

7.4.1 Tradeoffs in 3D Cache Organizations

The area utilization of modern, high-performance microprocessors is dominated by a variety of caches. The level-2/last-level cache already consumes about one half of the overall die area for many popular commodity chips. There are many other caches in the processor, such as the level-1 caches, translation look-aside buffers (TLBs), branch predictor history tables. In this section, we will focus on caches, but many of the ideas can be easily applied to the other SRAM-based memory arrays found in modern pipelines.

We first review the impact of the choice of granularity when it comes to 3D integration. Figure 7.8 shows several different approaches for applying 3D to the L2 cache. For this example, we assume that the L2 cache has been partitioned into eight banks. Figure 7.8a illustrates a traditional 2D layout. Depending on the location of the processor cores' L2 cache access logic, the overall worst-case routing distance (shown with arrows) can be as much as approximately 2x+4y, where x and y are the side-lengths of an L2 bank. Figure 7.8b shows the coarse-grained stacking approach similar to that described in Section 7.2. Note that while the overall footprint of the chip has been reduced by one half, the worst-case wire distance for accessing the farthest bit has not changed compared to the original 2D case.

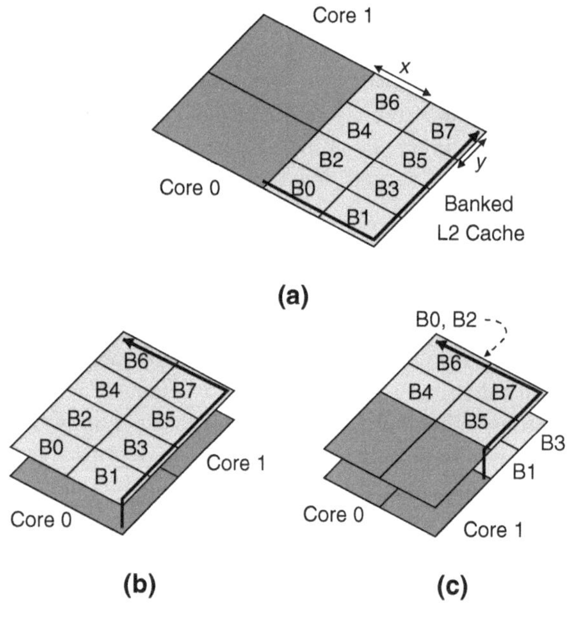

Fig. 7.8 A dual-core processor with an 8-banked L2 cache: (**a**) 2D version, (**b**) cache banks stacked on cores, and (**c**) banks stacked on banks.

At a slightly finer level of granularity, the L2 cache can be stacked on top of itself by rearranging the banks. Figure 7.8c shows a 3D bank-stacked organization. We have assumed that each of the processor cores has also been partitioned over two layers, for example, by stacking blocks on blocks as discussed in Section 7.3. In this example, the worst-case routing distance from the cores to the farthest bitcell has been reduced by $2y$. This wire length reduction translates directly into a reduction in the L2 cache access latency. An advantage of this organization is that the circuit layouts of the individual banks remain largely unchanged, and so even though the overall L2 cache has been partitioned across more than one layer, this approach does not require a radical redesign of the cache.

7.4.1.1 Three Dimension Splitting the Cache

While the long, global wires often contribute quite significantly to the latency of the L2 cache access, there are many wires *within* each bank that also greatly impact overall latency. The long global wires use the upper levels of metal which are typically engineered to facilitate the transmission of signals over longer distances. This may include careful selection of wire geometries (e.g., width-to-height aspect ratio), inter-wire spacing rules, as well as optimal placement and sizing of repeaters. The wires within a block may still have relatively long lengths, but intra-block routing usually employs the intermediate metal layers which are not as well optimized for long distances. Furthermore, the logic within blocks typically exhibits a higher density, which may make the optimal placement and sizing of repeaters nearly impossible. To deal with this, we can also consider *splitting* individual cache banks across multiple layers.

While the exact organization of a cache structure varies greatly from one implementation to the next in terms of cell design, number of sets, set associativity, tag sizes, line sizes, etc., the basic underlying structure and circuit topology are largely the same. Figure 7.9a illustrates a basic SRAM organization for reading a single bit. The address to be read is presented to the row decoder. The row decoder asserts one and only one (i.e., one hot) of its output wordlines. The activated wordline causes all memory cells in that row to output their values onto the bitlines. A column multiplexer, controlled by bits from the address, selects one of the bitline pairs (one column) and passes those signals to the sense amplifier. The sense amplifier speeds up the read access by quickly detecting any small differences between the bit lines. For a traditional cache, there may be multiple parallel arrays to implement both the data and tag portions of the cache. Additional levels of multiplexing, augmented with tag comparison logic, would be necessary to implement a set-associative cache. Such logic is not included in the figures in this section to keep the diagrams simple.

There are two primary approaches to split the SRAM array into a 3D organization [14,15], which we will discuss below in turn. We first consider splitting the cache by stacking columns on columns, as shown in Fig. 7.9b. There are effectively two primary ways to organize this column-stacked circuit. First, one can simply view each row as being split across the two layers. This implies that the original wordline now gets fanned out across both layers. This can still result in a faster wordline activation latency as the length of each line is now about half of its original length. Additional buffers/drivers may be needed to fully optimize the circuit. The output column multiplexer also needs to be partitioned across the two layers. The second column-stacked organization is to treat the array as if it were now organized with twice as many columns, but where each row now has half as many cells, as shown in Fig. 7.9c. This provides wordlines that are truly half of the original length (as opposed to two connected wordlines of half-length each), but it increases the row-decoder logic by one level.

The other natural organization is to stack rows on rows, thereby approximately halving the height of the SRAM array, as shown in Fig. 7.9d. This organization

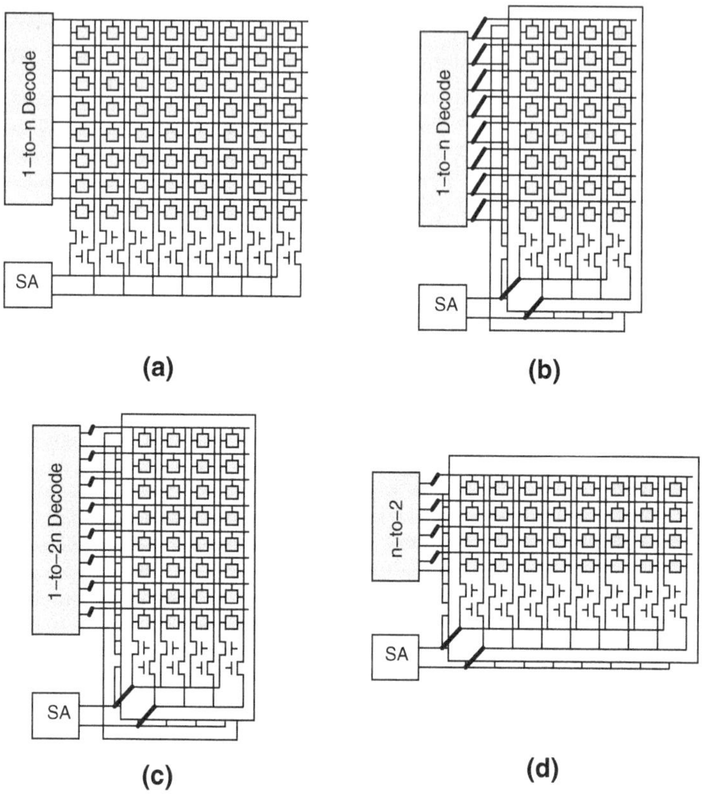

Fig. 7.9 SRAM array organizations: (**a**) original 2D layout, (**b**) 3D column-on-column layout with n split rows, (**c**) 3D column-on-column layout with $2n$ half-length rows, and (**d**) 3D row-on-row layout

requires splitting the row decoder across two layers to select a single row from either the top or the bottom layer of rows. Since only a single row will be selected, the stacked bitlines could in theory be tied together prior to going to the column multiplexer. For latency and power reasons, it will usually be better to treat the halved bitlines as separate multiplexer inputs (as illustrated) to isolate the capacitance between the bitlines. This requires the column multiplexer to handle twice as many inputs as in the baseline 2D case, but the overall cache read latency is less sensitive to column multiplexer latency since the setup of the multiplexer control inputs can be overlapped with the row-decode and memory cell access. In both row-stacked and column-stacked organizations, either the wordlines or the bitlines are reduced. In either case, this wire length reduction can translate into a simultaneous reduction in both latency and energy per access.

We now briefly present some experimental results to quantitatively assess the impact of 3D on cache organizations. These results are based on circuit-level simulations (SPICE) in a 65-nm model. Table 7.3 shows the latency of 2D and 3D

Table 7.3 Simulated latency results (in ns) of various 2D and 3D SRAM implementations in a 65-nm process.

Cache size (KB)	2D latency 1-layer	3D latency 2-layer	3D latency 4-layer
32	0.752	0.635 (−16%)	0.584 (−22%)
64	1.232	0.885 (−28%)	0.731 (−41%)
128	1.716	1.381 (−20%)	1.233 (−28%)
256	2.732	1.929 (−29%)	1.513 (−45%)
512	3.663	2.864 (−22%)	2.461 (−33%)
1024	5.647	3.945 (−30%)	3.066 (−46%)

caches for various sizes. The 3D caches make use of the column-on-column organization. For our circuit models, we found that the wordline latency had a greater impact than the bitline latency, and so stacking columns on columns (which reduces wordline lengths) results in faster cache access times. The general trend is that as the cache size increases, the relative benefit (% latency reduction) also increases. This is intuitive as the larger the cache, the longer the wires. The relative benefit does not increase monotonically because the organization at each cache size has been optimized differently to provide the lowest possible latency for the baseline 2D case.

Beyond splitting a cache across two layers, one can also distribute the cache circuits across four (or more) layers. Table 7.3 also includes simulation results for a four-layer 3D cache organization. For the four-layer version, we first split the SRAM arrays by stacking columns on columns as before, since the wordline latency dominated the bitline latency. After this reduction, however, the bitline latency is now a larger contributor to the overall latency than the remaining wordline component. Therefore to extend from two layers to four, we then stack half of the rows from each layer on top of the other half. The results demonstrate that further latency reductions can be achieved, although the benefit from going from two layers to four is less than that of going from one to two.

Although stacking columns on columns for a two-layer 3D cache provided the best latency improvements, the same was not true for energy reduction. We found that when energy is the primary objective function, stacking rows on rows provided more benefit. When the wordline accesses a row in the SRAM, all of the memory cells in that row attempt to toggle their respective bitlines. While the final column multiplexer only selects one of the bitcells to forward on to the sense amplifier, energy has been expended by *all* of the cells in the row to charge/discharge their bitlines. As a result, reducing the bitline lengths (via row-on-row stacking) directly reduces the amount of capacitance at the outputs of all of the bitcells. That is, the energy saving from reducing the bitline length is multiplied across all bitlines. In contrast, reducing the wordline length saves less energy because the row decoder only activates a single wordline per access. While these results may vary for different cache organizations and models, the general lesson that is important here is that different 3D organizations may be required depending on the exact design constraints and objectives. Designing a 3D circuit to minimize latency, energy, or area may all result in very different final organizations.

7.4.1.2 Dealing with TSVs

The 3D-split cache organizations described in this section require a larger number of TSVs. For example, when stacking columns on columns, the cache now effectively requires twice as many wordlines. This may either come in the form of split wordlines as shown in Fig. 7.9b, or twice as many true wordlines as shown in Fig. 7.9c. In either case, the cost in inter-layer connectivity is one TSV per wordline. Ideally, all of these TSVs should be placed in a single column as shown in Fig. 7.10a. There

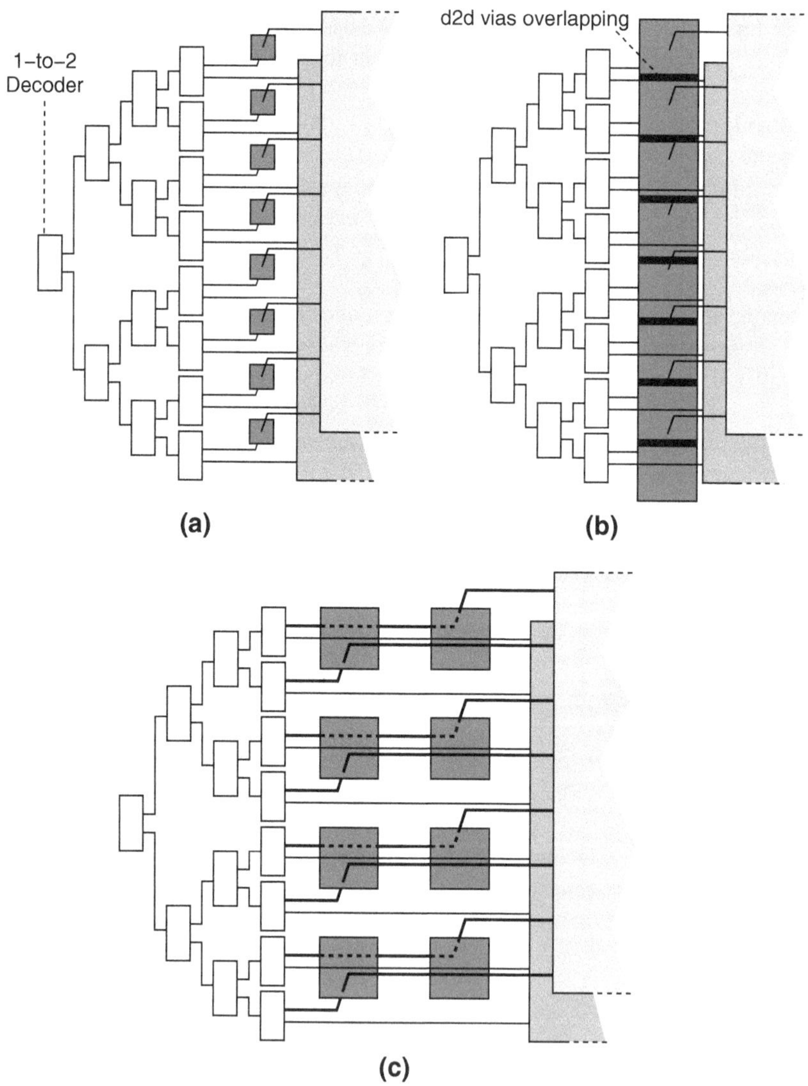

Fig. 7.10 Row-decoder detail for column-on-column 3D SRAM topology with (**a**) sufficiently small TSVs, (**b**) TSVs that are too large, and (**c**) an alternate layout that accommodates the larger TSVs

may be problems, however, if the pitch of the TSVs is greater than that of the wordlines. Fig. 7.10b shows that a larger TSV pitch would cause the TSVs to collide with each other.

Similar to the layout and placement of the regular vias used for connecting metal layers within a single 2D chip, the TSVs can be relocated to accommodate layout constraints, as shown in Fig. 7.10c. Similar to the TSV placement example for inter-block communications (Fig. 7.7 in Section 7.3.2), this may require some additional within-layer routing to get from the row-decoder output to the TSV and then back to the original placement of the wordlines. So long as this additional routing overhead (including the TSV) is significantly less than the wire reduction enabled by the 3D organization, the 3D cache will provide a net latency benefit. In a face-to-back 3D integration process, the TSVs must pass through the active silicon layer, which may in turn cause disruptions to the layout of the underlying transistors. In this case, additional white space may need to be allocated to accommodate the TSVs, which in turn increases the overall footprint of the cache block.

When communication is limited, such as the case when the number of signals exceeds the number of available TSVs, communication and computation may be traded. Figure 7.11a shows the row decoder logic for an SRAM with 16 wordlines split/duplicated across two layers. As a result, there is a need for 16 TSVs (one per wordline). Figure 7.11b shows another possible layout where we reduce the number of TSVs for the wordlines by one half, but at the cost of replicating the logic for the last layer of the row decoder. Figure 7.11c takes this one step further to reduce TSV requirements by one half again, but at the cost of adding more logic. The overall latency of either of these organizations will likely be similar, but increasing levels of logic replication will result in higher power costs. Nevertheless, when judiciously applied, such an approach provides the architect with one more technique to optimize the 3D design of a particular block.

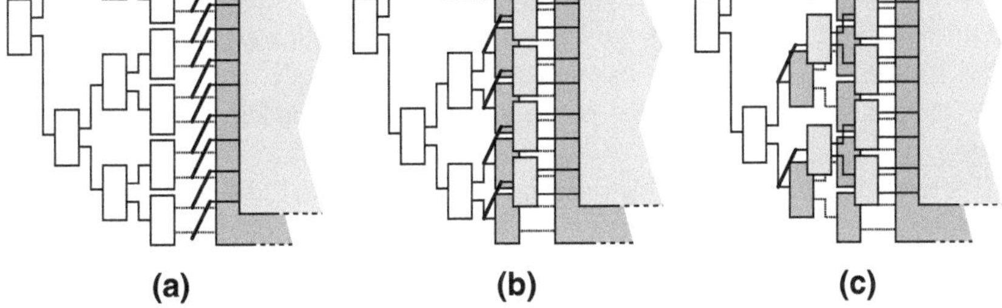

Fig 7.11 Row-decoder detail with **(a)** 16 TSVs, **(b)** 8 TSVs but one extra level of duplicated row-decoder logic, and **(c)** 4 TSVs with two levels of duplicated logic

7.4.2 Three Dimensional-Splitting Arithmetic Units

While caches and other SRAM structures dominate the vast majority of the silicon area of a modern high-performance microprocessor, there are many other logic components that are critical to performance in a variety of ways. The SRAM structures are very regular, and the different strategies for splitting the structures are intuitive. For other blocks that contain more logic and less regular structure, the splitting strategies may not be as obvious. In this section, we explore the design space for 3D-split arithmetic units. In particular, we focus on integer adders as they have an interesting mix of logic and wires, with some level of structure and regularity in layout, but not nearly as regular as the SRAM arrays.

7.4.3 Three-Dimensional Adders

While there are many styles of implementation for addition units, we only focus on the classic Look-ahead Carry Adder (LCA) in this section. Many of the techniques for 3D splitting can be extended or modified to deal with other implementation styles as well as other types of computation units such as multipliers and shifters. Figure 7.12a shows a simple structural view of an n=16-bit LCA. The critical path lies along the carry-propagate generation logic that starts from bit[0], traverses up the tree, and then back down the tree to bit[$n-1$].

There are several natural ways to partition the adder. Figure 7.12b shows an implementation that splits the adder based on its inputs. In this case, input x is placed on the bottom layer, and input y on the top layer. This also requires that the first level of the propagate logic be split across the two layers, requiring at least one TSV per bit. Depending on the relative sizes and pitches of the wires, TSVs, and the propagate logic, the overall width of the adder may be reduced. In the best case (as shown in the figure), this can result in a reduction in a halving of all wire lengths (in the horizontal direction) along the critical carry-propagate generation path. Note that after the first level of logic, all remaining circuitry resides on the top layer.

A second method for splitting the adder is by significance. We can place the least significant bits (e.g., $x[0{:}\frac{n}{2}-1]$) of both inputs on the bottom layer and most significant bits on the top layer. Figure 7.12c shows the schematic for this approach. Note that the two longest wires in the original 2D layout (those going to/coming from the root node) have now effectively been replaced by a single very short TSV, but all remaining wire lengths are left unchanged. Note that compared to the previous approach of partitioning by inputs, only the root node requires signals from both layers and so the total TSV requirements are independent of the size of the inputs (n).

There are many other possible rearrangements. Figure 7.12d shows a variation of the significance partitioning where the lower $\frac{n}{2}$ bits are placed on the right side of the circuit and the upper bits are on the left. As a result, a few of the intermediate wires have been replaced by TSVs, and the last-level wire lengths have also been reduced. All three 3D organizations can be viewed as different instantiations of

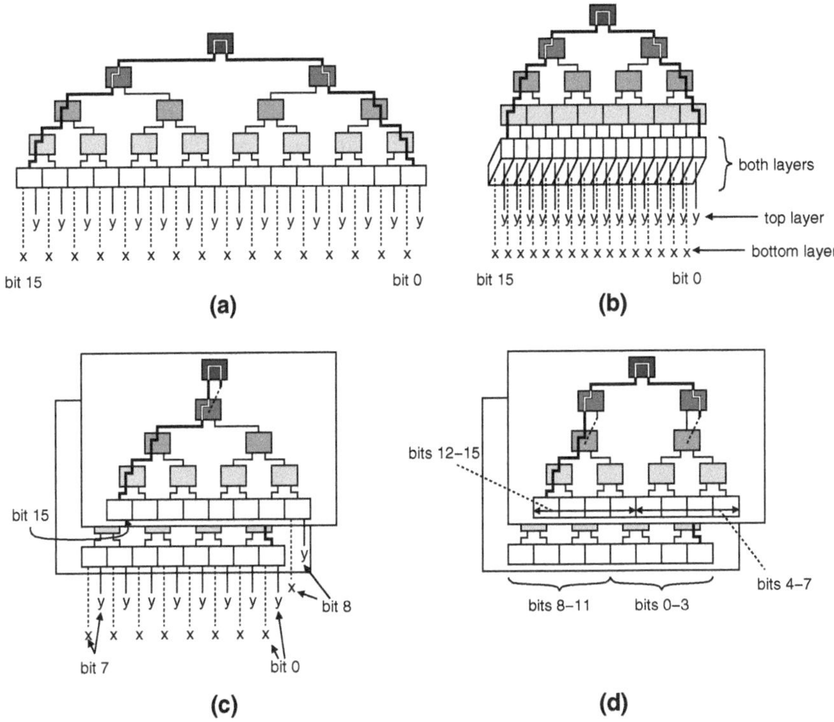

Fig. 7.12 **(a)** Two-dimensional Look-ahead Carry Adder (LCA) circuit, **(b)** 3D LCA adder with input partitioning, **(c)** 3D LCA with significance partitioning, and **(d)** 3D LCA with a mixed significance split

the same basic design where the changing parameter is the level of the tree that spans both layers. In the input-partitioned approach, the first level of the tree (at the leaves) spans both layers, whereas with significance partitioning, it is the root node that spans both layers. The configuration in Fig. 7.12d is like a hybrid of the two others: the top two levels of the tree (at the root) are structurally identical to the top of Fig. 7.12b, and the bottom three layers of the tree look very similar to the bottom levels of Fig. 7.12c. Such a layout could be useful for an adder that supports SIMD operations, where one addition occurs on the right and one occurs on the left. One example application of locating the logical addition operations in physically separate locations is to enable the localization of the wiring and control overhead for power or clock gating.

7.4.4 Interfacing Units

The optimal way to split a functional unit will depend on the design objectives, such as minimizing latency, power, or area footprint. The optimal way to split a *collection* of units may involve a combination of organizations where individual units are split in locally sub-optimal ways. Consider the three related blocks shown in Fig. 7.13a: a register file, an arithmetic unit, and a data cache. By themselves, it may be that splitting the register file by bit partitioning (least significant bits on lowest layer) results in the lowest latency, input partitioning (e.g., different ports on different layers) the data cache is its optimal configuration, and the ALU benefits the most from a hybrid organization such as that described at the end of the previous

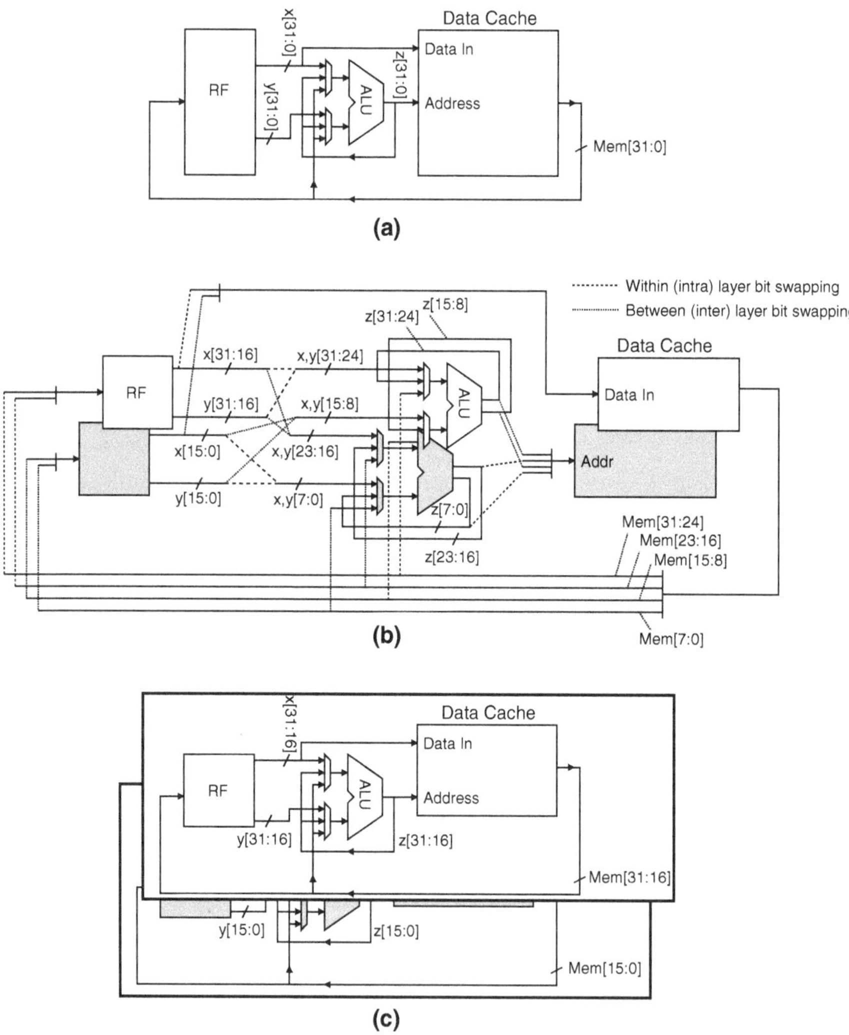

Fig. 7.13 (**a**) Two-dimensional organization of a register file (RF), arithmetic logic unit (ALU) and a data cache along with datapaths and bypasses, (**b**) 3D organization with each unit using different approaches for 3D splitting, and (**c**) 3D organization with all units using the same significance partitioning

section. This is not entirely surprising as each block has different critical paths with different characteristics, and therefore different techniques may be necessary to get the most benefit.

A processor consists of many interacting blocks, however, and the choice of splitting one block may have consequences on others. Consider the same three blocks, where values read from the register file are forwarded to the adder to compute an address, which are in turn then provided to the data cache, and finally the result from the cache gets written back to the register file. Figure 7.13b illustrates these blocks where each one has been split across two layers in a way that minimizes each block's individual latency. As a result, a very large number of TSVs are necessary because the *interface* between one block and the next is not the same. When the data operands come out of the register file, the least significant bits for both operands are on the bottom layer. The adder, however, requires the bits to be located on different layers. The output of the adder then in turn may not be properly aligned for the data cache. The final output from the data cache may need to use a bypass network to directly forward the result to both the adder and the register file, thereby requiring even more TSVs to deal with all of the different interfaces. All of these additional TSVs and the routing to get to and from the vias all increase wiring overhead and erode the original benefits of reducing wire through 3D organizations. As a result, the optimal overall configuration may involve simply using significance partitioning (for example) for all components, as shown in Fig. 7.13c. While this means that locally we are using sub-optimal 3D organizations for the adder and data cache, this still results in a globally optimal configuration.

The simple but important observation to be made is that the choice of 3D partitioning strategy for one block may have wide-reaching consequences for many other blocks in the processor. A significance partitioning of any data-path component will likely force *all* other data-path components to be split in a similar manner. The choice of a 3D organization for the instruction cache may in turn constrain the layout for the decode logic. For example, if the instruction cache delivers one half of its instructions to the bottom layer and the rest to the top layer, then the decode logic would be similarly partitioned with half of the decoders placed on each layer. These are not hard constraints, but the cost of dissimilar interfacing is additional TSVs to make the wires and signals match up.

7.5 Conclusions

From the computer architect's perspective, 3D integration provides two major benefits. First, physically organizing components in three dimensions can significantly reduce wire lengths. Second, devices from different fabrication technologies can be tightly integrated and combined in a 3D stack.

A statement as simple as "3D eliminates wire" can have many different interpretations and applications to microprocessor design. How can the computer architect leverage this reduction in wire lengths? While the previous sections have discussed specific techniques for 3D-integrated processor design, we now discuss some of the implications at a higher level. First, the techniques from the previous sections are not necessarily mutually exclusive. For example, one may choose to stack certain blocks on top of other blocks while splitting some units across multiple layers, and then integrating a complete last-level cache on a third layer. Different components of the processor have different design objectives and constraints, and different 3D-design strategies may be called upon to provide the best solution.

From the perspective of overall microarchitectural organization, the elimination of wire also presents several different options. As discussed in previous sections, wire elimination and refloorplanning can enable the elimination of entire pipeline stages. In other situations, two stages each with significant wire delays could be collapsed into a single stage. Besides the performance improvements that come with a shorter pipeline, there may be additional reductions in overall complexity that can also be achieved. For example, deep execution pipelines may require multiple levels of result bypassing to enable dependent instructions to execute in back-to-back cycles without stalling. The latency, area, power consumption, and other complexity metrics associated with conventional bypass network designs have been shown to scale super-linearly with many of the relevant parameters [13]. As such, eliminating one or more levels of bypass can provide substantial reductions in complexity.

The reduction of pipeline stages has some obvious benefits such as performance improvements, reduction in pipeline control logic, and reduction in power. The changes, however, may also lead to further opportunities for improving the overall microarchitecture of the pipeline. For example, there are a variety of buffers and queues in modern out-of-order, superscalar processors that buffer instructions for several cycles to tolerate various pipeline latencies. By reducing the overall pipeline length, it may be possible to reduce the sizes of some of these structures or in some cases even completely eliminate them. There are many microarchitectural techniques designed to tolerate wire delay, but if 3D greatly reduces the effects of many of these critical wires, the overall pipeline architecture may be relaxed to provide power, area, and complexity benefits.

Instead of eliminating pipeline stages, 3D may also be used to reduce the time spent per pipeline stage. This could result in a higher clock frequency and therefore improved performance, although perhaps at the cost of higher power. Note that this is different from traditional pipelining techniques for increasing clock speed. Traditionally, increasing processor frequency requires sub-dividing the pipeline into a larger number of shorter (lower latency) stages. With 3D, the total number of stages could be kept constant while reducing the latency per stage. In the regular pipeline, the architecture simply takes a fixed amount of work and breaks it into smaller pieces, whereas 3D actually reduces the total amount of work by removing wire delay.

Another potential option is to not use the eliminated wire for performance reasons but to convert the timing slack into power reduction. For example, gates and drivers on a critical timing path are often implemented using larger transistors with higher drive capacities to increase their speed. If the elimination of wire reduces the latency of the circuit, then the circuit designer could reduce the sizes of the transistors which in turn reduce their power consumptions. This may even provide opportunities to completely change the design style, converting from very fast dynamic/domino logic to lower power CMOS gates. In other blocks, transistors might be manufactured with *longer* channels which makes them slower but greatly reduces their leakage currents.

Earlier, we discussed how different 3D implementation styles (e.g., stacking vs. splitting) may be combined in a system to optimize different blocks in different ways. In a similar fashion, 3D may be applied in different ways across the various processor blocks to optimize for different objectives such as timing, area, or power. While we have mostly focused on issues such as wire delay, performance, and power, it is critical to also balance these objectives with the constraints of design complexity and the implied cost in redesigning new components, testing, and verification. In some situations, a finely partitioned 3D block may provide greater benefits, but the cost in terms of additional engineering effort and the impact on the overall project schedule and risk may force a designer to use more conservative organizations.

In this chapter, we have examined the application of 3D integration at several different levels of granularity, but we have not directly attempted to answer the question of what exactly *should* an architect do with 3D? At this time, we can only speculate on the answers, as the optimal answer will depend on many factors that are as yet still unknown. As discussed multiple times in this chapter, the exact organization of components will heavily depend on the exact dimensions and pitches of the TSVs provided by the manufacturing process. With future improvements in cooling technologies, computer architects may be able to pursue more aggressive organizations that focus more on eliminating wires than managing thermal problems. If cooling technologies do not progress as quickly, then the optimal 3D designs may look very different since the architect must more carefully manage the power densities of the processor.

Acknowledgments Much of the work and ideas presented in this chapter have evolved over several years in collaboration with many researchers, in particular Bryan Black and the other researchers we worked with at Intel including Bryan Black, Yuan Xie at Pennsylvania State University, and Kiran Puttaswamy while he was at Georgia Tech. Funding and equipment for this research have also been provided by the National Science Foundation, Intel Corporation, and the Center for Circuit and System Solutions (C2S2) which is funded under the Semiconductor Research Corporation's Focus Center Research Program.

References

1. B. Black, M. Annavaram, E. Brekelbaum, J. DeVale, L. Jiang, G. Loh, D. McCauley, P. Morrow, D. Nelson, D. Pantuso, P. Reed, J. Rupley, S. Shankar, J. P. Shen, and C. Webb. Die-stacking (3D) microarchitecture, *International Symposium on Microarchitecture*, pp. 469–479, 2006.
2. S. Mysore, B. Agarwal, N. Srivastava, S.-C. Lin, K. Banerjee, T. Sherwood. Introspective 3D chips, *Conference on Architectural Support for Programming Languages and Operating Systems*, pp. 264–273, 2006.

3. C. McNairy, R. Bhatia. Montecito: a dual-core, dual-thread Itanium processor, *IEEE Micro*, 25(2):10–20, 2005.
4. T. Austin. DIVA: A dynamic approach to microprocessor verification, *Journal of Instruction Level Parallelism*, 2:1–26, 2000.
5. N. Madan, R. Balasubramonian. Leveraging 3D technology for improved reliability, *International Symposium on Microarchitecture*, pp. 223–235, 2007.
6. G. Loh. 3D-stacked memory architectures for multi-core processors, *International Symposium on Computer Architecture*, pp. 453–464, 2008.
7. C. Liu, I. Ganusov, M. Burtscher, S. Tiwari. Bridging the processor-memory performance gap with 3D IC technology, *IEEE Design and Test*, 22(6):556–564, 2005.
8. G. L. Loi, B. Agarwal, N. Srivastava, S.-C. Lin, T. Sherwood. A thermally-aware performance analysis of vertically integrated (3-D) processor-memory hierarchy, *Design Automation Conference*, pp. 991–996, 2006.
9. T. Kgil, S. D'Souza, A. G. Saidi, N. Binkert, R. Dreslinksi, S. Reinhardt, K. Flautner, T. Mudge. PicoServer: using 3D stacking technology to enable a compact energy efficient chip multiprocessor, *Conference on Architectural Support for Programming Languages and Operating Systems*, pp. 117–128, 2006.
10. G. Schrom, P. Hazucha, J.-H. Hahn, V. Kursun, D. Gardner, S. Narendra, T. Karnik, V. De. Feasibility of monolithic and 3D-stacked DC-DC converters for microprocessors in 90 nm technology generation, *International Symposium on Low-Power Electronics and Design*, pp. 263–268, 2004.
11. G. Hinton, D. Sager, M. Upton, D. Boggs, D. Carmean, A. Kyler, P. Roussel. The microarchitecture of the Pentium 4 processor, *Intel Technology Journal*, Q1, 2001.
12. R. Kessler. The Alpha 21264 microprocessor, *IEEE Micro*, 19(2):24–36, 1999.
13. S. Palacharla. *Complexity-Effective Superscalar Processors*. PhD thesis, University of Wisconsin at Madison, 1998.
14. K. Puttaswamy, G. Loh. Implementing caches in a 3D technology for high performance processors, *International Conference on Computer Design*, pp. 525–532, 2005.
15. Y.-F. Tsai, Y. Xie, N. Vijaykrishnan, M. J. Irwin. Three-dimensional cache design using 3DCacti, *International Conference on Computer Design*, pp. 519–524, 2005.

Chapter 8
Three-Dimensional Network-on-Chip Architecture

Yuan Xie, Narayanan Vijaykrishnan, and Chita Das

Abstract On-chip interconnects are predicted to be a fundamental issue in designing multi-core chip multiprocessors (CMPs) and system-on-chip (SoC) architectures with numerous homogeneous and heterogeneous cores and functional blocks. To mitigate the interconnect crisis, one promising option is network-on-chip (NoC), where a general purpose on-chip interconnection network replaces the traditional design-specific global on-chip wiring by using switching fabrics or routers to connect IP cores or processing elements. Such packet-based communication networks have been gaining wide acceptance due to their scalability and have been proposed for future CMPs and SoC design. In this chapter, we study the combination of both three-dimensional integrated circuits and NoCs, since both are proposed as solutions to mitigate the interconnect scaling challenges. This chapter will start with a brief introduction on network-on-chip architecture and then discuss design space exploration for various network topologies in 3D NoC design, as well as different techniques on 3D on-chip router design. Finally, it describes a design example of using 3D NoC with memory stacked on multi-core CMPs.

Y. Xie (✉)
Pennsylvania State University, University Park, PA 16801, USA
e-mail: yuanxie@cse.psu.edu

Y. Xie et al. (eds.), *Three-Dimensional Integrated Circuit Design*,
Integrated Circuits and Systems, DOI 10.1007/978-1-4419-0784-4_8,

8.1 Introduction

As technology scales, integrating billions of transistors on a chip is now becoming a reality. For example, the latest Intel Xeon processor consists of 2.3 billion transistors [25]. At such integration levels, it is imperative to employ parallelism to effectively utilize the transistors. Consequently, modern superscalar microprocessors incorporate many sophisticated micro-architectural features, such as multiple instruction issue, dynamic scheduling, out-of-order execution, speculative execution, and dynamic branch prediction [26]. However, in order to sustain performance growth, future superscalar microprocessors must rely on even more complex architectural innovations. Circuit limitations and limited instruction-level parallelism will diminish the benefits afforded to the superscalar model by increased architectural complexity [26]. Increased issue widths cause a quadratic increase in the size of issue queues and the complexity of register files. Furthermore, as the number of execution units increases, wiring and interconnection logic complexity begin to adversely affect performance. These issues have led to the advent of chip multiprocessors (CMP) as a viable alternative to the complex superscalar architecture. CMPs are simple, compact processing cores forming a decentralized micro-architecture which scales more efficiently with increased integration densities. The 9-core cell processor [6], the 8-core Sun UltraSPARC T1 processor [13], the 8-core Intel Xeon processor [25], and the 64-core TILEPro64 embedded processor [1] all signal the growing popularity of such systems.

A fundamental issue in designing multi-core chip multiprocessor (CMP) architectures with numerous homogeneous and heterogeneous cores and functional blocks is the design of the on-chip communication fabric. It is projected that on-chip interconnects will be the prominent bottleneck [7] in terms of performance, energy consumption, and reliability as technology scales down further into the nano-scale regime, as discussed in Chapter 1. This is primarily because the scaling of wires increases resistance, and thus the wire delay and energy consumption, while tighter spacing affects signal integrity, and thus reliability. Therefore, design of scalable, high-performance, reliable, and energy-efficient on-chip interconnects is crucial to the success of multi-core/SoC design paradigm and has become an important research thrust.

Traditionally, bus-based interconnection has been widely used for networks with a small number of cores. However, bus-based interconnects will be performance bottlenecks as the number of cores increases. Consequently, they are not considered appropriate for future multi-core systems that have many cores. To overcome these limitations, one promising option is network-on-chip (NoC) [8, 4], where a general purpose on-chip interconnection network replaces the traditional design-specific global on-chip wiring, by using switching fabrics or routers to connect IP cores or processing elements (PEs). Typically, processor cores communicate with each other using a packet-switched protocol, which packetizes data and transmits it through an on-chip network. Much like traditional macro networks, NoC is very scalable. Figure 8.1 shows a conceptual view of the NoC idea, where many cores are connected by the on-chip network routers, rather than by an on-chip bus.

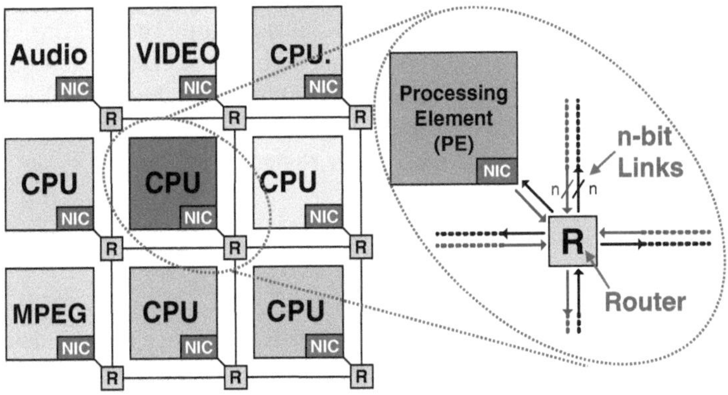

Fig. 8.1 A conceptual network-on-chip architecture: cores are connected to the on-chip network routers (R) via the network interface controllers (NICs)

Even though both 3D integrated circuits [32, 15, 29, 30] and NoCs [8, 4, 23] are proposed as alternatives for the interconnect scaling demands, the challenges of combining both approaches to design three-dimensional NOCs have not been addressed until recently [14, 10, 5, 34, 33, 16, 17, 21].

Chapter 7 presented the design of a single-core microprocessor using 3D integration (via splitting caches or function units into multiple layers) as well as dual-core processor designs using SRAM/DRAM memory stacking. However, all the designs discussed in Chapter 7 are not involved with network-on-chip architecture. In this chapter, we will focus on how to combine 3D integration with network-on-chip as the communication fabric among processor cores and memory banks.

In the following sections, we first give a brief introduction on NoC and then discuss various approaches to design 3D on-chip network topology and 3D router design. An example of memory stacking on top of chip multiprocessors (CMPs) using 3D NoC architecture will be presented.

8.2 A Brief Introduction on Network-on-Chip

Network-on-chip architecture has been proposed as a potential solution for the interconnect demands that arise with the nanometer era [8, 4]. In the network-on-chip architecture, a general purpose on-chip interconnection network replaces the traditional design-specific global on-chip wiring by the use of switching fabric or routers to connect IP cores or processing elements (PEs). The PEs communicate with each other by sending messages in packets through the routers. This is usually called packet-based interconnect.

A typical 2D NoC consists of a number of processing elements (PE) arranged in a grid-like mesh structure, much like a Manhattan grid. The PEs are interconnected through an underlying packet-based network fabric. Each PE interfaces to a network router through a network interface controller (NIC). Each router is, in turn, connected to four adjacent routers, one in each cardinal direction. The number of ports for a router is defined as the *radix* of the router.

8.2.1 NoC Topology

Network topology is a vital aspect of on-chip network design since it determines the power-performance metrics. For example, NoC topology determines zero-load latency, bisection bandwidth, router micro-architecture, routing complexity, channel lengths, and overall network power consumption.

The mesh topologies [4] (as shown in Fig. 8.1) have been popular for tiled CMPs because of the low complexity and planar 2D-layout properties. Such a simple and regular topology has a compact 2D layout. Other topologies such as concentrated meshes and the flattened butterfly could also be employed in the NoC design for various advantages.

For example, a concentrated mesh (Cmesh) [2] preserves the advantages of a mesh and works around the scalability problem by sharing the router between multiple processing elements. The number of nodes sharing a router is called the concentration degree of the network. Figure 8.2 shows the layout for Cmesh for 64

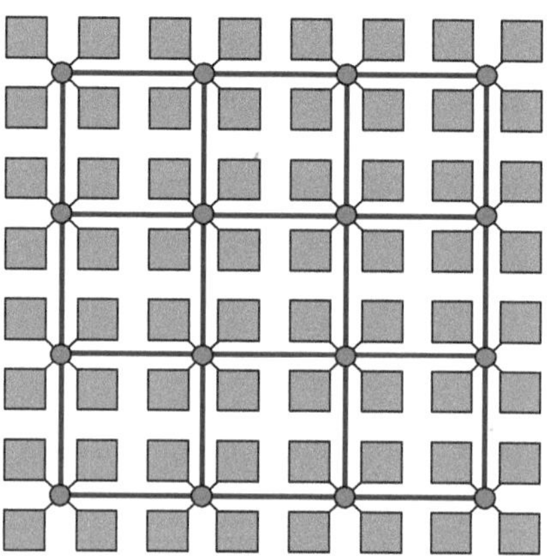

Fig. 8.2 A concentrated mesh network-on-chip topology

nodes. Such topology reduces the number of routers, resulting in reduced hop counts and thus yielding excellent latency savings over mesh. Cmesh has a radix (number of ports) of 8. It can also afford to have very wide channels (512+ bits) due to a slowly growing bisection.

Another example is the flattened butterfly topology [9], which reduces the hop count by employing both concentration and rich connectivity by using longer links to nonadjacent neighbors. The higher connectivity increases the bisection bandwidth and requires a larger number of ports (higher radix) in the router. This increased bisection bandwidth results in narrower channels. Figure 8.3 shows a possible layout for a flattened butterfly with 64 nodes. The rich connectivity trades off serialization latency for reducing the hop count. Such topology has a radix between 7 and 13, depending on the network size and small channel widths (128+ bits).

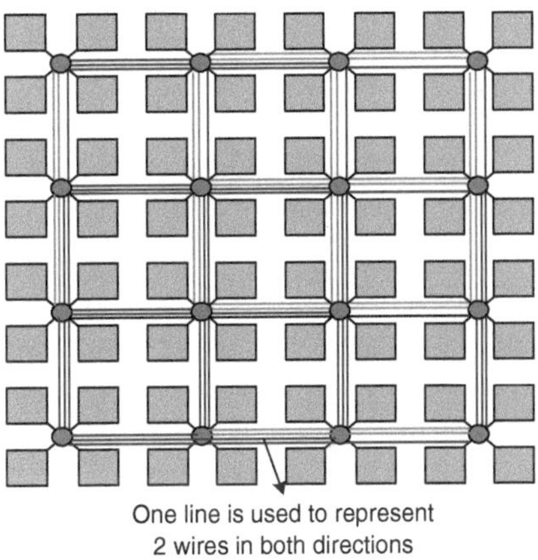

Fig. 8.3 A flattened butterfly topology with 64 nodes

8.2.2 NoC Router Design

A generic NoC router architecture is illustrated in Fig. 8.4. The router has P input and P output channels/ports (The number of ports is defined as radix). When P=5 (or radix is 5), it is a typical 2D NoC router for a mesh network, resulting in a 5 $\times$ 5 crossbar. When the network topology changes, the complexity of the router also changes. For example, a Cmesh network topology requires a router design with a radix of 8. The routing computation unit, RC, operates on the header *flit* (a flit is the smallest unit of flow control; one packet is composed of a number of flits) of an incoming packet and, based on the packet's destination, dictates the appropriate output physical channel/port (PC) and/or valid virtual channels (VC) within the selected output PC. The routing can be deterministic or adaptive. The virtual channel allocation unit (VA) arbitrates between all packets competing for access to

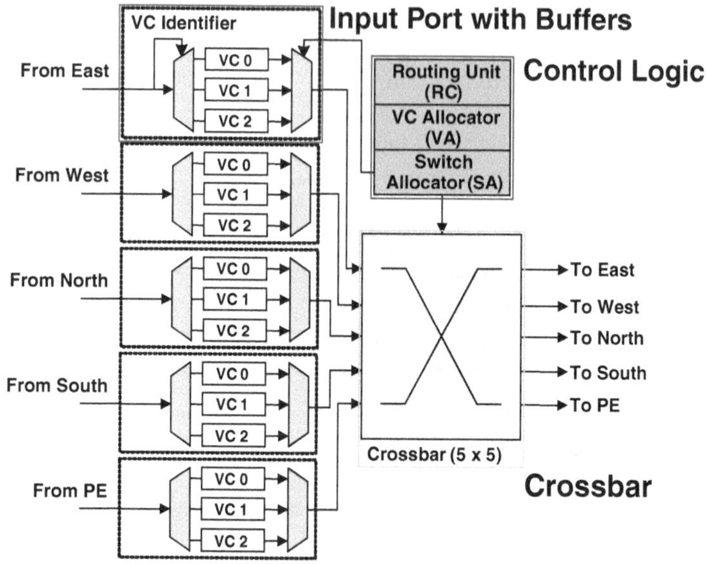

Fig. 8.4 A generic 2D network-on-chip router with five input ports and five output ports

the same output VCs and chooses a winner. The switch allocation unit (SA) arbitrates between all VCs requesting access to the crossbar. The winning flits can then traverse the crossbar and move on to their respective output links.

8.2.3 More Information on NoC Design

Network-on-chip design methodologies have gained a lot of industrial interest. For example, Tilera Corporation has built a 64-core embedded multi-core processor called TILE64 [1], which contains 64 full-featured, programmable cores that are connected by mesh-based NoC architecture. The Intel 80-core TeraFLOPS processor [31] comprises a network-on-chip architecture. The 80-core chip is arranged as an 8×10 array of PE cores and packet-switched routers, connecting with a mesh topology (similar to Fig. 8.1). Figure 8.5 shows the NoC block diagram for the processor. Each PE core contains two pipelined floating-point multiply accumulators (FPMAC), connecting to the router through an interface block (RIB). The router is a 5-port crossbar-based design with mesochronous interface (MSINT). The mesh NoC network provides a bisection bandwidth of 2 Terabits/s.

To learn more about the general background of network-on-chip architecture, one can refer to books [8, 4] and a few survey papers such as [19, 23, 12].

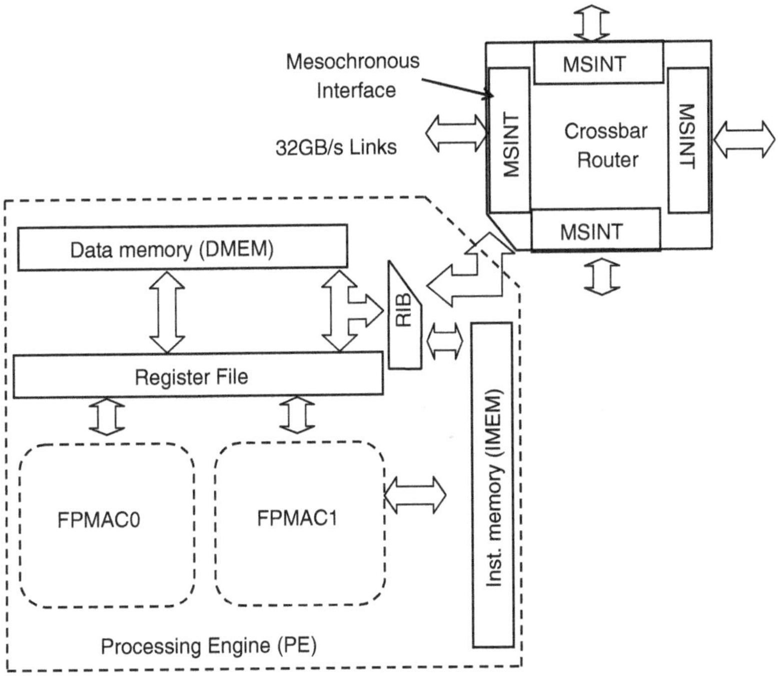

Fig. 8.5 NoC block diagram for Intel's 80-core TeraFLOPS processor

8.3 Three-Dimensional NoC Architectures

This section delves into the exploration of possible architectural designs for 3D NoC architectures. Expanding this 2D paradigm into the third dimension poses interesting design challenges. Given that on-chip networks are severely constrained in terms of area and power resources, while at the same time they are expected to provide ultra-low latency, the key issue is to identify a reasonable tradeoff between these contradictory design threads. In this section, we explore the extension of a baseline 2D NoC implementation into the third dimension while considering the aforementioned constraints.

8.3.1 Symmetric NoC Router Design

The natural and simplest extension to the baseline NoC router to facilitate a 3D layout is simply adding two additional physical ports to each router; one for Up and one for Down, along with the associated buffers, arbiters (VC arbiters and switch arbiters), and crossbar extension. We can extend a traditional NoC fabric to the third dimension by simply adding such routers at each layer (called a symmetric NoC, due to symmetry of routing in all directions). We call this architecture a 3D symmetric NoC, since both intra- and inter-layer movements bear identical characteristics: hop-by-hop traversal, as illustrated in Fig. 8.6. For example, moving from the bottom layer of a 4-layer chip to the top layer requires three network hops.

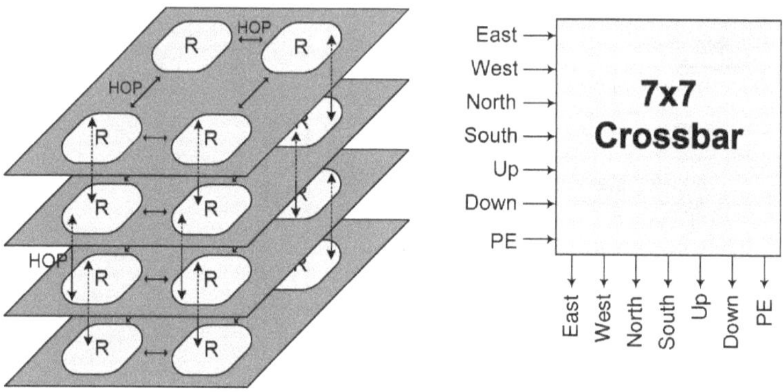

Fig. 8.6 A symmetric 3D network-on-chip router with two additional input/output ports (up and down), totally together needs seven input ports and seven output ports

This architecture, while simple to implement, has a few major inherent drawbacks.

- It wastes the beneficial attribute of a negligible inter-wafer distance in 3D chips (for example, in Chapter 2, we have seen that the thickness of a die could be as small as tens of microns). Since traveling in the vertical dimension is multi-hop, it takes as much time as moving within each layer. Of course, the average number of hops between a source and a destination does decrease as a result of folding a 2D design into multiple stacked layers, but inter-layer and intra-layer hops are indistinguishable. Furthermore, each flit must undergo buffering and arbitration at every hop, adding to the overall delay in moving up/down the layers.
- The addition of two extra ports necessitates a larger 7×7 crossbar, as shown in Fig. 8.6b. Crossbars scale upward very inefficiently, as illustrated in Table 8.1. This table includes the area and power budgets of all crossbar types investigated in this section based on synthesized implementations in 90-nm technology. Clearly, a 7×7 crossbar incurs significant area and power overhead over all other architectures. Therefore, the 3D symmetric NoC implementation is a somewhat naive extension to the baseline 2D network.
- Due to the asymmetry between vertical and horizontal links in a 3D architecture, there are several aspects, such as link bandwidth and buffer allocation, that will need to be customized along different directions in a 3D chip. Further, temperature gradients or process variation across the different layers of the 3D chip can cause identical router components to have different delays in different layers. As

Table 8.1 Area and power comparison of the crossbar switches implemented in 90-nm technology

Crossbar type	Area	Power (500 Mhz)
5×5 Crossbar	8523 μm^2	4.21 mW
6×6 Crossbar	11579 μm^2	5.06 mW
7×7 Crossbar	17289 μm^2	9.41 mW

an example, components operating at the chip layer farthest away from the heat sink will limit the highest frequency of the entire network.

8.3.2 Three-Dimensional NoC–Bus Hybrid Router Design

There is an inherent asymmetry in the delays of a 3D architecture between the fast vertical interconnects and the horizontal interconnects that connect neighboring cores due to differences in wire lengths (a few tens of microns in the vertical direction as compared to a few thousand of microns in the horizontal direction). The previous section argues that a symmetric NoC architecture with multi-hop communication in the vertical (inter-layer) dimension is not desirable.

Given the very small inter-layer distance, single-hop communication is, in fact, feasible. This technique revolves around the fact that vertical distance is negligible compared to intra-layer distances; a shared medium can provide single-hop traversal between any two layers. This realization opens the door to a very popular shared medium interconnect, the bus. The NoC router can be hybridized with a bus link in the vertical dimension to create a 3D NoC–bus hybrid structure as shown in Fig. 8.7. This hybrid system provides both performance and area benefits. Instead of an unwieldy 7 × 7 crossbar, it requires a 6 × 6 crossbar (Fig. 8.7), since the bus

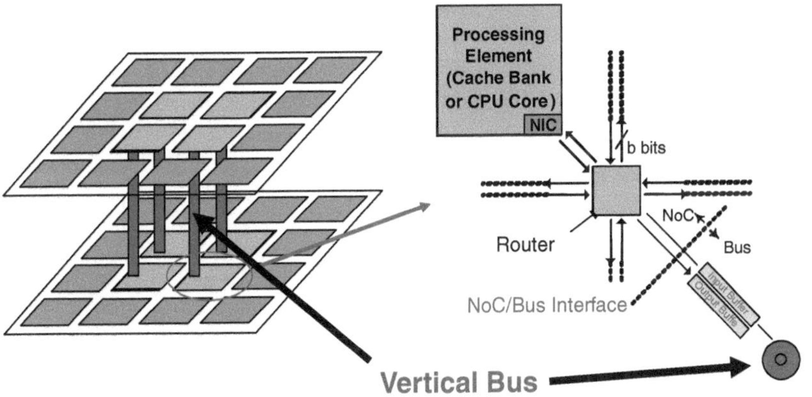

Fig. 8.7 A hybrid 3D NoC/bus architecture. The router has one additional input/output ports to connect with the vertical bus

adds a single additional port to the generic 2D 5 × 5 crossbar. The additional link forms the interface between the NoC domain and the bus (vertical) domain. The bus link has its own dedicated queue, which is controlled by a central arbiter. Flits from different layers wishing to move up/down should arbitrate for access to the shared medium. Figure 8.8 illustrates the view of the vertical via structure. This schematic depicts the usefulness of the large via pads between the different layers; they are deliberately oversized to cope with misalignment issues during the fabrication process. Consequently, it is the large via pads which ultimately limit vertical via density in 3D chips.

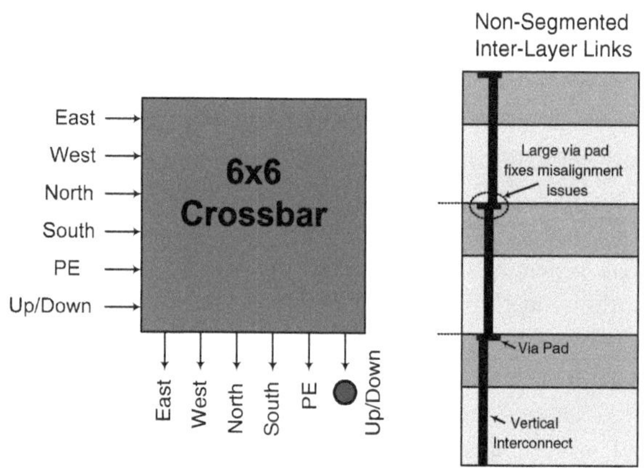

Fig. 8.8 The router has one additional input/output port to connect with the vertical bus, and therefore it needs six input ports and six output ports. The bus is formed by 3D vias connecting multiple layers

Despite the marked benefits over the 3D symmetric NoC router, the bus approach also suffers from a major drawback: it does not allow concurrent communication in the third dimension. Since the bus is a shared medium, it can only be used by a single flit at any given time. This severely increases contention and blocking probability under high network load. Therefore, while single-hop vertical communication does improve performance in terms of overall latency, inter-layer bandwidth suffers. More details on the 3D NoC–bus hybrid architecture can be found in [14].

8.3.3 True 3D Router Design

Moving beyond the previous options, we can envision a true 3D crossbar implementation, which enables seamless integration of the vertical links in the overall router operation. Figure 8.9 illustrates such a 3D crossbar layout. It should be noted at this point that the traditional definition of a crossbar – in the context of a 2D physical layout – is a switch in which each input is connected to each output through a single connection point. However, extending this definition to a physical 3D structure would imply a switch of enormous complexity and size (given

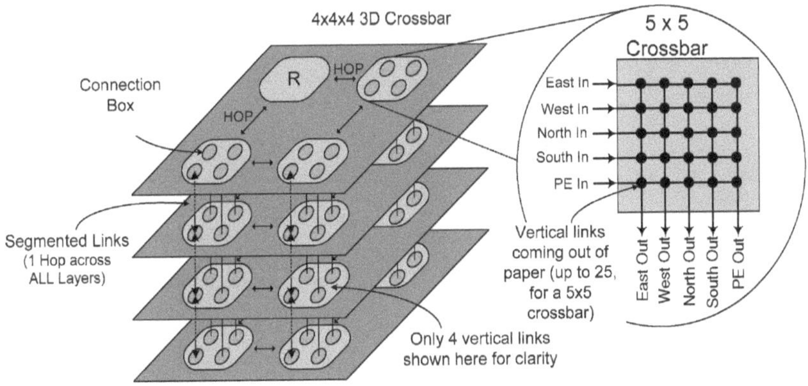

Fig. 8.9 The true 3D router design

the increased numbers of input- and output-port pairs associated with the various layers). Therefore, we chose a simpler structure which can accommodate the interconnection of an input to an output port through more than one connection points. While such a configuration can be viewed as a multi-stage switching network, we still call this structure a crossbar for the sake of simplicity. The vertical links are now embedded in the crossbar and extend to all layers. This implies the use of a 5 × 5 crossbar, since no additional physical channels need to be dedicated for inter-layer communication.

As shown in Table 8.1, a 5 × 5 crossbar is significantly smaller and less power-hungry than the 6 × 6 crossbar of the 3D NoC–bus hybrid and the 7 × 7 crossbar of the 3D symmetric NoC. Interconnection between the various links in a 3D crossbar would have to be provided by dedicated connection boxes at each layer. These connecting points can facilitate linkage between vertical and horizontal channels, allowing flexible flit traversal within the 3D crossbar. The internal configuration of such a connection box (CB) is shown in Fig. 8.10. The vertical link segmentation also affects the via layout as illustrated in Fig. 8.10. While this layout is more complex than that shown in Fig. 8.8, the area between the offset vertical vias can still be utilized by other circuitry, as shown by the dotted ellipse in Fig. 8.10. Hence, the 2D crossbars of all layers are physically fused into one single three-dimensional crossbar. Multiple internal paths are present, and a traveling flit goes through a number of switching points and links between the input and output ports. Moreover, flits reentering another layer do not go through an intermediate buffer; instead, they directly connect to the output port of the destination layer. For example, a flit can move from the western input port of layer 2 to the northern output port of layer 4 in a single hop.

However, despite this encouraging result, there is an opposite side to the coin which paints a rather bleak picture. Adding a large number of vertical links in a 3D crossbar to increase NoC connectivity results in increased path diversity. This translates into multiple possible paths between source and destination pairs. While this

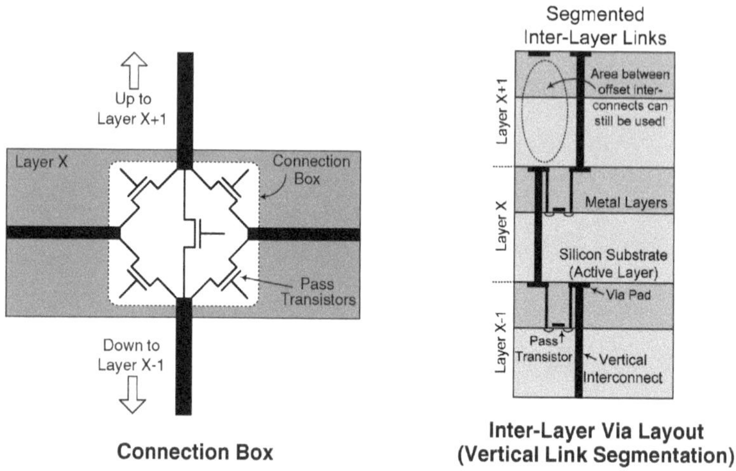

Fig. 8.10 Side view of the inter-layer via structure in 3D crossbar for the true 3D router design

increased diversity may initially look like a positive attribute, it actually leads to a dramatic increase in the complexity of the central arbiter, which coordinates inter-layer communication in the 3D crossbar. The arbiter now needs to decide between a multitude of possible interconnections and requires an excessive number of control signals to enable all these interconnections. Even if the arbiter functionality can be distributed to multiple smaller arbiters, then the coordination between these arbiters becomes complex and time consuming. Alternatively, if dynamism is sacrificed in favor of static path assignments, the exploration space is still daunting in deciding how to efficiently assign those paths to each source–destination pair. Furthermore, a full 3D crossbar implies 25 (i.e., 5×5) connection boxes per layer. A four-layer design would therefore require 100 CBs! Given that each CB consists of six transistors, the whole crossbar structure would need 600 control signals for the pass transistors alone! Such control and wiring complexity would most certainly dominate the whole operation of the NoC router. Pre-programming static control sequences for all possible input–output combinations would result in an oversize table/index; searching through such a table would incur significant delays, as well as area and power overhead. The vast number of possible connections hinders the otherwise streamlined functionality of the switch. Note that the prevailing tendency in NoC router design is to minimize operational complexity in order to facilitate very short pipeline lengths and very high frequency. A full crossbar with its overwhelming control and coordination complexity poses a stark contrast to this frugal and highly efficient design methodology. Moreover, the redundancy offered by the full connectivity is rarely utilized by real-world workloads, and is, in fact, design overkill [10].

8.3.4 3D Dimensionally-Decomposed NoC Router Design

Given the tight latency and area constraints in NoC routers, vertical (inter-layer) arbitration should be kept as simple as possible. Consequently, a true 3D router design, as described in the previous section, is not a realistic option. The design complexity can be reduced by using a limited amount of inter-layer links. This section describes a modular 3D decomposable router (called row–column–vertical (RoCoVe) Router) [10].

In a typical 2D NoC router, the 5 × 5 crossbar has five inputs/outputs that correspond to the four cardinal directions and the connection from the local PE. The crossbar is the major contributor to the latency and area of a router. It has been shown [11] that through the use of a preliminary switching process known as guided flit queuing, incoming traffic can be decomposed into two independent streams: (a) East–West traffic (i.e., packet movement in the X dimension) and (b) North–South traffic (i.e., packet movement in the Y dimension). Such segregation of traffic flow allows the use of smaller crossbars and the isolation of the two flows in two independent router submodules, which are called *Row Module* and *Column Module* [11].

With the same idea of traffic decomposition, the traffic flow in 3D NoC can be decomposed into three independent streams, with a third traffic flow in the Z dimension (i.e., inter-layer communication). An additional module is required to handle all traffic in the third dimension, and this module is called *Vertical Module.* In addition, there must be links between vertical module and row/column modules to allow the movement of packets from the vertical module to the row module and the column module. Consequently, such a dimensionally decomposed approach allows for a much smaller crossbar design (4 × 2) resulting in a much faster and power-efficient 3D NoC router design. The architectural view of this 3D dimensionally-decomposed NoC router design is shown in Fig. 8.11. More details can be found in [10].

8.3.5 Multi-layer 3D NoC Router Design

All the 3D router design options discussed earlier (symmetric 3D router, 3D NoC–bus hybrid router, true 3D router, and 3D dimensionally decomposed router) are based on the assumption that the processing element (PE) (which could be a processor core or a cache bank) itself is still a 2D design. In Section 7.4, a fine-granularity design of a microprocessor is introduced such that one can split a PE across multiple layers. For example, 3D cache design and 3D functional units are described in Section 7.4. Three-dimensional block design and the floorplanning algorithms are also discussed in Chapter 4. Consequently, a PE in the NoC architecture can be implemented with such a fine-granularity approach. Although such a multi-layer stacking of a PE is considered aggressive in the current technology, it could

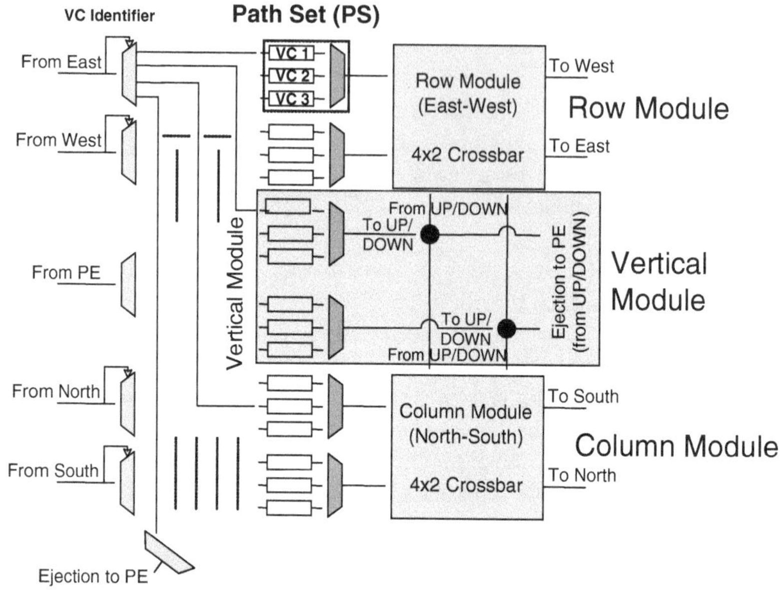

Fig. 8.11 Architectural detail of the 3D-dimensionally decomposed NoC router NoC router design

be possible as 3D technology matures with smaller TSV pitches (as discussed in Section 7.4).

With such a multi-layer stacking of processing elements in the NoC architecture, it is necessary to design a multi-layer 3D router that will span across multiple layers of a 3D chip. Logically, such a NoC architecture with multi-layer PEs and multi-layer routers is identical to the traditional 2D NoC case with the same number of nodes, albeit the smaller area of each PE and router and the shorter distance between routers. Consequently, the design of a multi-layer router requires no additional functionality as compared to a 2D router, and only requires distribution of the functionality across multiple layers.

The router modules can be classified into two categories – separable and nonseparable based on the ability to systematically split the module into smaller sub-modules across layers with the inter-layer wiring constraints and the need to balance areas across layers [5]. Input buffers, crossbar, and inter-router links are classified as separable modules, while arbitration logic and routing logic are classified as nonseparable since they cannot be systematically broken into subsets. The saving in chip area can be used for enhancing the router capability, for example, adding express paths between nonadjacent PEs to reduce the average hop count, and help to boost performance and reduce power. Furthermore, because a large portion of the communication traffic consists of short flits and frequent patterns, it is possible to dynamically shut down some layers of the multi-layer router to reduce the power consumption.

8.3.6 3D NoC Topology Design

Until now, all the router designs discussed so far are based on the mesh-based NoC topology. As described in Section 8.2, there exists various NoC topologies, such as the concentrated mesh or the flattened butterfly topologies, all of which have advantages and disadvantages. By employing different topologies rather than the mesh topology, the router designs discussed above could also have different variants. For example, in 2D concentrated mesh topology, the router itself has a radix of 8 (i.e., an 8-port router, with four to local PEs and the others to four cardinal directions). With such topology, the 3D NoC–bus hybrid approach would result in a 9-port router design. Such high-radix router designs are power-hungry with degraded performance, even though the hop count between PEs is reduced. Consequently, a topology-router co-design method for 3D NoC is desirable, so that the hop count between any two PEs and the radix of the 3D router design is as small as possible. Xu et al. [33] proposed a 3D NoC topology with a low diameter and low radix router design. The level 2D mesh is replaced with a network of long links connecting nodes that are at least m mesh-hops away, where m is a design parameter. In such a topology, long-distance communications can leverage the long physical wire and vertical links to reach their destination, achieving low total hop count while the radix of the router is kept low. For application-specific NoC architecture, Yan and Lin [34] also proposed a 3D NoC synthesis algorithm called ripup-reroute-and-router-merging (RRRM) that is based on a rip-up and reroute formulation for routing flows and a router merging procedure for network optimization to reduce the hop count.

8.3.7 Impact of 3D Technology on NoC Designs

Chapter 2 discussed the 3D integration technology options. In this section, the impact of various 3D integration approaches on NoC design is discussed.

Since TSV vias contend with active device area, they impose constraints on the number of such vias per unit area. Consequently, the NoC design should be performed holistically in conjunction with other system components such as the power supply and the clock network that will contend for the same interconnect resources.

The 3D integration using TSV (through silicon via) can be classified into one of the two following categories: (1) *monolithic approach* and (2) *stacking approach*. The first approach involves a sequential device process, where the front-end processing (to build the device layer) is repeated on a single wafer to build multiple active device layers before the back-end processing builds interconnects among devices. The second approach (which could be wafer-to-wafer, die-to-wafer, or die-to-die stacking) processes each active device layer separately using conventional fabrication techniques. These multiple device layers are then assembled to build up 3D ICs using bonding technology. Dies can be bonded face-to-face (F2F) or face-to-back (F2B). The microbump in face-to-face wafer bonding does not go through a thick buried Si layer and can be fabricated with a higher pitch density. In stacking bonding, the dimension of the TSVs is not expected to scale at the same rate as the feature size because alignment tolerance and thinned die/wafer height during bonding pose limitation on the scaling of the vias.

The TSV (or micropad) size, length, and the pitch density, as well as the bonding method (face-to-face or face-to-back bonding, SOI-based 3D or bulk CMOS-based 3D) can have a significant impact on the 3D NoC topology design. For example, the relatively large size of TSVs can hinder partitioning a design at very fine granularity across multiple device layers and make the true 3D router design less possible. On the other hand, the monolithic 3D integration provides more flexibility in the vertical 3D connection because the vertical 3D via can potentially scale down with feature size due to the use of local wires for connection. Availability of such technologies makes it possible to partition the design at a very fine granularity. Furthermore, face-to-face bonding or SOI-based 3D integration may have a smaller via pitch size and higher via density than face-to-back bonding or bulk CMOS-based integration. This influence of the 3D technology parameters on the NoC topology design will be thoroughly studied, and suitable NoC topologies for different 3D technologies will be identified with respect to the performance, power, thermal, and reliability optimizations.

8.4 Chip Multiprocessor Design with 3D NoC Architecture

In the previous section, various router designs and topology explorations for 3D NoC architecture are discussed. In this section, we use the 3D NoC–bus hybrid architecture as an example to study the chip multiprocessor design with memory stacking that employs 3D NoC architecture and to evaluate the benefits of such an architecture design [14].

The integration of multiple cores on a single die is expected to accentuate the already daunting memory bandwidth problem. Supplying enough data to a chip with a massive number of on-die cores will become a major challenge for performance scalability. Traditional on-chip memory will not suffice due to the I/O pin limitation. According to the ITRS projection, the number of pins on a package will not continue to grow rapidly enough for the next decade to overcome this problem. Consequently, it is anticipated that memory stacking on top of multi-core would be one of the early commercial uses of 3D technology.

The adoption of CMPs and other multi-core systems is expected to increase the sizes of both L2 and L3 caches in the foreseeable future. However, diminutive feature sizes exacerbate the impact of interconnect delay, making it a critical bottleneck in meeting the performance and power consumption budgets of a design. Hence, while traditional architectures have assumed that each level in the memory hierarchy has a single, uniform access time, increases in interconnect delays will render access times in large caches dependent on the physical location of the requested cache line. That is, access times will be transformed into variable latencies based on the distance traversed along the chip.

The concept of nonuniform cache architectures (NUCA) [3] has been proposed based on the above observation. Instead of a large uniform monolithic L2 cache, the L2 space in NUCA is divided into multiple banks, which have different access latencies according to their locations relative to the processor. These banks are connected through a mesh-based interconnection network. Cache lines are allowed to migrate within this network for the purpose of placing more frequently accessed data in the cache banks closer to the processor. Several recent proposals extend the NUCA concept to CMPs. An inherent problem of NUCA in CMP architectures is the management of data shared by multiple cores. Proposed solutions to this problem include data replication and data migration. Still, large access latencies and high power consumption stand as inherent problems for NUCA-based CMPs.

The introduction of three-dimensional (3D) circuits provides an opportunity to reduce wire lengths and increase memory bandwidth. Consequently, this technology can be useful in reducing the access latencies to the remote cache banks of a NUCA architecture.

Section 7.2 discusses the design of stacking SRAM or DRAM L2 cache for dual-core processors without the need of using NoC architecture or the concept of NUCA architecture. In this section, we consider the design of a 3D topology for a NUCA that combines the benefits of network-on-chip and 3D technology to reduce L2 cache latencies in CMP-based systems. This section provides new insights on network topology design for 3D NoCs and addresses issues related to data management in L2, taking into account network traffic and thermal issues.

8.4.1 The 3D L2 Cache Stacking on CMP Architecture

As previously mentioned in Chapter 1, one of the advantages of 3D chips is the very small distance between the layers. In Chapter 2, we have seen that the distance between two layers is on the order of tens of microns, which is negligible compared to the distance travelled between two network-on-chip routers in 2D network-on-chip architecture. (For example, 1500 μm on average for a 64 KB cache bank implemented in 65-nm technology.) This characteristic makes communication in the vertical (inter-layer) direction very fast as compared to the horizontal (intra-layer). In this section, we introduce the architecture of stacking a large L2 cache on top of a CMP processor, where a fast access to the stacked L2 cache from CMP processor is enabled by the 3D technology.

As discussed in Section 8.3, a straightforward 3D NoC router design is the symmetric 3D NoC router, which increases the design complexity (using a 7×7 crossbar) and results in multi-hop communication between nonadjacent layers. A NoC–bus hybrid design not only reduces the design complexity (using a 6×6 crossbar), but also results in single-hop communication among the layers because of the short distance between them. In this section, a NoC–bus hybrid architecture is described, which uses dynamic time-division multiple access (dTDMA) buses as "Communication Pillars" between the wafers, as shown in Fig. 8.7. These vertical bus pillars provide single-hop communication between any two layers and can be interfaced to a traditional NoC router for intra-layer traversal using minimal hardware, as will be shown later. Due to technological limitations and router complexity issues (to be discussed later), not all NoC routers can include a vertical bus, but the ones that do form gateways to the other layers. Therefore, those routers connected to vertical buses have a slightly modified architecture.

8.4.2 The dTDMA Bus as a Communication Pillar

The dTDMA bus architecture [24] eliminates the transactional character commonly associated with buses and instead employs a bus arbiter which dynamically grows and shrinks the number of time slots to match the number of active clients. Single-hop communication and transaction-less arbitrations allow for low and predictable latencies. Dynamic allocation always produces the most efficient time slot configuration, making the dTDMA bus nearly 100% bandwidth efficient. Each pillar node requires a compact transceiver module to interface with the bus, as shown in Fig. 8.12.

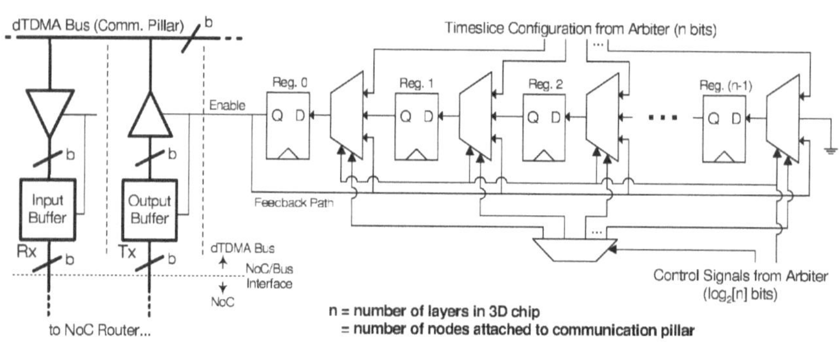

Fig. 8.12 Transceiver module of a dTDMA bus

The dTDMA bus interface (Fig. 8.12) consists of a transmitter and a receiver connected to the bus through a tri-state driver. The tri-state drivers on each receiver and transmitter are controlled by independently programmed fully tapped feedback shift registers. Because of its very small size, the dTDMA bus interface is a minimal addition to the NoC router.

The presence of a centralized arbiter is another reason why the number of vertical buses, or pillars, in the chip should be kept low. An arbiter is required for each pillar with control signals connecting all layers. The arbiter should be placed in the middle layer of the chip to keep wire distances as uniform as possible. Naturally, the number of control wires increases with the number of pillar nodes attached to the pillar, i.e., the number of layers present in the chip. The arbiter and all the other components of the dTDMA bus architecture have been implemented in Verilog HDL and synthesized using commercial 90-nm TSMC libraries. The area occupied by the arbiter and the transceivers is much smaller compared to the NoC router, thus fully justifying the decision to use this scheme as the vertical gateway between the layers. The area and power numbers of the dTDMA components and a generic 5-port (North, South, East, West, local node) NoC router (all synthesized in 90-nm technology) are shown in Table 8.2. Clearly, both the area and power overheads due to the addition of the dTDMA components are orders of magnitude smaller than the overall budget. Therefore, using the dTDMA bus as the vertical interconnect is of minimal area and power impact. The dTDMA bus is observed to be better than a symmetric 3D router design for the vertical direction as long as the number of device layers is < 9 (bus contention becomes an issue beyond that).

Table 8.2 Area and power overhead of dTDMA bus

Component	Power	Area (mm^2)
Generic NoC router (5-port)	119.55 mW	0.3748
dTDMA bus Rx/Tx (2 per client)	97.39 μW	0.00036207
dTDMA bus arbiter (1 per bus)	204.98 μW	0.00065480

Table 8.3 Area overhead of inter-wafer wiring for different via pitch sizes

	Inter-wafer area (due to dTDMA bus wiring)			
Bus width	10 μm	5 μm	1 μm	0.2 μm
128 bits(+42 control)	62500 μm^2	15625 μm^2	625 μm^2	25 μm^2

As discussed in Chapter 2, the parasitics of TSVs have a small effect on power and delay, because of their small size. The density of the inter-layer vias determines the number of pillars which can be employed. Table 8.3 illustrates the area occupied by a pillar consisting of 170 wires (128-bit bus + 3 × 14 control wires required in a 4-layer 3D SoC) for different via pitch sizes. In face-to-back 3D implementations, the pillars must pass through the active device layer, implying that the area occupied by the pillar translates into wasted device area. This is the reason why the number of inter-layer connections must be kept to a minimum. However, as via density increases, the area occupied by the pillars becomes smaller and negligible compared to the area occupied by the NoC router (see Tables 8.2 and 8.3). However, as previously mentioned in Chapter 2, via densities are still limited by via pad sizes, which are not scaling as fast as the actual via sizes. As shown in Table 8.3, even at a pitch of 5 μm, a pillar induces an area overhead of around 4% to the generic 5-port NoC router, which is not overwhelming. These results indicate that, for the purposes of our 3D architecture, adding extra dTDMA bus pillars is feasible.

Via density, however, is not the only factor limiting the number of pillars. Router complexity also plays a key role. As previously mentioned, adding an extra vertical link (dTDMA bus) to an NoC router will increase the number of ports from 5 to 6, and since contention probability within each router is directly proportional to the number of competing ports, an increase in the number of ports increases the contention probability. This, in turn, will increase congestion within the router, since more flits will be arbitrating for access to the router's crossbar. Thus, arbitrarily adding vertical pillars to the NoC routers adversely affects the performance of each pillar router. Hence, the number of high-contention routers (pillar routers) in the network increases, thereby increasing the latency of both intra-layer and inter-layer communication.

On the other hand, there is a minimum acceptable number of pillars. In this work, we place each CPU on its own pillar. If multiple CPUs were allowed to share the same pillar, there would be fewer pillars, but such an organization would give rise to other issues such as contentions.

8.4.3 3D NoC–Bus Hybrid Router Architecture

Section 8.2 provided a brief description of the 3D NoC–bus hybrid router design. A detailed design is described in this section.

A generic NoC router consists of four major components: the routing unit (RT), the virtual channel allocation unit (VA), the switch allocation unit (SA), and the crossbar (XBAR). In the mesh topology, each router has five physical channels (PC): North, South, East, and West, and one for the connection with the local processing element (CPU or cache bank). Each physical unit has a number of virtual channels (VC) associated with it. These are first-in-first-out (FIFO) buffers which hold flits from different pending messages. In our implementation, we used 3 VCs per PC, each 1 message deep. Each message was chosen to be 4 flits long. The width of the router links was chosen to be 128 bits. Consequently, a 64B cache line can fit in a packet (i.e., 4 flits/packet $\times$ 128 bits/flit $=$ 512 bits/packet $=$ 64 B/packet).

The most basic router implementations are 4-stage ones, i.e., they require a clock cycle for each component within the router. In our L2 architecture, low network latency is of utmost importance, thereby necessitating a faster router. Lower latency router architectures have been proposed which parallelize the RT, VA, and SA using a method known as speculative allocation [22]. This method predicts the winner of the VA stage and performs SA based on that. Moreover, a method known as look-ahead routing can also be used to perform routing one step ahead (perform the routing of node i+1 at node i). These two modifications can significantly improve the performance of the router. Two-stage, and even single-stage [20], routers are now possible which parallelize the various stages of operation. In our proposed architecture, we use a single-stage router to minimize latency.

Routers connected to pillar nodes are different, as an interface between the dTDMA pillar and the NoC router must be provided to enable seamless integration of the vertical links with the 2D network within the layers. The modified router is shown in Fig. 8.7. An extra physical channel (PC) is added to the router, which corresponds to the vertical link. The extra PC has its own dedicated buffers and is indistinguishable from the other links to the router operation. The router only sees an additional physical channel.

8.4.4 Processors and L2 Cache Organization

Figure 8.13 illustrates the organization of the processors and L2 caches in our design. Similar to CMP-DNUCA [3], we separate cache banks into multiple clusters. Each cluster contains a set of cache banks and a separate tag array for all the cache lines within the cluster. Some clusters have processors placed in the middle of them, while others do not. All the banks in a cluster are connected through a network-on-chip for data communication, while the tag array has a direct connection to the local processor in the cluster. Note that each processor has its own private L1 cache and an associated tag array for L2 cache banks within its local cluster. For a cluster without a local processor, the tag array is connected to a customized logic block which is responsible for receiving a cache line request, searching the tag array and forwarding the request to the target cache bank. This organization of processors and caches can be scaled by changing the size and/or number of the clusters.

8.4.5 Cache Management Policies

Based on the organization of processors and caches given in the previous section, we developed our cache management policies, consisting of a cache line search policy, a cache placement and replacement policy, and a cache line migration policy, all of which are detailed in the following sections.

8.4.5.1 Search Policy

Our cache line search strategy is a two-step process. In the first step, the processor searches the local tag array in the cluster to which it belongs and also sends requests to search the tag array of its neighboring clusters. All the vertically neighboring clusters receive the tag that is broadcast through the pillar. If the cache line is not found in either of these places, then the processor multicasts the requests to the remaining clusters. If the tag match fails in all the clusters, then it is considered an L2 miss. On a tag match in any of the clusters, the corresponding data is routed to the requesting processor through the network-on-chip.

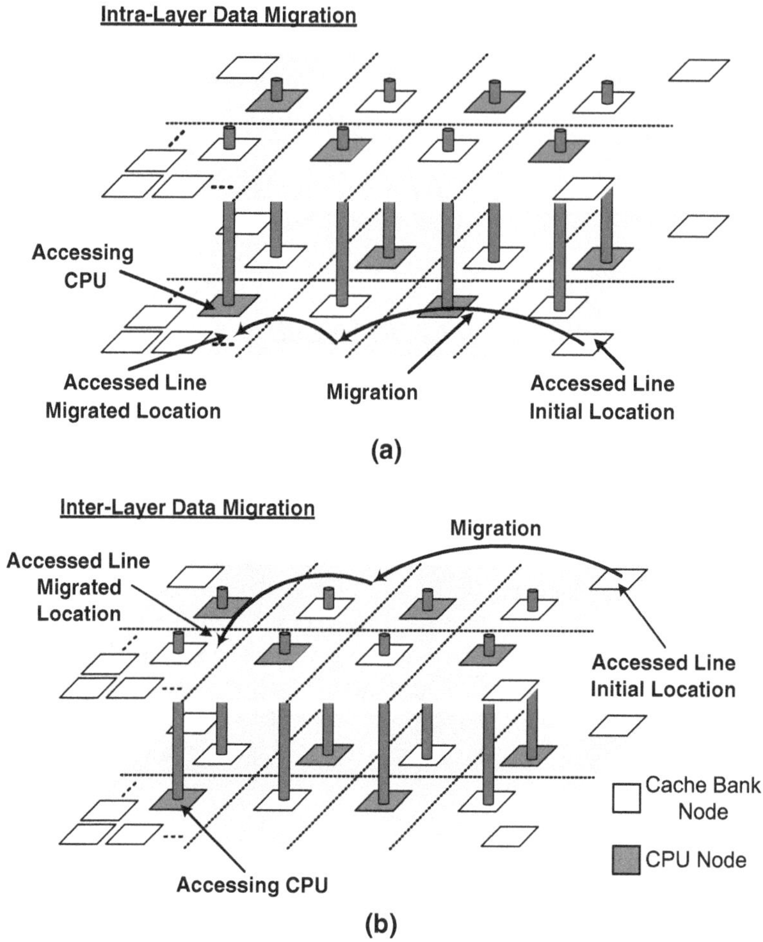

Fig. 8.13 Intra-layer and inter-layer data migration in the 3D L2 architecture. *Dotted lines* denote clusters

8.4.5.2 Placement and Replacement Policy

We use cache placement and replacement policies similar to those of CMP-DNUCA [3]. Initially a cache line is placed according to the low-order bits of its cache tag; that is, these bits determine the cluster in which the cache line will be placed initially. The low-order bits of the cache index indicate the bank in the cluster into which the cache line will be placed. The remaining bits of the cache index determine the location in the cache bank. The tag entry of the cluster is also updated when the cache line is placed. The placement policy can only be used to determine the initial location of a cache line, because when cache lines start migrating, the lower order bits of the cache tag can no longer indicate the cluster location. Finally, we use a pseudo-LRU replacement policy to evict a cache line to service a cache miss.

8.4.5.3 Cache Line Migration Policy

Similar to prior approaches, our strategy attempts to migrate data closer to the accessing processor. However, our policy is tailored to the 3D architecture and migrations are treated differently based on whether the accessed data lie in the same or different layer as the accessing processor. For data located within the same layer, the data is migrated gradually to a cluster closer to the accessing processor. When moving the cache lines to a closer cluster, we skip clusters that have processors (other than the accessing processor) placed in them since we do not want to affect their local L2 access patterns, and move the cache lines to the next closest cluster without a processor. Eventually, if the data is accessed repeatedly by only a single processor, it migrates to the local cluster of the processor. Figure 8.13a illustrates this intra-layer data migration.

For data located in a different layer, the data is migrated gradually closer to the pillar closest to the accessing processor (see Fig. 8.13b). Since clusters accessible through the vertical pillar communications are considered to be in local vicinity, we never migrate the data across the layers. This decision has the benefit of reducing the frequency of cache line migrations, which in turn reduces power consumption.

To avoid false misses (misses caused by searches for data in the process of migration), we employ a lazy migration mechanism as in CMP-DNUCA [3].

8.4.6 *Methodology*

We simulated the 3D CMP architecture by using Simics [18] interfaced with a 3D NoC simulator. A full-system simulation of an 8-processor CMP architecture running Solaris 9 was performed. Each processor uses in-order issue and executes the SPARC ISA. The processors have private L1 caches and share a large L2 cache. The default configuration parameters for processors, memories, and network-in-memory are given in Table 8.4. Some of the parameters in this table are modified for studying different configurations. The shown cache bank and tag array access latencies are extracted using the well-known cache simulator Cacti [27].

To model the latency of the 3D, hybrid NoC/bus interconnect, we developed a cycle-accurate simulator in C based on an existing 2D NoC simulator [12]. For this work, the 2D simulator was extended to three dimensions, and the dTDMA bus was integrated as the vertical communication channel. The 3D NoC simulator produces, as output, the communication latency for cache access.

In our cache model, private L1 caches of different processors are maintained coherent by implementing a distributed directory-based protocol. Each processor has a directory tracking the states of the cache lines within its L1 cache. L1 access events (such as read misses) cause state transitions and updates to directories, based on the MESI protocol. The traffic due to L1 cache coherence is taken into account in our simulation.

We simulated nine SPEC OMP benchmarks [28] with our simulation platform. For each benchmark, we marked an initialization phase in the source code. The

Table 8.4 Default system configuration parameters (L2 cache is organized as 16 clusters of size 16×64 KB)

Processor parameters	
Number of processors	8
Issue width	1
Memory parameters	
L1 (split I/D)	64 KB, 2-way, 64 B line, 3-cycle, write-through
L2 (unified)	16 MB (256×64 KB), 16-way, 64 B line, 5-cycle bank access
Tag array (per cluster)	24 KB, 4-cycle access
Memory	4 GB, 260 cycle latency
Network parameters	
Number of layers	2
Number of pillars	8
Routing scheme	Dimension-Order
Switching scheme	Wormhole
Flit size	128 bits
Router latency	1 cycle

cache model is not simulated until this initialization completes. After that, each application runs 500 million cycles for warming up the L2 caches. We then collected statistics for the next 2 billion cycles following the cache warm-up period.

8.4.7 Results

We first introduce the schemes compared in our experiments. We refer to the scheme with perfect search from [3] as CMP-DNUCA. We name our 2D and 3D schemes as CMP-DNUCA-2D and CMP-DNUCA-3D, respectively. Note that our 2D scheme is just a special case of our 3D scheme discussed in the paper, with a single layer. Both of these schemes employ cache line migration. To isolate the benefits due to 3D technology, we also implemented our 3D scheme without cache line migration, which is called CMP-SNUCA-3D.

Our first set of results give the average L2 hit latency numbers under different schemes. The results are presented in Fig. 8.14. We observe that our 2D scheme (CMP-DNUCA-2D) generates competitive results with the prior 2D approach (CMP-DNUCA [3]). Our 2D scheme shows slightly better IPC results for several benchmarks because we place processors not on the edges of the chip, as in CMP-DNUCA, but instead surround them with cache banks as shown in Fig. 8.13. Our results with 3D schemes reiterate the expected benefits from the increase in locality. It is interesting to note that CMP-SNUCA-3D, which does not employ migration, still outperforms the 2D schemes that employ migration. On the average, L2 cache latency reduces by 10 cycles when we move from CMP-DNUCA-2D to CMP-SNUCA-3D. Further gains are also possible in the 3D topology using data

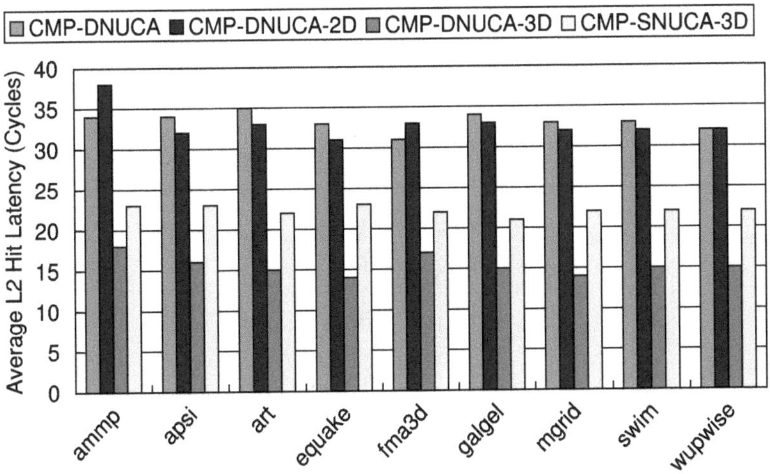

Fig. 8.14 Average L2 hit latency values under different schemes

migration. Specifically, CMP-DNUCA-3D reduces average L2 latency by seven cycles as compared to the static 3D scheme. Further, we note that even when employing migration, as shown in Fig. 8.15, 3D exercises it much less frequently compared to 2D, due to the increased locality. The reduced number of migrations in turn reduces the traffic on the network and the power consumption. These L2 latency savings translate to IPC improvements commensurate with the number of L2 accesses. Fig. 8.16 illustrates that the IPC improvements brought by CMP-DNUCA-3D (CMP-SNUCA-3D) over our 2D scheme are up to 37.1% (18.0%). The IPC

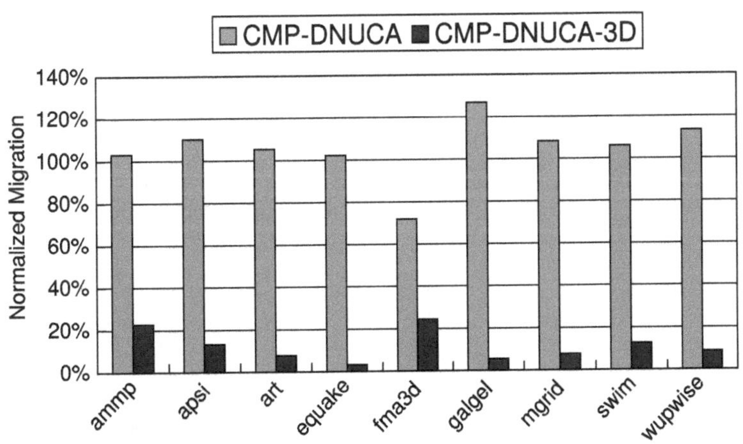

Fig. 8.15 Number of block migrations for CMP-DNUCA and CMP-DNUCA-3D, normalized with respect to CMP-DNUCA-2D

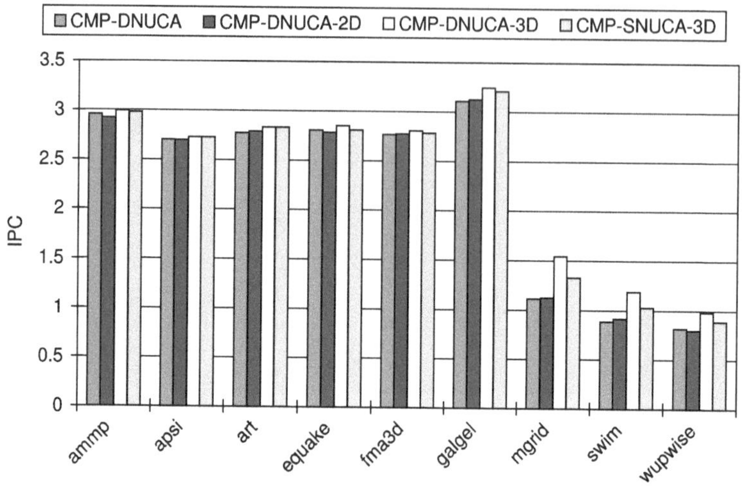

Fig. 8.16 IPC values under different schemes

improvements are higher with mgrid, swim, and wupwise since these applications exhibit a higher number of L2 accesses.

We next study the impact of larger cache sizes on our savings using CMP-DNUCA-2D and CMP-DNUCA-3D. When we increase the size of the L2 cache, we increase the size of each cluster while maintaining the 16-way associativity. Figure 8.17 shows the average L2 latency results with 32 MB and 64 MB L2 caches for four representative benchmarks (art and galgel with low L1 miss rates and mgrid and swim with high L1 miss rates). We observe that L2 latencies increase with the large cache sizes, albeit at a slower rate with the 3D configuration (on average seven cycles for 2D vs. five cycles for 3D), indicating that 3D topology is a more scalable option when we move to larger L2 sizes.

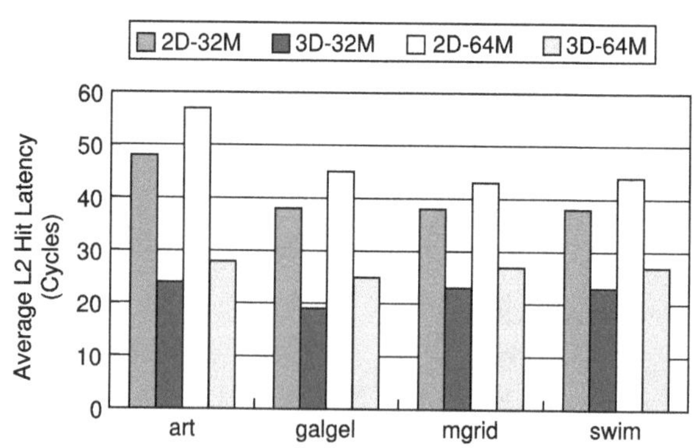

Fig. 8.17 Average L2 hit latency values under different schemes

Next we make experiments by modifying some of the parameters in the underlying 3D topology. The results with the CMP-DNUCA-3D scheme using different numbers of pillars to capture the effect of the different inter-layer via pitches are given in Fig. 8.18. As the number of pillars reduces, the contention for the shared resource (pillar) increases to service inter-layer communications. Consequently, average L2 latency increases by 1 to 7 cycles when we move from 8 to 2 pillars. Also, when the number of layers increases from 2 to 4, the L2 latency decreases by 3 to 8 cycles, primarily due to the reduced distances in accessing data, as illustrated in Fig. 8.19 for the CMP-SNUCA-3D scheme.

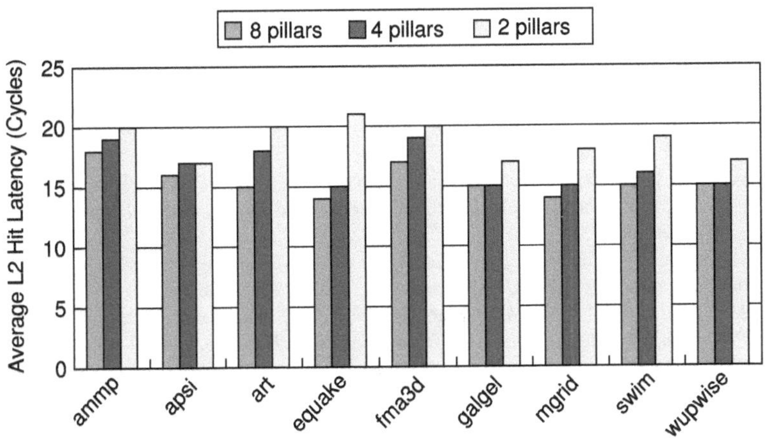

Fig. 8.18 Impact of the number of pillars (the CMP-DNUCA-3D scheme)

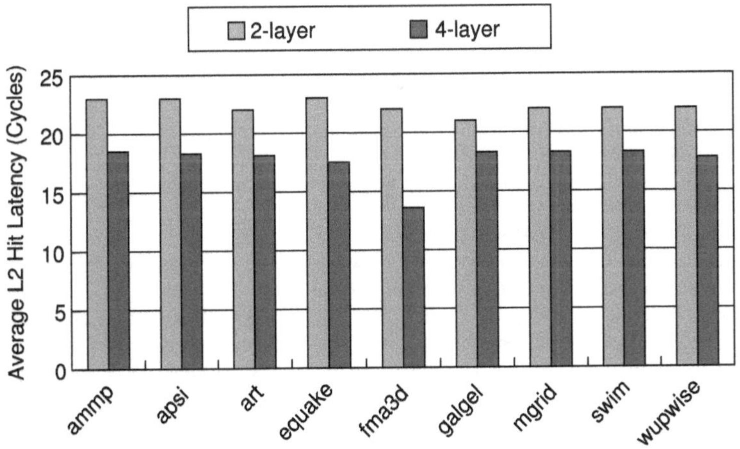

Fig. 8.19 Impact of the number of layers (the CMP-SNUCA-3D scheme)

8.5 Conclusion

Three-dimensional circuits and networks-on-chip (NoC) are two emerging trends for mitigating the growing complexity of interconnects. In this chapter, we describe various approaches to designing 3D NoC architecture and demonstrate that combining on-chip networks and 3D architectures can be a promising option for designing future chip multiprocessors.

Acknowledgments Much of the work and ideas presented in this chapter have evolved over several years of working with our colleagues and graduate students, in particular Professor Mahmut Kandemir, Dr. Mazin Yousif from Intel, Chrysostomos Nicopoulos, Thomas Richardson, Feihui Li, Jongman Kim, Dongkook Park, Reetuparna Das, Asit Mishra, and Soumya Eachempati. The research was supported in part by NSF grants, EIA-0202007, CCF-0429631, CNS-0509251, CCF-0702617, CAREER 0093085, and a grant from DARPA/MARCO GSRC.

References

1. A. Agarwal, L. Bao, J. Brown, B. Edwards, M. Mattina, C. Miao, C. Ramey, and D. Wentzlaff. Tile processor: Embedded multicore for networking and multimedia. In *Proceedings of Hot Chips Symposium*, 2007.
2. J. Balfour and W. J. Dally. Design tradeoffs for tiled CMP on-chip networks. In *Proceedings of International conference on Supercomputing*, pp. 187–198, 2006.
3. B. M. Beckmann and D. A. Wood. Managing wire delay in large chip-multiprocessor caches. In *Proceedings of International Symposium on Microarchitecture*, pp. 319–330, 2004.
4. G. De Micheli and L. Benini. *Networks on Chips*. Morgan Kaupmann, San Francisco, CA, 2006.
5. P. Dongkook, S. Eachempati, R. Das, A. K. Mishra, Y. Xie, N. Vijaykrishnan, and C. R. Das. Mira: A multi-layered on-chip interconnect router architecture. In *Proceedings of International Symposium on Computer Architecture*, pp. 251–261, 2008.
6. M. Gschwind, P. Hofstee, B. Flachs, M. Hopkins, Y. Watanabe, and T. Yamazaki. A compiler enabling and exploiting the cell broadband processor architecture. *IBM Systems Journal Special Issue on Online Game Technology*, 45(1), 2006.
7. R. Ho, K. Mai, and M. Horowitz. The future of wires. *Proceedings of the IEEE*, 89(4):490–504, April 2001.
8. A. Jantsch and H. Tenhunen. *Networks on Chip*. Kluwer Academic Publishers, Boston, 2003.
9. J.Kim, J. Balfour, and W. J. Dally. Flattened butterfly topology for onchip networks. In *Proceedings of International Symposium on Microarchitecture*, pp. 172–182, 2007.
10. J. Kim, C. Nicopoulos, D. Park, R. Das, Y. Xie, N. Vijaykrishnan, and C. Das. A novel dimensionally-decomposed router for on-chip communication in 3D architectures. In *Proceedings of International Symposium on Computer Architecture*, pp. 138–149, 2007.
11. J. Kim, C. Nicopoulos, D. Park, V. Narayanan, M. S. Yousif, and C. Das. A gracefully degrading and energy-efficient modular router architecture for on-chip networks. In *Proceedings of International Symposium on Computer Architecture*, pp. 4–15, 2006.
12. J. Kim, D. Park, C. Nicopoulos, N. Vijaykrishnan, and C. Das. Design and analysis of an NoC architecture from performance, reliability and energy perspective. In *Proceedings of Symposium on Architecture for Networking and Communications Systems*, pp. 173–182, October 2005.
13. P. Kongetira, K. Aingaran, and K. Olukotun. Niagara: A 32-way multithreaded SPARC processor. *IEEE MICRO*, 25(2):21–29, 2005.
14. F. Li, C. Nicopoulos, T. Richardson, Y. Xie, V. Narayanan, and M. Kandemir. Design and management of 3D chip multiprocessors using network-in-memory. In *Proceedings of International Symposium on Computer Architecture*, pp. 130–141, 2006.

15. G. H. Loh. 3D-stacked memory architectures for multi-core processors. In *Proceedings of International Symposium on Computer Architecture*, pp. 453–464, 2008.
16. I. Loi, F. Angiolini, and L. Benini. Developing mesochronous synchronizers to enable 3D NoCs. In *Proceedings of Design, Automation and Test in Europe Conference*, pp. 1414–1419, 2008.
17. I. Loi, S. Mitra, T. H. Lee, S. Fujita, and L. Benini. A low-overhead fault tolerance scheme for tsv-based 3D network on chip links. In *Proceedings of International Conference on Computer-Aided Design*, pp. 598–602, 2008.
18. P. Magnusson, M. Christensson, J. Eskilson, D. Forsgren, and G. Hallberg. Simics: A full system simulation platform. *IEEE Computer*, 35(2):50–58, February 2002.
19. R. Marculescu, U. Y. Ogras, L. S. Peh, N. E. Jerger, and Y. Hoskote. Outstanding research problems in NoC design: System, microarchitecture, and circuit perspectives. *IEEE Transactions on Computer-Aided Design of Integrated Circuits and Systems*, 28(1):3–21, January 2009.
20. R. Mullins, A. West, and S. Moore. Low-latency virtual-channel routers for on-chip networks. In *Proceedings of International Symposium on Computer Architecture*, p. 188, June 2004.
21. V. F. Pavlidis and E. G. Friedman. 3-D topologies for networks-on-chip. *IEEE Transactions on Very Large Scale Integration (VLSI) Systems*, 15(10):1081–1090, 2007.
22. L. Peh and W. Dally. A delay model and speculative architecture for pipelined routers. In *Proceedings of International Symposium on High Performance Computer Architecture*, pp. 255–266, January 2001.
23. A. Pullini, F. Angiolini, S. Murali, D. Atienza, G. De Micheli, and L. Benini. Bringing NoCs to 65 nm. *IEEE Micro*, 27(5):75–85, 2007.
24. T. Richardson, C. Nicopoulos, D. Park, V. Narayanan, Y. Xie, C. Das, and V. Degalahal. A hybrid SoC interconnect with dynamic TDMA-based transaction-less buses and on-chip networks. In *Proceedings of International Symposium on VLSI Design*, pp. 657–664, 2006.
25. S. Rusu, S. Tam, H. Muljono, J. Stinson, D. Ayers, J. Chang, R. Varada, M. Ratta, and S. Kottapalli. A 45 nm 8-core enterprise xeo processor. In *Proceedings of International Solid-State Circuits Conference*, February 2009.
26. J. Shen and M. Lipasti. *Modern Processor Design: Fundamentals of Superscalar Processors*. McGraw-Hill, Boston, 2005.
27. P. Shivakumar and N. Jouppi. Cacti 3.0: An integrated cache timing, power and area model. In *Technical Report, Compaq Computer Corporation*, August 2001.
28. Standard Performance Evaluation Corporation. SPEC OMP. http://www.spec.org.
29. G. Sun, X. Dong, Y. Xie, J. Li, and Y. Chen. A novel architecture of the 3D stacked mram l2 cache for CMPs. In *Proceedings of International Symposium on High Performance Computer Architecture*, pp. 239–249, 2009.
30. B. Vaidyanathan, W. Hung, F. Wang, Y. Xie, N. Vijaykrishnan, and M. Irwin. Architecting microprocessor components in 3D design space. In *Proceedings of International Conference on VLSI Design*, pp. 103–108, 2007.
31. S. R. Vangal, J. Howard, G. Ruhl, S. Dighe, H. Wilson, J. Tschanz, D. Finan, A. Singh, T. Jacob, S. Jain, V. Erraguntla, C. Roberts, Y. Hoskote, N. Borkar, and S. Borkar. An 80-tile sub-100-w teraFLOPS processor in 65-nm cmos. *IEEE Journal of Solid-State Circuits*, 43(1):29–41, 2008.
32. Y. Xie, G. H. Loh, B. Black, and K. Bernstein. Design space exploration for 3D architectures. *ACM Journal of Emerging Technology of Computer Systems*, 2(2):65–103, 2006.
33. Y. Xu, Y. Du, B. Zhao, X. Zhou, Y. Zhang, and J. Yang. A low-radix and low-diameter 3D interconnection network design. In *Proceedings of International Symoposium on High Performance Computer Architecture*, pp. 30–41, 2009.
34. S. Yan and B. Lin. Design of application-specific 3D networks-on-chip architectures. In *Proceedings of International Conference of Computer Design*, pp. 142–149, 2008.

Chapter 9
PicoServer: Using 3D Stacking Technology to Build Energy Efficient Servers

Taeho Kgil, David Roberts, and Trevor Mudge

Abstract With power and cooling increasingly contributing to the operating costs of a datacenter, energy efficiency is the key driver in server design. One way to improve energy efficiency is to implement innovative interconnect technologies such as 3D stacking. Three-dimensional stacking technology introduces new opportunities for future servers to become low power, compact, and possibly mobile. This chapter introduces an architecture called Picoserver that employs 3D technology to bond one die containing several simple slow processing cores with multiple memory dies sufficient for a primary memory. The multiple memory dies are composed of DRAM. This use of 3D stacks readily facilitates wide low-latency buses between processors and memory. These remove the need for an L2 cache allowing its area to be re-allocated to additional simple cores. The additional cores allow the clock frequency to be lowered without impairing throughput. Lower clock frequency means that thermal constraints, a concern with 3D stacking, are easily satisfied. PicoServer is intentionally simple, requiring only the simplest form of 3D technology where die are stacked on top of one another. Our intent is to minimize risk of introducing a new technology (3D) to implement a class of low-cost, low-power, compact server architectures.

9.1 Introduction

Datacenters are an integral part of today's computing platforms. The success of the internet and the continued scalability in Moore's law have enabled internet service providers such as Google and Yahoo to build large-scale datacenters with millions of servers. For large-scale datacenters, improving energy efficiency becomes a critical

T. Kgil (✉)
Intel, 2111 NE 25th Ave, Hillsboro, OR, 97124, USA
e-mail: taeho.kgil@intel.com

Y. Xie et al. (eds.), *Three-Dimensional Integrated Circuit Design*, Integrated Circuits and Systems, DOI 10.1007/978-1-4419-0784-4_9,

task. Datacenters based on off-the-shelf general purpose processors are unnecessarily power hungry, require expensive cooling systems, and occupy a large space. In fact, the cost of power and cooling these datacenters will likely contribute to a significant portion of the operating cost. Our claim can be confirmed in Fig. 9.1 which breaks down the annual operating cost for datacenters. As Fig. 9.1 clearly shows, the cost in power and cooling servers is increasingy contributing to the overall operating costs of a datacenter.

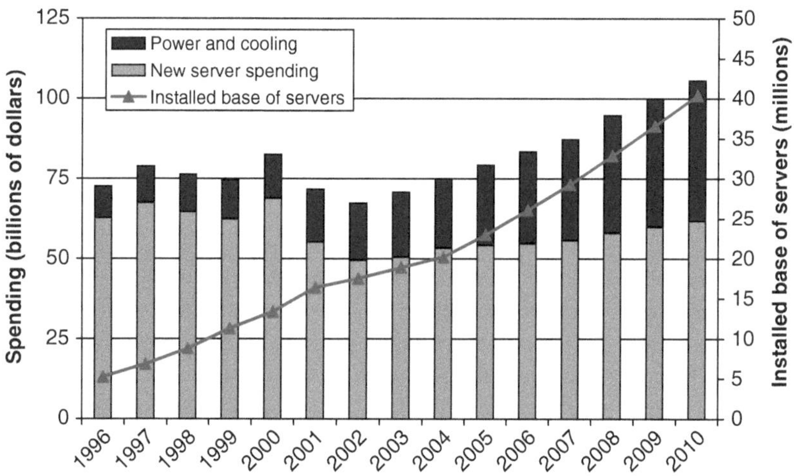

Fig. 9.1 IDC estimates for annual cost spent (1) power and cooling servers and (2) purchasing additional servers [52]

One avenue to designing energy efficient servers is to introduce innovative interconnect technology. Three-dimensional stacking technology is an interconnect technology that enables new chip multiprocessor (CMP) architectures that significantly improve energy efficiency. Our proposed architecture, PicoServer,[1] employs 3D technology to bond one die containing several simple slow processor cores with multiple DRAM dies that form the primary memory. In addition, 3D stacking enables a memory processor interconnect that is of both very high bandwidth and low latency. As a result the need for complex cache hierarchies is reduced. We show that the die area normally spent on an L2 cache is better spent on additional processor cores. The additional cores mean that they can be run slower without affecting throughput. Slower cores also allow us to reduce power dissipation and with its thermal constraints, a potential roadblock to 3D stacking. The resulting system is ideally suited to throughput applications such as servers. Our proposed architecture is intentionally simple and requires only the simplest form of 3D technology where die is stacked on top of one another. Our intent is to minimize risk of realizing a

[1] This chapter is based on the work in [32] and [29]

class of low-cost, low-power, compact server architectures. Employing PicoServers can significantly lower power consumption and space requirements.

Server applications handle events on a per-client basis, which are independent and display high levels of thread-level parallelism. This high level of parallelism makes them ill-suited for traditional monolithic processors. CMPs built from multiple simple cores can take advantage of this thread-level parallelism to run at a much lower frequency while maintaining a similar level of throughput and thus dissipating less power. By combining them with 3D stacking we will show that it is possible to cut power requirements further. Three-dimensional stacking enables the following key improvements:

- **High-bandwidth buses between DRAM and L1 caches that support multiple cores – thousands of low-latency connections with minimal area overhead between dies are possible.** Since the interconnect buses are on-chip, we are able to implement wide buses with a relatively lower power budget compared to inter-chip implementations.
- **Modification in the memory hierarchy due to the integration of large capacity on-chip DRAM.** It is possible to remove the L2 cache and replace it with more processing cores. The access latency for the on-chip DRAM[2] is also reduced because address multiplexing and off-chip I/O pad drivers [47] are not required. Further, it also introduces opportunities to build nonuniform memory architectures with a fast on-chip DRAM and relatively slower off-chip secondary system memory.
- **Overall reduction in system power primarily due to the reduction in core clock frequency.** The benefits of 3D stacking stated in items 1 and 2 allow the integration of more cores clocked at a modest frequency – in our work 500-1000 MHz – on-chip while providing high throughput. Reduced core clock frequency allows their architecture to be simplified; for example, by using shorter pipelines with reduced forwarding logic.

The potential drawback of 3D stacking is thermal containment (see Chapter 3). However, this is not a limitation for the type of simple, low-power cores that we are proposing for the PicoServer, as we show in Section 9.4.5. In fact the ITRS projections of Table 9.2 predict that systems consuming just a few watts do not even reequire a heat sink.

The general architecture of a PicoServer is shown in Fig. 9.2. For the purposes of this work we assume a stack of five to nine dies. The connections are by vias that run perpendicular to the dies. The dimensions for a 3D interconnect via vary from 1 to 3 μm with a separation of 1 to 6 μm. Current commercial offerings

[2] We will refer to die that is stacked on the main processor die as "on-chip," because they form a 3D chip.

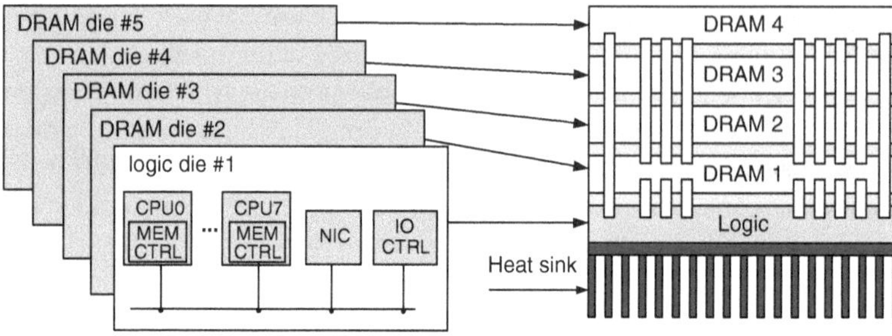

Fig. 9.2 A diagram depicting the PicoServer: a CMP architecture connected to DRAM using 3D stacking technology with an on-chip network interface controller (NIC) to provide low-latency high-bandwidth networking

can support 1,000,000 vias per cm^2 [26]. This is far more than we need for PicoServer. These function as interconnect and thermal pipes. For our studies, we assume that the logic-based components – the microprocessor cores, the network interface controllers (NICs), and peripherals – are on the bottom layer and conventional capacity-oriented DRAMs occupy the remaining layers. To understand the design space and potential benefits of this new technology, we explored the tradeoffs of different bus widths, number of cores, frequencies, and the memory hierarchy in our simulations. We found bus widths of 1024 bits with a latency of two clock cycles at 250 MHz to be reasonable in our architecture. In addition, we aim for a reasonable area budget constraining the die size area to be below 80 mm^2 at 90 nm process technology. Our 12-core PicoServer configuration which occupies the largest die area is conservatively estimated to be approximately 80 mm^2. The die areas for our 4- and 8-core PicoServer configurations are, respectively 40 mm^2 and 60 mm^2.

We also extend our analysis of PicoServer and show the impact of integrating Flash onto a PicoServer architecture. We provide a qualitative analysis of two configurations that integrate (1) Flash as a discrete component and (2) directly stacks Flash on top of our DRAM + logic die stack. Both configurations leverage the benefits of 3D stacking technology. The first configuration is driven by bigger system memory capacity requirements, while the second configuration is driven by small form factor.

This chapter is organized as follows. In the next section we provide background for this work by describing an overview of server platforms, 3D stacking technology, and trends in DRAM technology. In Section 9.3, we outline our methodology for the design space exploration. In Section 9.4, we provide more details for the PicoServer architecture and evaluate various PicoServer configurations. In Section 9.5, we present our results in the PicoServer architecture for server benchmarks and compare our results to conventional architectures that do not employ 3D stacking. These architectures are CMPs without 3D stacking and conventional high-performance desktop architectures with Pentium 4-like characteristics. A summary and concluding remarks are given in Section 9.6.

9.2 Background

This section provides an overview of the current state of server platforms, 3D stacking technology, and DRAM technology. We first show how servers are currently deployed in datacenters and analyze the behavior of current server workloads. Next, we explain the state of 3D stacking technology and how it is applied in this work. Finally, we show the advances in DRAM technology. We explain the current and future trends in DRAM used in the server space.

9.2.1 Server Platforms

9.2.1.1 Three-Tier Server Architecture

Today's datacenters are commonly built around a 3-tier server architecture. Figure 9.3 shows a 3-tier server farm and how it might handle a request for service. The first tier handles a large bulk of the requests from the client (end user). Tier 1 servers handle web requests. Because Tier 1 servers handle events on a per-client basis, they are independent and display high levels of thread-level parallelism. For requests that require heavier computation or database accesses, they are forwarded to Tier 2 servers. Tier 2 servers execute user applications that interpret script languages and determine what objects (typically database objects) should be accessed. Tier 2 servers generate database requests to Tier 3 servers. Tier 3 servers receive database queries and return the results to Tier 2 servers.

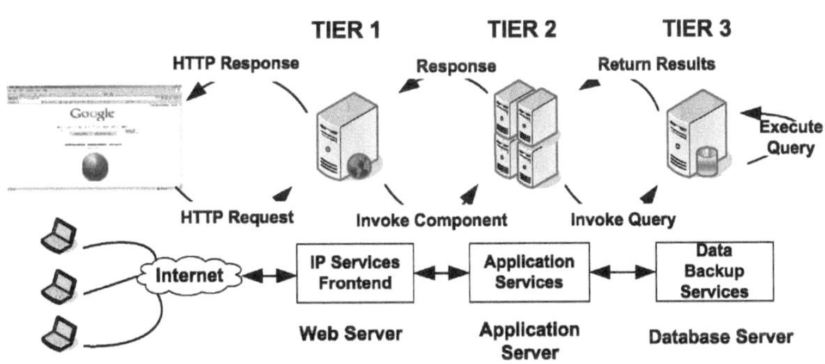

Fig. 9.3 A typical 3-tier server architecture. Tier 1 – web server, Tier 2 – application server, Tier 3 – database server

For example, when a client requests a Java Servlet Page (JSP web page), it is first received by the front end server – Tier 1. Tier 1 recognizes a Java Servlet Page that must be handled and initiates a request to Tier 2 typically using remote message interfaces (RMI). Tier 2 initiates a database query on the Tier 3 servers, which in turn generate the results and send the relevant information up the chain all the way to Tier 1. Finally, Tier 1 sends the generated content to the client.

Three-tier server architectures are commonly deployed in today's server farms, because they allows each level to be optimized for its workload. However, this strategy is not always adopted. *Google* employs essentially the same machines at each level, because economies of scale and manageability issues can outweigh the advantages. We will show that, apart from the database disk system in the third tier, the generic PicoServer architecture is suitable for all tiers.

9.2.1.2 Server Workload Characteristics

Server workloads display a high degree of thread-level parallelism (TLP) because connection-level parallelism through client connections can be easily mapped to thread-level parallelism (TLP). Table 9.1 shows the behavior of commercial server workloads. Most of the commercial workloads display high TLP and low instruction-level parallelism (ILP) with the exception of decision support systems. Conventional general-purpose processors, however, are typically optimized to exploit ILP. These workloads suffer from a high cache miss rate regularly stalling the machine. This leads to low instructions per cycle (IPC) and poor utilization of processor resources. Our studies have shown that except for computation intensive workloads such as PHP application servers, video-streaming servers, and decision support systems, out-of-order processors have an IPC between 0.21 and 0.54 for typical server workloads, i.e., at best modest computation loads with an L2 cache of 2 MB. These workloads do not perform well because much of the requested data has been recently DMAed from the disk to system memory, invalidating cached data that leads to cache misses. Therefore, we can generally say that single-thread-optimized out-of-order processors do not perform well on server workloads. Another interesting property of most server workloads is the appreciable amount of time spent in kernel code, unlike SPECCPU benchmarks. This kernel code is largely involved in

Table 9.1 Behavior of commercial workloads adapted from [38]

Attribute	Web99	JBOB(JBB)	TPC-C	SAP 2T	SAP 3T DB	TPC-H
Application category	Web server	Server java	OLTP*	ERP†	ERP†	DSS‡
Instruction-level parallelism	low	low	low	med	low	high
Thread-level parallelism	high	high	high	high	high	high
Instruction/data working-set	large	large	large	med	large	large
Data sharing	low	med	high	med	high	med
I/O bandwidth	high (network)	low	high (disk)	med (disk)	high (disk)	med (disk)

∗ OLTP : online transaction processing
† ERP : enterprise resource planning
‡ DSS : decision support system

interrupt handling for the NIC or disk driver, packet transmission, network stack processing, and disk cache processing.

Finally, a large portion of requests are centered around the same group of files. These file accesses request access to memory and I/O. Due to the modest computation requirements, memory and I/O latency are critical to high performance. Therefore, disk caching in the system memory plays a critical part in providing sufficient throughput. Without a disk cache, the performance degradation due to the hard disk drive latency would be unacceptable.

To perform well on these classes of workloads an architecture should naturally support multiple threads to respond to independent requests from clients. Thus intuition suggests that a CMP or SMT architecture should be able to better utilize the processor die area.

9.2.1.3 Conventional Server Power Breakdown

Figure 9.4 shows the power breakdown of a server platform available today. This server uses a chip multiprocessor implemented with many simple in-order cores to reduce power consumption. The power breakdown shows that 1/4 is consumed by the processor, 1/4 is consumed by the system memory, 1/4 is consumed by the power supply, and 1/5 is consumed by the I/O interface. Immediately we can see that using a relatively large amount of system memory results in the consumption of a substantial fraction of power. This is expected to increase as the system memory clock frequency increases and system memory size increases. We also find that despite using simpler cores that are energy efficient, a processor would still consume a noticeable amount of power. The I/O interface consumes a large amount of power due to the high I/O supply voltage required in off-chip interfaces. The I/O supply voltage is likely to reduce as we scale in the future but would not scale as fast as the core supply voltage. Therefore, there are many opportunities to further reduce power by integrating system components on-chip. And finally, we find that the power supply displays some inefficiency. This is due to the multiple levels of voltage it has to support. A reduced number of power rails will dramatically improve the power supply efficiency. Three-dimensional stacking technology has the potential to

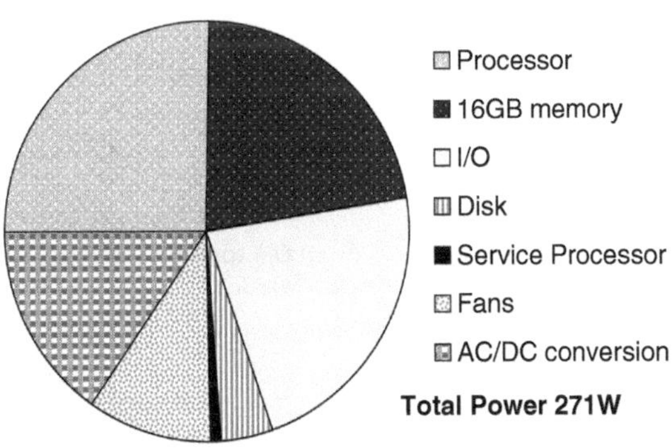

Fig. 9.4 Power breakdown of T2000 UltraSPARC executing SpecJBB

(1) reduce the power consumed by the processor and the I/O interfaces by integrating additional system components on-chip and also (2) improve power supply efficiency by reducing the number of power rails implemented in a server.

9.2.2 Three-Dimensional Stacking Technology

This section provides an overview of 3D stacking technology. In the past there have been numerous efforts in academia and industry to implement 3D stacking technology [17, 40, 37, 44, 57]. They have met with mixed success. This is due to the many challenges that need to be addressed. They include (1) achieving high yield in bonding die stacks; (2) delivering power to each stack; and (3) managing thermal hotspots due to stacking multiple dies. However, in the past few years strong market forces in the mobile terminal space have accelerated a demand for small form factors with very low power. In response, several commercial enterprises have begun offering reliable low-cost die-to-die 3D stacking technologies.

In 3D stacking technology, dies are typically bonded as face-to-face or face-to-back. Face-to-face bonds provide higher die-to-die via density and lower area overhead than face-to-back bonds. The lower via density for face-to-back bonds result from the through silicon vias (TSVs) that have to go through silicon bulk. Figure 9.5 shows a high-level example of how dies can be bonded using 3D stacking technology. The bond between layer 1 (starting from the bottom) and 2 is face-to-face, while the bond between layer 2 and 3 is face-to-back. Using the bonding techniques in 3D stacking technology opens up the opportunity of stacking heterogeneous dies together. For example, architectures that stack DRAM and logic are manufactured from different process steps. References [43, 24, 16] demonstrate the benefits of stacking DRAM on logic. Furthermore, with the added third dimension from the vertical axis, the overall wire interconnect length can be reduced and wider bus width can be achieved at lower area costs. The parasitic capacitance and resistance for 3D vias are negligible compared to global interconnect. We also note that the size and pitch of 3D vias only adds a modest area overhead. Three-dimensional via pitches are equivalent to 22λ for 90-nm technology, which is about the size of a 6T SRAM cell. They are also expected to shrink as this technology becomes mature.

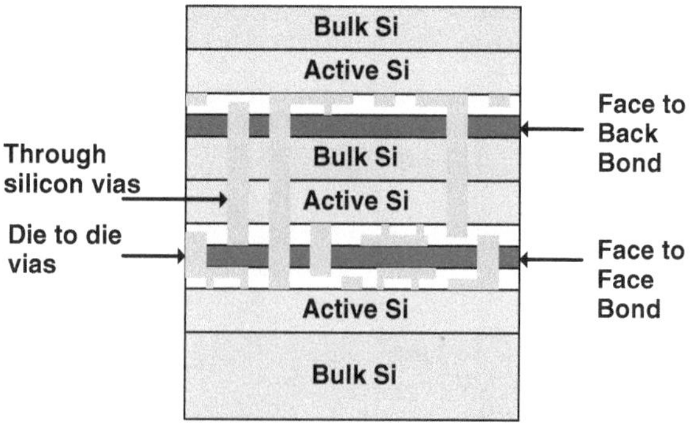

Fig. 9.5 Example of a three-layer 3D IC

The ITRS roadmap in Table 9.2 predicts deeper stacks being practical in the near future. The connections are by vias that run perpendicular to the dies. As noted earlier, the dimensions for a 3D interconnect via vary from 1 to 3 μm with a separation of 1 to 6 μm. Current commercial offerings can support 1,000,000 vias per cm^2 [26].

Table 9.2 ITRS projection [12] for 3D stacking technology, memory array cells, and maximum power budget for power aware platforms. ITRS projections suggest DRAM density exceeds SRAM density by 15–18× entailing large capacity of DRAM can be integrated on-chip using 3D stacking technology as compared to SRAM

	2007	2009	2011	2013	2015
Low-cost/handheld #die/stack	7	9	11	13	14
SRAM density Mbits/cm^2	138	225	365	589	948
DRAM density Mbits/cm^2 at production	1,940	3,660	5,820	9,230	14,650
Maximum power budget for cost-performance systems (W)	104	116	119	137	137
Maximum power budget for low-cost/handheld systems with battery (W)	3.0	3.0	3.0	3.0	3.0

Table 9.3 Three-dimensional stacking technology parameters [26, 13, 44]

	Face-to-back	Face-to-face	RPI	MIT 3D FPGA
Size	1.2μ × 1.2μ	1.7μ × 1.7 μ	2μ × 2μ	1μ × 1μ
Minimum pitch	<4μ	2.4μ	N/A	N/A
Feed through capacitance	2–3 fF	≈0	N/A	2.7 fF
Series resistance	<0.35Ω	≈0	≈0	≈0

Overall yield using 3D stacking technology is a product of the yield of each individual die layer. Therefore, it is important that the individual die is designed with high yield in mind. Memory stacking is a better choice than logic-to-logic stacking. Memory devices typically show higher yield, because fault tolerance fits well with their repetitive structure. For example, re-fusing extra bitlines to compensate for defective cells and applying single bit error correction logic to memory boost yield. Several studies including [48] show that DRAM yields are extremely high suggesting chips built with a single logic layer and several DRAM layers generate yields close to the logic die.

9.2.3 DRAM Technology

This section overviews the advances in DRAM technology in the server space. DRAM today is offered in numerous forms usually determined by the application space. In particular, for server platforms, DDR2/DDR3 DRAM has emerged as the primary solution for system memory. FBDIMM DRAM that delivers higher throughput and higher capacity than DDR2/DDR3 is emerging as an alternative, but higher power, solution. RLDRAM and NetRAM [55, 7] are also popular DRAM choices for network workloads in the server space.

The common properties for these memories are high-throughput and low latency. In the server space, DRAM must meet the high-throughput and low-latency demands to deliver high performance. These demands can only be achieved at the price of increasing power consumption in the DRAM I/O interface and the DRAM arrays. As a result, power has increased to a point where the I/O power and DRAM power contribute to a significant amount of overall system power (as we showed in Section 9.2.1.3). Industry has addressed this concern by reducing the I/O supply voltage and introducing low-power versions of the DDR2 interface both at the price of sacrificing throughput and latency. We will show that DRAM stacked using 3D stacking technology can be implemented to deliver high-throughput and low-latency DRAM interfaces while consuming much less power.

9.3 Methodology

This section describes our methodology in evaluating the benefits of 3D stacking technology. The architectural aspects of our studies were obtained from a microarchitectural simulator called M5 [15] that is able to run Linux and evaluate full system-level performance. We model the benefits gained using 3D stacking technology in this full system simulator. We also model multiple servers connected to multiple clients in M5. The client requests are generated from user-level network application programs. We measure server throughput – network bandwidth or transactions per second – to estimate performance. Our die area estimations are derived from previous publications and developed models for delay and power [12, 26, 50, 2, 13, 17]. DRAM timing and power values were obtained from IBM and Micron technology datasheets [3]. A detailed description of our methodology is described in the following sections.

9.3.1 Simulation Studies

9.3.1.1 Full System Architectural Simulator

To evaluate the performance of PicoServer we used the M5 full system simulator. M5 boots an unmodified Linux kernel on a configurable architecture. Multiple systems are defined in the simulator to model the clients and servers and connected via an ethernet link model. The server side executes Apache – a web server, Fenice – a video-streaming server, mySQL – a database server, and NFS – a file server. The client side executes benchmarks that generate representative requests for dynamic and static web page content, video stream requests, database queries, and network file commands respectively. For comparison purposes we defined a Pentium 4-like system [53] and a chip multiprocessor-like system similar to [36]. We also looked at several configurations using 3D stacking technology on these platforms. We assume that with 3D stacking technology, wider bus widths can be implemented with lower power overhead. Table 9.4 shows the configurations used in our simulations.

9.3.1.2 Server Benchmarks

We use several benchmarks that directly interact with client requests. We used two web content handling benchmarks, SURGE [14] and SPECweb99 [10], to measure web server performance. Both benchmarks request filesets of more than a 1 GB. A web script handling benchmark SPECweb2005 [9] using PHP is selected to represent script workloads. A video-streaming benchmark, Fenice [6], that uses the RTSP protocol along with the UDP protocol is chosen to measure behavior for on-demand workloads. For a file-sharing benchmark we use an NFS server and stressed it with dbench. Finally, we executed two database benchmarks to measure database performance for Tier 2 and 3 workloads.

SURGE The SURGE benchmark represents client requests for static web content. We modified the SURGE fileset and used a zipf distribution to generate reasonable client requests. Based on the zipf distribution a static web page which is approximately 12 KB in file size is requested 50% of the time in our client requests. We configured the SURGE client to have 20 outstanding client requests.

SPECweb99 To evaluate a mixture of static web content and simple dynamic web content, we used a modified version of SURGE to request SPECweb99 filesets (behavior illustrated in Table 9.1). We used the default configuration for SPECweb99 to generate client requests. Seventy percent of client requests are for static web content and 30% are for dynamic web contents.

SPECweb2005 Scripting languages are a popular way to describe web pages. SPECweb2005 offers three types of benchmarks: a banking benchmark that emulates the online banking activity of a user, an E-commerce benchmark that emulates the online purchase activity, and a support benchmark that emulates the online stream activity. All benchmarks require a dynamic web page to be generated from a script interpreter. We use a PHP interpreter to measure the behavior of Tier 2 Servers. The client requests are generated from methods described for SPECweb99 and SURGE clients.

Fenice On-demand video serving is also an important workload for Tier 1 servers. For copyright protection and live broadcasts, the RTSP protocol is commonly used for real-time video playback. Fenice is an open source streaming project [6] that provides workloads supporting the RTSP protocol. We modified it to support multithreading. Client requests were generated with a modified version of nemesi, a RTSP supporting MPEG player. Nemesi is also from the open source streaming project. We generated multiple client requests that fully utilized the server CPUs for a high-quality 16 Mbps datastream of 720 × 480 resolution MPEG2 frames.

dbench This benchmark is commonly used to stress NFS daemons. In our tests we used the in-kernel NFS daemon which is multithreaded and available in standard Linux kernels. We generated NFS traffic using dbench on the client side that stressed

Table 9.4 Commonly used simulation configurations. System memory latencies are generated from DDR2 DRAM models. We assume cores clocked at lower clock frequencies (below 1 GHz) have higher L1 cache associativity. L2 cache unloaded latency for single-core and multicore configurations differs due to longer global interconnect lengths in multicore platforms [39]

	OO4-small baseline with and w/out 3D stacking	OO4-large baseline/with and w/out 3D stacking	Conventional CMP MP4/8 w/out 3D stacking	PicoMP4/8/12-500 MHz/1000 MHz*
Operating frequency	4 GHz	4 GHz	1 GHz	500 MHz/1 GHz
Number of processors	1	1	4/8	4/8/12
Processor type	out-of-order	out-of-order	in-order	in-order
Issue width	4	4	1	1
L1 cache	2 way 16 KB	2 way 128 KB	4 way 16 KB per core	4 way 16 KB per core
L2 cache	8 way 256 KB 7.5 ns unloaded latency	8 way 2 MB 7.5 ns unloaded latency	8 way 2 MB 16 ns unloaded latency	N/A
Memory bus width	64 bit@400 MHz/1024 bit@250 MHz	64 bit@400 MHz/1024 bit@250 MHz	64 bit@250 MHz	1024 bit@250 MHz
System memory	512 MB DDR2 DRAM	512 MB DDR2 DRAM	512 MB DDR2 DRAM	128 MB – 512 MB DDR2 DRAM

*PicoServer platform using 3D stacking technology. The core clock frequency of PicoServer is typically 500 MHz. PicoServer configurations with 1 GHz core clock frequency are later used to show the impact of 3D stacking technology.

the file server. dbench generates workloads that both read and write to the file server while locking these files so that a different client could not access it simultaneously.

OLTP On-line transaction processing is a typical workload executed on Tier 2 and 3 servers (behavior illustrated in Table 9.1). The TPC council has described in detail benchmarks for OLTP. We used a modified version of TPC-C made available by the Open Source Development Lab (OSDL) called DBT2 [5]. DBT2 generates transaction orders. Our database server is MySQL 5.0. We use the InnoDB storage engine that supports transactions and provides a reasonable amount of scalability for multicores. We generated a 1 GB warehouse which is typically used for small-scale computation intensive databases. We chose a small working-set size due to simulation time limitations. We selected a buffer pool size accordingly.

DSS Decision support system is another typical workload used to evaluate Tier 2 and 3 servers. We used TPC-H, the current version of a DSS workload. Again a modified version of TPC-H available by OSDL (DBT3) [5] is used in this study. We loaded the TPC-H database onto mySQL and used the defined TPC-H queries to measure performance. The query cache is disabled to prevent speedup in query time due to caching. To reduce our simulation time to a reasonable amount, we only performed and measured the time for a Q22 query out of the many TPC-H queries. Q22 query takes a modest amount of time to execute and displays the behaviors illustrated in Table 9.1.

9.3.2 Estimating Power and Area

Power and area estimation at the architectural level is difficult to do with great accuracy. To make a reasonable estimation and to show general trends, we resorted to industry white papers, datasheets, and academia publications on die area, and we compared our initial analytical power models with real implementations and widely used cycle-level simulation techniques. We discuss this further in the next subsections.

9.3.2.1 Processors

We relied to a large extent on figures reported in [20, 1, 51] for an ARM processor to estimate processor power and die area. The ARM is representative of a simple in-order 32-bit processor that would be suitable for the PicoServer. Due to the architectural similarities with our PicoServer cores, we extrapolated the die area and power consumption for our PicoServer cores at 500 MHz from published data in [20, 1, 51]. Table 9.5 lists these estimates along with values listed in [1, 51] and a Pentium 4 core for comparison. An analysis of the expected die area per core was also conducted. We collected several die area numbers available from ARM, MIPS, PowerPC, and other comparable scalar in-order processors. We also synthesized several 32-bit open source cores that are computationally comparable to a single PicoServer core. We synthesized them using the Synopsys physical compiler toolset.

Table 9.5 Published power consumption values for various microprocessors [20, 1, 51, 53]

	Pentium 4 90 nm	ARM11 130 nm	Xscale 90 nm*	PicoServer MP 90 nm†
L1 cache	16 KB	16 KB	32 KB	16 KB
L2 cache	1 MB	N/A	N/A	N/A
Total power (W)	89–103 W	250 mW @ 550 MHz	850 mW @ 1.5 GHz	190 mW @ 500 MHz
Total die area (mm^2)	112	5–6	6–7	4–5

*Die area for a 90 nm Xscale excludes L2 cache [51]
†For the PicoServer core, we estimated our power to be in the range of an ARM11, Xscale

The power values listed in Table 9.5 include static power. Our estimates for a 500 MHz PicoServer core are conservative compared to the ARM core values, especially with respect to [51]. Given that the Xscale core consumes 850 mW at 1.5 GHz and 1.3 V, a power consumption of 190 mW at 500 MHz for the 90 nm PicoServer core is conservative when applying the 3× scaling in clock frequency and the additional opportunities to scale voltage. For power consumption at other core clock frequencies, for example 1 GHz, we generated a power vs. frequency plot. It followed a cubic law [23]. We assumed a logic depth of 24 FO4 (fan out of 4) logic gates and used the 90-nm PTM process technology [51].

Support for 64 bit in a PicoServer core seems inevitable in the future. We expect the additional area and power overhead for 64-bit support in a PicoServer core to be modest when we look at the additional area and power overhead for 64-bit support in commercially available cores like MIPS and Xeon. As for the L2 cache, we referred to [56] and scaled the area and power numbers generated from actual measurements. We assumed the power numbers in [56] were generated when the cache access rate was 100%. Therefore, we scaled the L2 cache power by size and access rate while assuming leakage power would consume 30% of the total L2 cache power.

9.3.2.2 Interconnect Considering 3D Stacking Technology

For the purposes of this study, we adopted the data published in [12, 26, 50] as typical of 3D stacking interconnects. In general, we found die-to-die interconnect capacitance to be below 3fF. We also verified this with extracted parasitic capacitance values from 3D Magic, a tool recently developed at MIT. The extracted capacitance was found to be 2.7fF, which agrees with the results presented in [26]. By comparison with 2D on-chip interconnect, a global interconnect wire was estimated to have capacitance of 400fF per millimeter, based on [27]. Therefore, we can assume that the additional interconnect capacitance in 3D stacking vias is negligible. As for the number of I/O connections that are possible between dies, a figure of 10,000 connects per square millimeter is reported [26]. Our needs are much less. From our studies, we need roughly 1100 I/O connections: 32 bits for our address bus, 1024 bits for the data bus, and some additional control signals. For estimating the interconnect capacitance on our processor and peripheral layer, we again referred to [27] to generate analytical and projected values. We selected a wire length of 12 mm to account for 1.3 times the width/height of a 80 mm^2 die and scaled the wire length accordingly for smaller die sizes. We assumed we would gain a 33% reduction in wire capacitance compared to a 2D on-chip implementation from projections on interconnect wire length reduction shown in [22]. Based on these initial values, we calculated the number of repeaters required to drive the interconnect range at 250–400 MHz from hspice simulations. We found we needed only a maximum of two to three repeaters to drive this bus since the frequency of this wide on-chip bus was relatively slow.

We measured the toggle rate and access rate of these wires and calculated interconnect power using the well-known dynamic power equation. Table 9.6 shows the expected interconnect capacitance for 1024 bits in the case of 2D on-chip, 3D stacking, and 2D off-chip implementations. Roughly speaking, on-chip implementations have at most 33% capacitance of an off-chip implementation. Furthermore, because the supply voltages in I/O pads – typically 1.8–2.5 V are generally higher than the core supply voltage, we find the overall interconnect power for an off-chip implementation consumes an order of magnitude more power than an on-chip one. With modest toggle rates, small to modest access rates for typical configurations found in our benchmarks, and modest bus frequency – 250 MHz, we conclude that inter-die interconnect power contributes very little to overall power consumption.

Table 9.6 Parasitic interconnect capacitance for on-chip 2D, 3D, and off-chip 2D for a 1024 bit bus.

	130 nm	90 nm
On-chip 2D 12 mm	5.6 nF	5.4 nF
On-chip 3D 8 mm	3.7 nF	3.6 nF
Off-chip 2D	16.6 nF	16.6 nF

9.3.2.3 DRAM

We made DRAM area estimates for the PicoServer using the data in [45]. Currently, it is reasonable to say that 80 mm^2 of chip area is required for 64 MB of DRAM in 90 nm technology.

Conventional DRAM is packaged separately from the processor and is accessed through I/O pad pins and wires on a PCB. However, for our architecture, DRAM exists on-chip and connects to the processor and peripherals through a 3D stacking via. Therefore, the pad power consumed by the packages, necessary for driving signals off-chip across the PCB, is avoided in our design. Using the Micron DRAM spreadsheet calculator [3], modified to omit pad power, and profile data from M5 including the number of cycles spent on DRAM reads, writes, and page hit rates, we generated an average power for DRAM. We compared the estimated power from references on DRAM and especially with the DRAM power values generated from the SunFire T2000 Server Power Calculator [11]. The Micron spreadsheet uses actual current measurements for each DRAM operation – read, write, refresh, bank precharge, etc. We assumed a design with a 1.8-V voltage supply.

9.3.2.4 Network Interface Controller – NIC

Network interface controller power was difficult to model analytically due to lack of information on the detailed architecture of commercial NICs. For our simulations, we looked at the National Semiconductor 82830 gigabit ethernet controller. This chip implements the MAC layer of the ethernet card and interfaces with the physical layer (PHY) using the gigabit media-independent interface (GMII). We analyzed the datasheet and found the maximum power consumed by this chip to be 743 mW [4]. This power number is for 180–nm technology. We assumed maximum power is consumed when all the input and output pins were active. We then derated this figure based on our measured usage. In addition, we assumed static power at 30% of the maximum chip power. We believe our power model is conservative considering the significant improvements made on NICs since [4].

9.4 PicoServer Architecture

Table 9.7 shows the latency and bandwidth achieved for conventional DRAM, XDR DRAM, L2 cache, and on-chip DRAM using 3D stacking technology. With a 1024-bit wide bus, the memory latency and bandwidth achieved in a 3D stacking on-chip DRAM are comparable to an L2 cache and XDR DRAM. This suggests an L2 cache is not needed if stacking is used. Furthermore, the removal of off-chip drivers in conventional DRAM reduces access latency by more than 50% [47]. This strengthens our argument that on-chip DRAM can be as effective as an L2 cache. Another example that strengthens our case is that DRAM vendors are producing and promoting DRAM implementations with reduced random access latency [57, 7]. Therefore, our PicoServer architecture does not have an L2 cache and the on-chip DRAM is connected through a shared bus architecture to the L1 caches of each core. The role of this on-chip DRAM is as a primary system memory.

The PicoServer architecture is comprised of single-issue in-order processors that together create a chip multiprocessor which is a natural match to applications with a high level of TLP [36]. Each PicoServer CPU core is clocked at a nominal value

Table 9.7 Bandwidth and latency suggest on-chip DRAM can easily provide enough memory bandwidth compared to an L2 cache noted in [39, 56]. Average access latency for SDRAM and DDR2 DRAM is estimated to be $t_{RCD}+t_{CAS}$ where t_{RCD} denotes RAS to CAS delay and t_{CAS} denotes CAS delay. For, XDRAM, $t_{RAC\text{-}R}$ is used where $t_{RAC\text{-}R}$ denotes the read access time

	SDRAM	DDR2 DRAM	XDR DRAM	L2 cache @1.2 GHz	On-chip DRAM 3D IC
Bandwidth (GB/sec)	1.0	5.2	31.3	21.9	31.3
Average access latency (ns)	30	25	28	16	25*

*Average access latency with no 3D stacking aware optimizations. On-chip DRAM latency expected to reduce by more than 50% [47] when 3D stacking optimizations are applied.

of 500 MHz and has an instruction and data cache, with the data caches using a MESI cache coherence protocol. Our studies showed the majority of bus traffic is generated from cache miss traffic, not cache coherence. This is due to the properties of the target application space and the small L1 caches – 16 KB instruction and 16 KB data per core. With current densities, the capacity of the on-chip DRAM stack in PicoServer is hundreds of megabytes. In the near future this will rise to several gigabytes as noted in the Table 9.2. Other components such as the network interface controller (NIC), DMA controller, and additional peripherals that are required in implementing a full system are integrated on the CPU die.

9.4.1 Core Architecture and the Impact of Multithreading

PicoServer is composed of simple-single issue in-order cores with a five-stage pipeline. A 32-bit architecture is assumed for each core. Branch prediction is still useful in a server workload. Each core has a hybrid branch predictor with a 1 KB history table. Our studies showed the accuracy of the branch predictor for server workloads is about 95%.

Each core also includes architectural support for a shared memory protocol and a memory controller that is directly connected to DRAM. The memory controller responds to shared bus snoops and cache misses. On a request to DRAM, the memory controller delivers the address, data for memory writes, or cpu ID for memory reads. The cpu ID is needed for return routing of read data. Our estimated die area for a single core is 4–5 mm^2 (shown in Table 9.5).

Despite some benefits that can be obtained from multithreading (described in later paragraphs), we assume no support for multithreading due to the limitation in our simulation environment. Without significant modification to a commodity Linux kernel, it is difficult to scale server applications to more than 16 cores or threads. For this reason our study of multithreading examined a single core with multiple threads. We extrapolated this to the multicore case to show how many threads would be optimal when we leverage 3D stacking technology. Multithreading has the potential to improve overall throughput by switching thread contexts during lengthy stalls to memory.

To study the impact of multithreading on PicoServer, we assume multithreading support that includes an entire thread context – register file, store buffer, and interrupt trap unit. An additional pipeline stage is required to schedule threads. We assumed a die area overhead of supporting four threads to be about 50%. Although [39] predicted a 20% die area overhead to support four threads in the Niagara core, our cores are much smaller – 5 mm^2 vs. 16 mm^2. Register and architectural state die area estimates from [20, 51] take up a larger percentage of the total die area. Therefore, we assessed a greater area overhead for PicoServer cores.

In the multithreading study we varied the number of threads that can be supported and access latency to memory from a single core and measured the network bandwidth (a metric for throughput) delivered by this core. We did our analysis running SURGE because it displayed the highest L1 cache miss rate which implies it would benefit the most from multithreading. Our metrics used in this study are total network bandwidth and network bandwidth/mm^2. We varied the cache size to see the impact of threading.

Figures 9.6 and 9.7 show our simulated results. From these we are able to conclude threading indeed helps improve overall throughput, however, only to a limited extent when considering the area overhead and the impact of 3D stacking. Three-dimensional stacking reduces the access latency to memory by simplifying the core to memory interface and reducing the transfer latency. Three-dimensional stacked memory can be accessed in tens of cycles which correspond to the plots shown in Figs. 9.6b and 9.7b. The latter plot suggests that if area efficiency and throughput are taken together, limiting to only two threads appears optimal. We also find that the memory and I/O traffic increases as we add additional threads to the core. Therefore, a system must be able to deliver sufficient I/O and memory bandwidth to accommodate the additional threads. Otherwise, threading will be detrimental to overall system throughput.

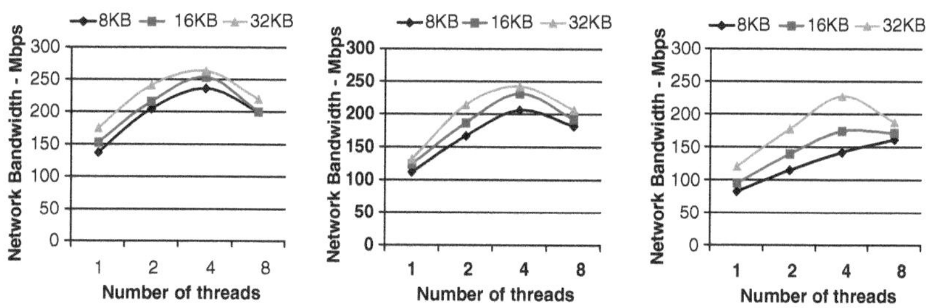

(a) memory latency = 1 (b) memory latency = 10 (c) memory latency = 100

Fig. 9.6 Impact of multithreading for varying memory latency on SURGE for varying four-way set associative cache sizes (8 KB, 16 KB, 32 KB) and varying number of threads. We assume the core is clocked at 500 MHz

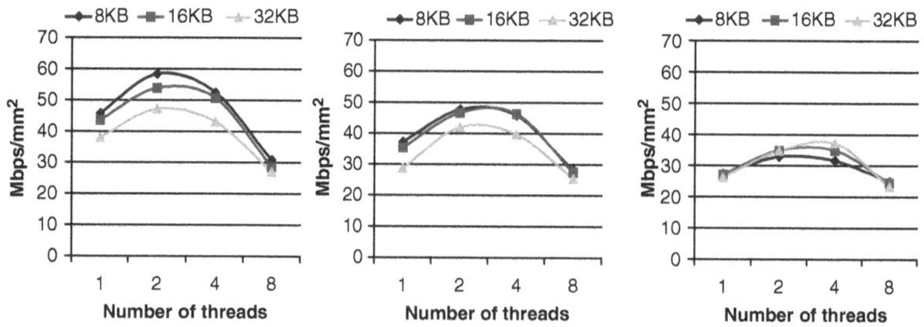

(a) memory latency = 1 (b) memory latency = 10 (c) memory latency = 100

Fig. 9.7 Impact of multithreading for Mbps/mm^2 when varying memory latency on SURGE. The same setup and assumptions in Fig. 9.6 are applied

9.4.2 Wide Shared Bus Architecture

PicoServer adopts a simple wide shared bus architecture that provides high memory bandwidth and fully utilizes the benefits of 3D stacking technology. Our bus architecture was determined from SURGE runs on M5. We limited our tests to SURGE because it generates a representative cache miss rate per core on our benchmarks. To explore the design space of our bus architecture, we first ran simulations for varying bus widths on a single-shared bus ranging from 128 to 2048 bits. We varied the cacheline size as well to match the bus width (varied it from 16 to 256 bytes). Network bandwidth (a metric for throughput) is measured to determine the impact of bus width on the PicoServer. As shown in Fig. 9.8a, a relatively wide data bus is necessary to achieve scalable network performance to satisfy the outstanding cache miss requests. This is because of the high bus contention on the shared data bus for high bus traffic that is generated for narrow bus widths as shown in Fig. 9.8b, c. As we decrease the bus width, the bus traffic increases, resulting in a superlinear increase in latency. Reducing bus utilization implies reduced bus arbitration latency, thus improving network bandwidth. Wide bus widths also help speed up NIC DMA transfers by allowing a large chunk of data to be copied in one transaction. A 1024-bit bus width seems reasonable for our typical PicoServer configurations of 4, 8, and 12 cores. More cores cause network performance to saturate unless wider buses are employed. We also looked at interleaved bus architectures, but found that with our given L1 cache miss rates, a 1024-bit bus is wide enough to handle the bus requests. For architectures and workloads that generate higher bus requests as a result of increasing the number of cores to 16 or more, or by having L1 caches with higher miss rates – more than 10% – then interleaving the bus becomes more effective. An interleaved bus architecture increases the number of outstanding bus requests, thus addressing the increase in the number of bus requests.

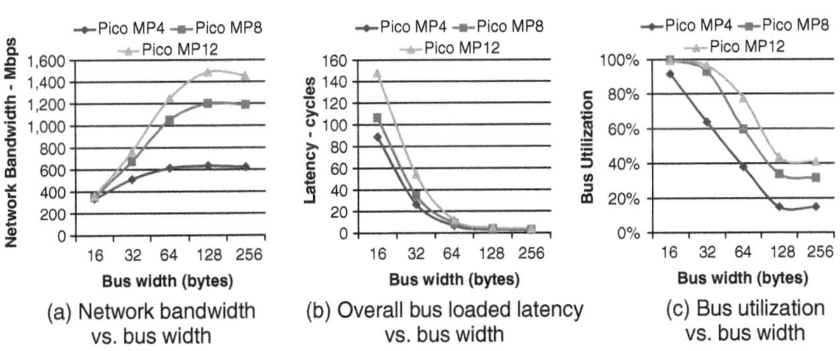

Fig. 9.8 Network performance on SURGE for various shared bus architectures with an L1 cache of 16 KB (each for I and D). We assumed a CPU clock frequency of 500 MHz for these experiments. Our bus architecture must be able to handle high bandwidths as the number of processors increases

9.4.3 On-chip DRAM Architecture

9.4.3.1 Role of On-chip DRAM

Based on the logic die area estimates, we projected the DRAM die size for a 4-core, 8-core, and 12-core PicoServers to be 40, 60, and 80 mm^2, respectively. Table 9.8 shows the on-chip memory alternatives for PicoServers. For example, to obtain a total DRAM size of 256 MB, we assume DRAM is made up of a stack of four layers. For Tier 3 servers we employ eight layers because they rely heavily on system memory size. With current technology – 90 nm, it is feasible to create a four-layer stack containing 256 MB of physical memory for a die area of 80 mm^2. Although a large amount of physical memory is common in server farms (4–16 GB) today, we believe server workloads can be scaled to fit into smaller systems with smaller physical memory based on our experience with server workloads and discussions with data center experts [41]. From our measurements on memory usage for server applications shown in Fig. 9.9, we found for many of the server applications (except TPC-C and TPC-H) that a modest amount – around 64 MB – of system memory is occupied by the user application, data, and the kernel OS code. The remainder of the memory is either free or used as a disk cache. When we consider that much of the user memory space in TPC-C and TPC-H is allocated as user-level cache, this is even true for TPC-C and TPC-H. Considering the fact that 256 MB can be integrated on-chip for four die layers, a large portion of on-chip DRAM can be used as a disk cache. Therefore, for applications that require small/medium filesets, an on-chip DRAM of 256 MB is enough to handle client requests.

For large filesets, there are several options to choose from. First, we could add additional on-chip DRAM by stacking additional DRAM dies, as in the eight-layer case. From the ITRS roadmap in Table 9.2, recall that the number of stacked dies we assume is conservative. With aggressive die stacking, we could add more die stacks to improve on-chip DRAM capacity – ITRS projects more than 11 layers in the next 2–4 years. This is possible because our power density in the logic layer is quite small – less than 5 W/cm^2. Another alternative is to add a secondary system memory which functions as a disk cache. For the workloads we considered in this study, we found that the access latency of this secondary system memory could be as

Table 9.8 Projected on-chip DRAM size for varying process technologies. Area estimates are generated based on Semiconductor SourceInsight 2005 [45]. Die size of 80 mm^2 is similar to that of a Pentium M at 90 nm

	130 nm	110 nm	90 nm	80 nm
DRAM stack of four layers each layer 40 mm^2	64 MB	96 MB	128 MB	192 MB
DRAM stack of eight layers each layer 40 mm^2	128 MB	192 MB	256 MB	384 MB
DRAM stack of four layers each layer 60 mm^2	96 MB	144 MB	192 MB	288 MB
DRAM stack of eight layers each layer 60 mm^2	192 MB	288 MB	384 MB	576 MB
DRAM stack of four layers each layer 80 mm^2	128 MB	192 MB	256 MB	384 MB
DRAM stack of eight layers each layer 80 mm^2	256 MB	384 MB	512 MB	768 MB

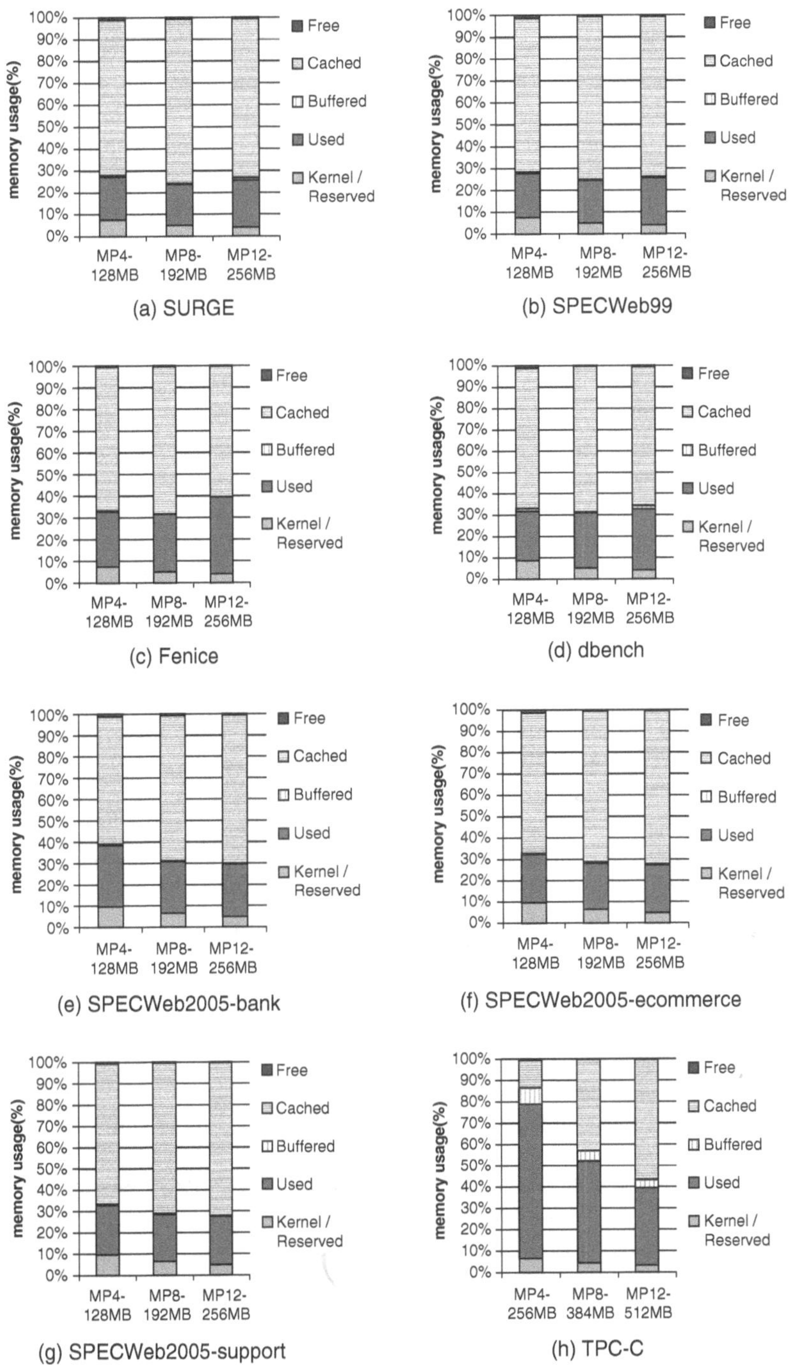

Fig. 9.9 Breakdown in memory for server benchmarks (SURGE, SPECWeb99, Fenice, dbench, SPECWeb2005, TPC-C); TPC-H is excluded because it displayed memory usage similar to TPC-C

slow as hundreds of micro seconds without affecting throughput. An access latency as slow as hundreds of micro seconds implies that Flash memory that consumes less active/standby power can be used as secondary system memory – shown in Section 9.4.6. This idea has been explored in [30, 31]. Therefore, for workloads requiring large filesets, we could build a nonuniform memory architecture with fast on-chip DRAM and relatively slower off-chip secondary system memory. The fast on-chip DRAM would primarily hold code, data, and a small disk cache while the slow system memory would function as a large disk cache device.

9.4.3.2 On-Chip DRAM Interface

To maximize the benefits of 3D stacking technology, the conventional DRAM interface needs to be modified for PicoServer's 3D stacked on-chip DRAM. Conventional DDR2 DRAMs are designed assuming a small pin count and use address multiplexing and burst mode transfer to make up for the limited number of pins. With 3D stacking technology, there is no need to use narrow interfaces and address multiplexing with the familiar two-phase commands, RAS then CAS. Instead, the additional logic required for latching and muxing narrow address/data can be removed. The requested addresses can be sent as a single command while data can be driven out in large chunks. Further, conventional off-chip DRAMs are offered as DIMMs made up of multiple DDR2 DRAM chips. The conventional off-chip DIMM interface accesses multiple DDR2 DRAM chips per request. For 3D stacked on-chip DRAM, only one subbank needs to be accessed per request. As a result 3D stacked on-chip DRAM consumes much less power per request than off-chip DRAM. Figure 9.10 shows an example of a read operation without multiplexing. In particular, it shows that RAS and CAS address requests are combined into a single address request. DRAM vendors already provide interfaces that do not require address multiplexing such as reduced latency DRAM from Micron [7] and NetDRAM [55] from Samsung. This suggests the interface required for 3D

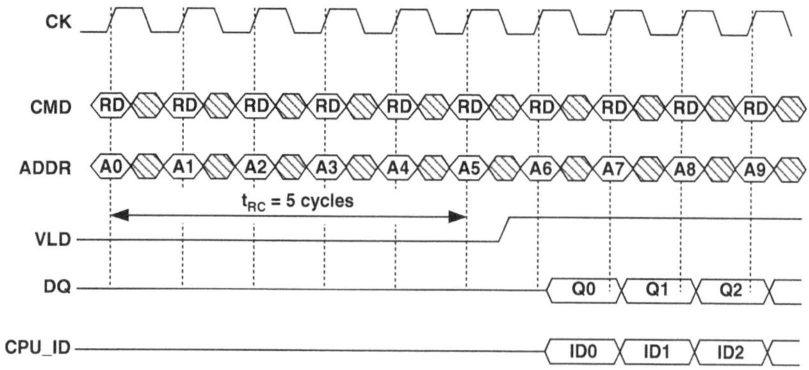

Fig. 9.10 On-chip DRAM read timing diagram without address multiplexing

stacked on-chip DRAM can be realized with only minor changes to existing solutions. Additional die area made available through the simplification of the interface can be used to speed up the access latency to on-chip DRAM. By investing more die area to subbank the on-chip DRAM, latencies as low as 10 ns can be achieved.[3]

9.4.3.3 Impact of On-Chip DRAM Refresh on Throughput

DRAM periodically requires each DRAM cell to be refreshed. Retention time of each DRAM cell is typically defined as 64 ms for industry standard temperature and decreases to 32 ms in hotter environments. Based on our thermal analysis presented in Section 9.4.5, our maximum junction temperature was well under the industry standard temperature constraints. As a result, we assumed a 64 ms refresh cycle per cell. However, refresh circuits are commonly shared among multiple DRAM cell arrays to reduce the die area overhead, thus reducing the average DRAM refresh interval to approximately 7.8125 μs and requiring approximately 200 ns to complete. Roughly speaking, this implies a DRAM bank cannot be accessed for a duration of hundreds of CPU clock cycles every ten thousands of CPU clock cycles. To measure the impact of refresh cycles, we modeled the refresh activity of DRAM on M5 and observed the CPI overhead. The access frequency to on-chip DRAM is directly correlated to the amount of L1 cache misses observed. We found that for a 5% L1 cache miss rate and 12 cores clocked at 500 MHz (PicoMP12-500 MHz running SURGE), this would incur a CPI refresh overhead of 0.03 CPI. This is because many of the L1 cache misses do not coincide with a refresh command, executed resulting in only a minimal performance penalty.

9.4.4 The Need for Multiple NICs on a CMP Architecture

A common problem of servers with large network pipes is handling bursty behavior in the hundreds of thousands of packets that can arrive each second. Interrupt coalescing is one method of dealing with this problem. It works by starting a timer when a noncritical event occurs. Any other noncritical events that occur before the timer expires are coalesced into one interrupt, reducing the total number. Even with this technique, however, the number of interrupts received by a relatively low-frequency processor, such as one of the PicoServer cores, can overwhelm it. In our simulations we get around this difficulty by having multiple NICs, one for each of a subset of the processors. For an eight-chip multiprocessor architecture with one NIC and on-chip DRAM, we found the average utilization per processor to be below 60%, as one processor could not manage the NIC by itself. To fully utilize each processor in our multiple processor architecture, we inserted one NIC for every two processors.

---

[3] In this study, we took a conservative approach and did not apply the latency reduction due to additional subbanking. We only applied latency optimizations resulting from the removal of the drivers of off-chip signals

For example, a four-processor architecture would have two NICs, an eight-processor architecture would have four NICs, and so forth.

Although our simulation environment does not support it, a more ideal solution would have a smarter single NIC that could route interrupts to multiple CPUs, each with separate DMA descriptors and TX/RX queues. This could be one NIC either with multiple interface IP addresses or an intelligent method of load-balancing packets to multiple processors. Such a NIC would need to keep track of network protocol states at the session level. There have been previous studies of intelligent load-balancing on NICs to achieve optimal throughput on platforms [21]. TCP splicing and handoff are also good examples of intelligent load balancing at higher network layers [46].

9.4.5 Thermal Concerns in 3D Stacking

A potential concern with 3D stacking technology is heat containment. To address this concern, we investigated the thermal impact of 3D stacking on the PicoServer architecture. Because we could not measure temperature directly on a real 3D stacked platform, we modeled the 3D stack with the grid model in Hotspot version 3.1 [28]. Mechanical thermal simulators such as FLOWTHERM and ANSYS were not considered in our studies due to the limited information we could obtain about the mechanical aspects of the 3D stacking process. However, Hotspot's RC equivalent heat flow model is adequate to show trends and potential concerns in 3D stacking. Because our work is aimed at answering whether 3D stacking can provide an advantage in the server space, instead of describing the details in heat transfer, we present general trends.

The primary contributors to heat containment in 3D stacking technology are the interface material (SiO_2) and the free air interface between silicon and air as can be seen in Table 9.9. Silicon and metal conduct heat much more efficiently. We first configured our PicoServer architecture for various scenarios by: (1) varying the amount of stacked dies; (2) varying the location of the primary heat-generating die – the logic die in the stack; and (3) varying the thickness of the SiO_2 insulator that is typically used in-between stacked dies. Our baseline configuration has a logic die directly connected to a heat sink and assumes a room temperature of 27°C. Hotspot requires information on properties of the material and power density to generate steady-state temperatures. We extracted 3D stacking properties from [37, 44, 57]

Table 9.9 Thermal parameters for commonly found materials in silicon devices

	Thermal conductivity (W/m·K)	Heat capacity (J/m^3·K)
Si	148	1.75×10^6
SiO_2	1.36	1.86×10^6
Cu	385	3.86×10^6
Air at 25°C	0.026	1.2×10^3

and assigned power density at the component level based on area and power projections for each component. Components were modeled at the platform-level – processor, peripheral, global bus interconnect, etc. We generated the maximum junction temperature in the PicoServer architecture shown in Fig. 9.11.

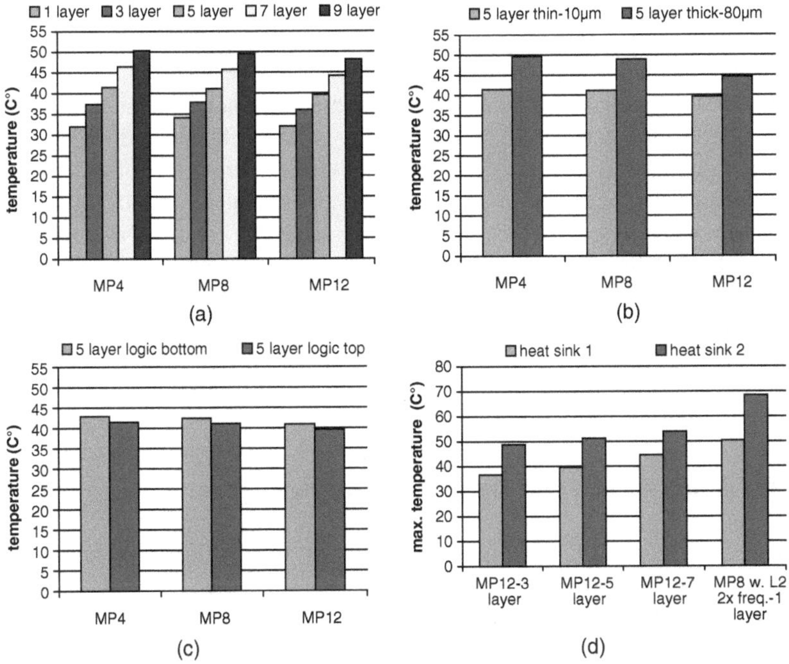

Fig. 9.11 Maximum junction temperature for sensitivity experiments on Hotspot: (**a**) varying the number of layers; (**b**) varying 3D interface thickness; (**c**) varying location of logic die; and (**d**) maximum junction temperature for heat sink quality analysis. A core clock frequency of 500 MHz is assumed in calculating power density. We varied the size of on-chip memory based on the number of layers stacked. One layer assumes no on-chip memory at all

Figure 9.11a shows the sensitivity to the number of stacked layers. We find roughly a 2–3°C increase in maximum junction temperature for each additional layer stacked. Interestingly, maximum junction temperature reduces as we increase the die area. We believe this is due to our floorplan and package assumptions. Further analysis is needed, and we leave it for future research. Figure 9.11b shows the sensitivity to the 3D stacking dielectric interface. We compared the effect of the SiO_2 thickness (the interface material) for 10 and 80 μm. In [17, 37, 44, 57] we find the maximum thickness of the interface material does not exceed 10 μ m for 3D stacking. The 80 μm point is selected to show the impact of heat containment as the thickness is increased substantially. It results in a 6°C increase in junction temperature. While notable, this is not a great change given the dramatic change in material thickness. We assumed the increase in dielectric interface thickness did not increase bus latency because the frequency of our on-chip bus was relatively slow.

Figure 9.11c shows the sensitivity to placement in the stack – top or bottom layer. We find the primary heat generating die is not sensitive to the location of the heat sink.

We also conducted an analysis of the impact of heat sink quality. We varied the heat sink configuration to model a high-cost heat sink (heat sink 1) and a low-cost heat sink (heat sink 2). Figure 9.11d shows the impact of 3D stacking technology on heat sink quality. It clearly suggests that a low-cost heat sink can be used on platforms using 3D stacking technology.

The above results suggest that heat containment is not a major limitation for the PicoServer architecture. The power density is relatively low. It does not exceed 5 W/cm^2. As a result, the maximum junction temperature does not exceed 50°C. Three-dimensional vias can also act as heat pipes, which we did not take into account in our analysis however, this can be expected to improve the situation. An intelligent placement would assign the heat-generating layer (the processor layer) adjacent to the heat sink resulting in a majority of the heat being transferred to the heat sink. There is independent support for our conclusions in [19, 25].

9.4.6 Impact of Integrating Flash onto PicoServer

This section examines the architectural impact of directly attaching NAND Flash devices onto a PicoServer architecture, and it is also a case study on 3D-stacked Flash devices integrated onto PicoServer. Flash is emerging as an attractive memory device to integrate onto a server platform primarily due to the rapid rate of density improvement. There are two primary usage models for Flash in the server space: (1) a solid state disk (SSD) and (2) a memory device. It is widely believed that Flash integration improves overall server throughput while consuming lower system memory and disk drive power. For example, when Flash is used as a memory device and assigned the role of a disk cache, the higher density of Flash allows us to implement a larger cache with a higher cache hit rate that consumes lower power than DRAM. Higher cache hit rates reduce accesses to disk which results in improved system performance and reduction in disk power.

However, integrating Flash onto a server platform is not straightforward. There are two key challenges that Flash needs to address to successfully integrate onto a server platform; (1) improving transfer latency to/from Flash and (2) providing sufficient memory (RAM) to efficiently manage Flash. NAND Flash displays higher overall access latency (see Table 9.10) than typical memory devices such as DRAM primarily due to the high transfer latency (low-bandwidth narrow 8- or 16-bit interface) for typical off-the-shelf Flash devices. Although the page read latency into the internal buffer for a SLC NAND Flash page is approximately 25 μs, the transfer latency to read several KBs from a NAND Flash chip is substantially higher. One way to reduce transfer latency is leveraging 3D stacking. It enables the use of a wider bus that accesses a larger amount of data per cycle thus reducing transfer latency.

Table 9.10 ITRS 2007 roadmap for memory technology

	2007	2009	2011	2013	2015
NAND Flash-SLC* (μm^2/bit)	0.0130	0.0081	0.0052	0.0031	0.0021
NAND Flash-MLC* (μm^2/bit)	0.0065	0.0041	0.0013	0.0008	0.0005
DRAM cell density(μm^2/bit)	0.0324	0.0153	0.0096	0.0061	0.0038
Flash write/erase cycles – SLC/MLC†	1E+05/ 1E+04	1E+05/ 1E+04	1E+06/ 1E+04	1E+06/ 1E+04	1E+06/ 1E+04
Flash data retention (years)	10–20	10–20	10–20	20	20

*SLC – single-level cell, MLC – multi-level cell
†write/erase cycles for MLC Flash estimated from prior work [34]

The latency reduction may allow more critical data to be moved from DRAM onto Flash for greater energy savings.

Additionally, the amount of memory (RAM) required to efficiently manage a NAND Flash subsystem scales with capacity. While NAND Flash may still be managed with a small amount of memory, these types of subsystems display a limited read/write bandwidth and accelerate Flash wear out. To meet the memory (RAM) requirements that deliver a high bandwidth, high longevity, and high-capacity NAND Flash subsystem, DRAM is commonly integrated onto the NAND Flash subsystem that stores the Flash manageability code and data. However, the cost of implementing a dedicated DRAM device that is required in a NAND Flash subsystem such as a SSD is appreciable and inefficient. Consolidating the entire code and data available in a server platform into a single DRAM module saves cost and improves efficiency. The system integration benefit of 3D stacking allows Flash manageability code and data to reside in DRAM that is shared with other components in the system (shown in Fig. 9.12).

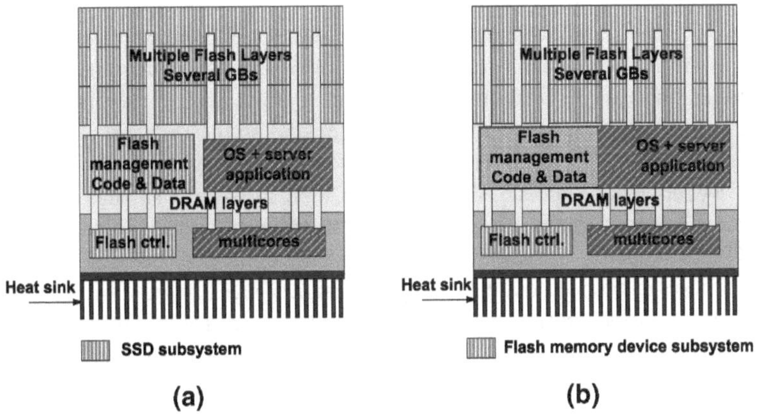

Fig. 9.12 Three-dimensional stacked Flash architecture for (**a**) SSD usage model and (**b**) memory device usage model

For the solid state disk (SSD) usage model, 3D stacking can be used as a way to implement energy efficient on-chip interfaces that can replace the conventional hard disk drive interface. Figure 9.12a shows this approach. Logic for the SSD controller is placed on a die using the same process technology as the CPU cores. The low-level details of Flash management (error checking, wear leveling, and buffer management) are isolated from the processors, providing a simple interface. The memory requirements for Flash manageability are easily satisfied with the integrated on-chip DRAM. A 3D-stacked SSD could provide: (1) lower power, (2) higher random-access throughput, (3) lower latency, and (4) greater physical robustness than disk drives. SSDs consist of a controller and buffer RAM in addition to Flash. In a 3D-stacked PicoServer, exposing an SSD interface in hardware could allow drivers to work without modification. Other benefits over external SSDs include a higher bandwidth interface, the option of data transfer directly into main memory or processor cache, and the ability to allow the PicoServer cores to control the Flash using application-specific algorithms. It has been shown that using a combination of SSD and conventional disk-based storage together can provide higher performance due to the different characteristics of each device [35]. Disk is most effective for high-bandwidth, high-density storage while Flash provides lower latency (especially when reading) and more IOPs (I/O per second). Algorithms dynamically place read-dominated data on SSD while write-dominated data is moved to the HDD for an overall performance improvement.

For the memory device usage model, there have been several proposals that span a wide range of application spaces, including disk buffers [33], disk caches [29, 30, 31], and code/data storage [42, 49]. Figure 9.12b shows how the physical structure is very similar to that of a 3D-stacked SSD except that low-level control of data transfers between Flash is given to the processor cores. This approach increases software complexity, however, achieves higher bandwidth and more efficient management than the SSD usage model. This is because Flash manageability is performed by the processor that has more computation capacity than a Flash controller. When Flash is used as a disk buffer, it is used as a staging buffer between DRAM and disk. Energy savings can be achieved by spinning down the disk for longer periods of time while data drains from the Flash buffer. This scheme also reduces the amount of DRAM in the system and saves 30–40% of the system energy. Similarly, Flash disk caches are simple to implement in an operating system, extending the standard DRAM-based disk cache. Data-intensive server applications such as web servers require large disk caches. By replacing some DRAM with Flash, there are energy savings due to lower idle power and performance gains from having more total cache capacity (due to higher Flash density). Three-dimensional stacking enhances this scheme by satisfying requests to cached data more quickly. Cached pages migrate from secondary Flash storage to the primary page cache in DRAM, so a wide 3D-stacked interconnect completes these data transfers faster and with lower bus energy. There are additional benefits the memory device usage model provides. With small enhancements in the microarchitecture of the Flash controller, code that typically is loaded from NAND Flash and residing in DRAM may directly reside in NAND Flash. The next code block prediction technique described in [42] and "demand-based paging" architecture outlined in [49] have shown the potential benefits. Three dimensional stacking helps this scheme by reducing the energy cost and delay while swapping pages between DRAM and Flash. For the same DRAM and Flash capacities, total energy will be reduced further because program execution completes more quickly, burning less idle energy in DRAM and other system components such as the power supply.

Based on the findings from [29, 30, 31], we present a case study of 3D stacked Flash as a memory device integrated onto a PicoServer architecture.

We first provide an analysis of the OS-managed disk cache behavior running a server workload. Figure 9.13 shows the disk cache access behavior in system memory in a web server. It shows that file access behavior in server workloads displays a short-tailed distribution where a large portion of the disk cache is infrequently accessed and only a small portion of the disk cache is frequently accessed. Further, Fig. 9.14 shows server throughput for varying access latencies to the infrequently accessed files in the disk cache. We are able to observe constant throughput for access latencies of tens to hundreds of microseconds. This is because we can hide

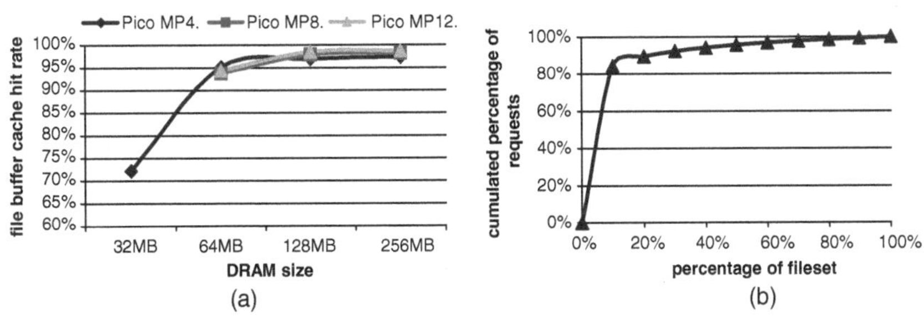

Fig. 9.13 (a) Disk cache access behavior on the server side for client requests. We measured for 4, 8, 12 PicoServer configurations and varied the DRAM size. (b) A typical cumulative distribution function of a client request behavior. Ninety percent of requests are for 20% of the web content files.

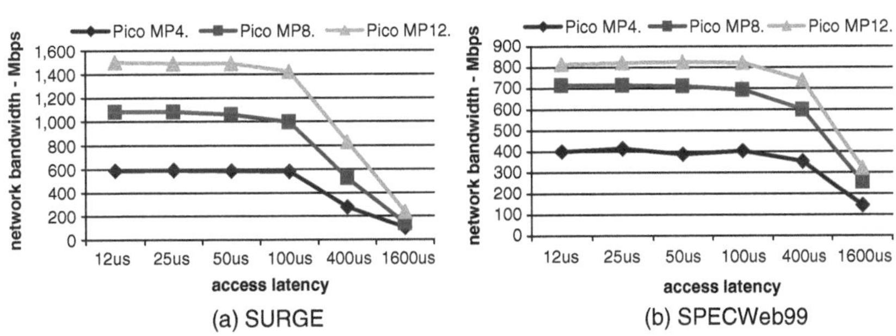

Fig. 9.14 Measured network bandwidth for full system simulation while varying access latency to a secondary disk cache. We assumed a 128 MB DRAM with a slower memory of 1 GB. We measured bandwidth for 4, 8, 12 PicoServer 500 MHz configurations. The secondary disk cache can tolerate access latencies of hundreds of microseconds while providing equal network bandwidth

the lengthy access latency to infrequently accessed files in the disk cache through the multithreaded nature of a server workload. One way to take advantage of this behavior is to integrate Flash as a second-level disk cache that replaces a large portion of DRAM allocated for a disk cache.

Because NAND Flash consumes much less power than DRAM and is more than 4× denser than DRAM (shown in Tables 9.10 and 9.11), a server that integrates Flash in its memory system is expected to be more energy efficient and has larger system memory capacity. Bigger disk cache sizes reduce disk cache miss rate and allow the HDD to be spun down longer. As we show in Table 9.11, HDDs in idle mode consume a significant amount of power.

Table 9.11 Performance, power consumption and cost for DRAM, NAND-based SLC/MLC Flash and HDD.

	Active power	Idle power	Read latency	Write latency	Erase latency
1 Gb DDR2 DRAM	878 mW	80 mW†	55 ns	55 ns	N/A
1 Gb NAND-SLC	27 mW	6 μ W	25 μs	200 μs	1.5 ms
4 Gb NAND-MLC	N/A	N/A	50 μs	680 μs	3.3 ms
HDD‡	13.0 W	9.3 W	8.5 ms	9.5 ms	N/A

† DRAM idle power in active mode. Idle power in powerdown mode is 18 mW
‡ Data for 750 GB hard disk drive [8]

The benefits that are achieved by integrating Flash onto a server can also be applied to PicoServer. Figures 9.15 and 9.16 show two such configurations that integrate Flash onto PicoServer using 3D stacking technology. The first configuration shown in Figs. 9.15a and 9.16a integrates Flash as a discrete component. It stacks eight die layers of Flash using 3D stacking technology to create a separate Flash chip – a discrete component. This Flash chip is connected to PicoServer through a PCB board layout. An off-chip I/O interface (typically PCI express) that is capable of delivering bandwidth of hundreds of megabytes per second is implemented between PicoServer and the discrete Flash chip. Flash is managed (wear-leveling and Flash command interface) by the OS running on the in-order PicoServer cores. The memory capacity of the discrete Flash chip is not constrained by the die area of PicoServer and allows us to stack more dies than PicoServer due to the lower power consumption of Flash. Therefore, discrete Flash chips with tens of gigabytes can be integrated onto a PicoServer architecture. Server workloads with large filesets and modest I/O bandwidth requirements will benefit most from this configuration. One potential drawback to the discrete Flash chip configuration is the idle power consumption of the PCI express interface. Idle power management must be performed on the PCI express I/O pins when they are not active.

The second configuration shown in Figs. 9.15b and 9.16b directly integrates Flash to PicoServer (discussed earlier in the section). It stacks four additional Flash die layers directly on top of the on-chip DRAM using 3D stacking technology. The Flash dies connect to other components in PicoServer through the wide shared bus.

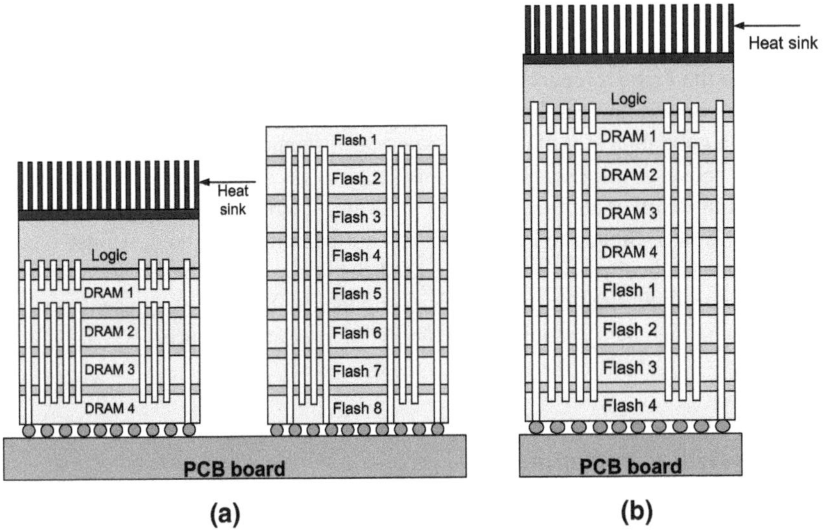

Fig. 9.15 (**a**) Eight-die-stacked Flash integrated as discrete component (PCB layout) in PicoServer (**b**) four-die-stacked Flash directly integrated to PicoServer

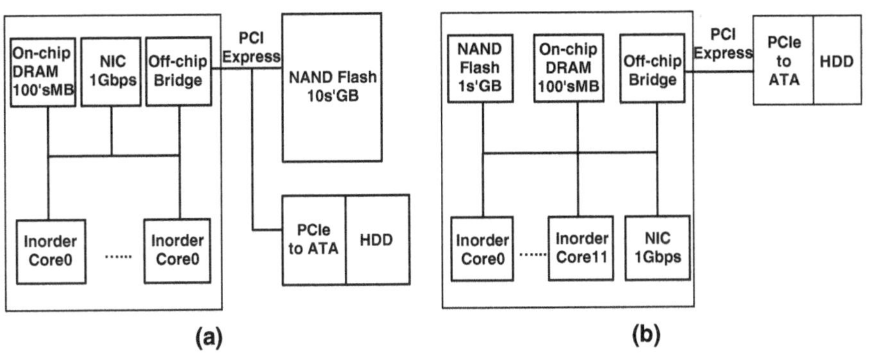

Fig. 9.16 (**a**) High-level block diagram of eight-die-stacked Flash integrated as discrete component (PCB layout) in PicoServer, (**b**) high-level block diagram of four-die-stacked Flash directly integrated to PicoServer

The wide on-chip shared bus interface delivers tens of gigabytes per second bandwidth. Flash is also managed – wear-leveling, Flash command interface – by the OS running on top of the in-order PicoServer cores. Flash capacity is limited by the die area of the on-chip DRAM and logic components in PicoServer. As a result, the Flash capacity is expected to be several gigabytes in size. We expect server workloads that require small filesets and high I/O bandwidth will benefit most from this configuration.

9.5 Results

To evaluate the PicoServer architecture two metrics are important – throughput and power. Throughput that can be measured as network bandwidth or transactions per seconds is a good indicator of overall system performance because it is a measure of how many requests were serviced. In this section, we compare various PicoServer configurations to other architectures first in terms of achievable throughput and then in terms of power. Since the PicoServer has not been implemented, we use a combination of analytical models and published data to make a conservative estimate about the power dissipation of various components. Finally we present a pareto chart showing the energy efficiency of the PicoServer architecture.

9.5.1 Overall Performance

Figures 9.17 and 9.18 show the throughput for some of our Tier 1–3 workload runs. Each bar shows the contribution to throughput in three parts: (1) a baseline with no L2 cache and a narrow (64 bit) bus; (2) the baseline but with an L2 cache; or (3) the baseline with wide bus, no L2 cache, and 3D stacking for DRAM. Hence, we are able to make comparisons that differentiate the impact of 3D stacking technology with the impact of having an L2 cache. Figure 9.17 shows that 3D stacking technology alone improves overall performance equal to or more than having an L2 cache. A fair comparison for a fixed number of cores, for example, would be a Pico MP4-1000 MHz vs. a conventional CMP MP4 without 3D-1000 MHz. In general, workloads that generated modest to high cache miss rates (SURGE, SPECweb99, SPECweb2005 and dbench), showed dramatic improvement from adopting 3D stacking technology. Fenice shows less dramatic improvements, because it involves video stream computations that generate lower cache miss rates. Interestingly, the script language Tier 2 benchmark – SPECweb2005, performed well against the OO4 configurations that have been expressly designed for single-threaded performance.

For OO4 configurations, we combine the impact of having an L2 cache and 3D stacking since the L2 cache latency on a uniprocessor is likely to be smaller than the access latency to a large-capacity DRAM, making it less appealing to only have a high-bandwidth on-chip DRAM implemented from 3D stacking. We find that 3D stacking improves performance by 15% on OO4 configurations. When we compare an OO4 architecture without 3D stacking with our PicoServer architecture, a PicoServer MP8 operating at 500 MHz performs better than a 4 GHz OO4 processor with a small L1 and L2 cache of 16 KB and 256 KB, respectively. For a similar die area comparison, we believe comparing PicoServer MP8 and a OO4-small architecture is a fair comparison, because the OO4-large requires additional die area for a 128 KB L1 cache and a 2 MB L2 cache.

If we assume that the area occupied by the L2 cache in our conventional CMP MP4/8 without 3D stacking technology is replaced with additional processing cores – a benefit made possible by using 3D stacking technology – a comparison in

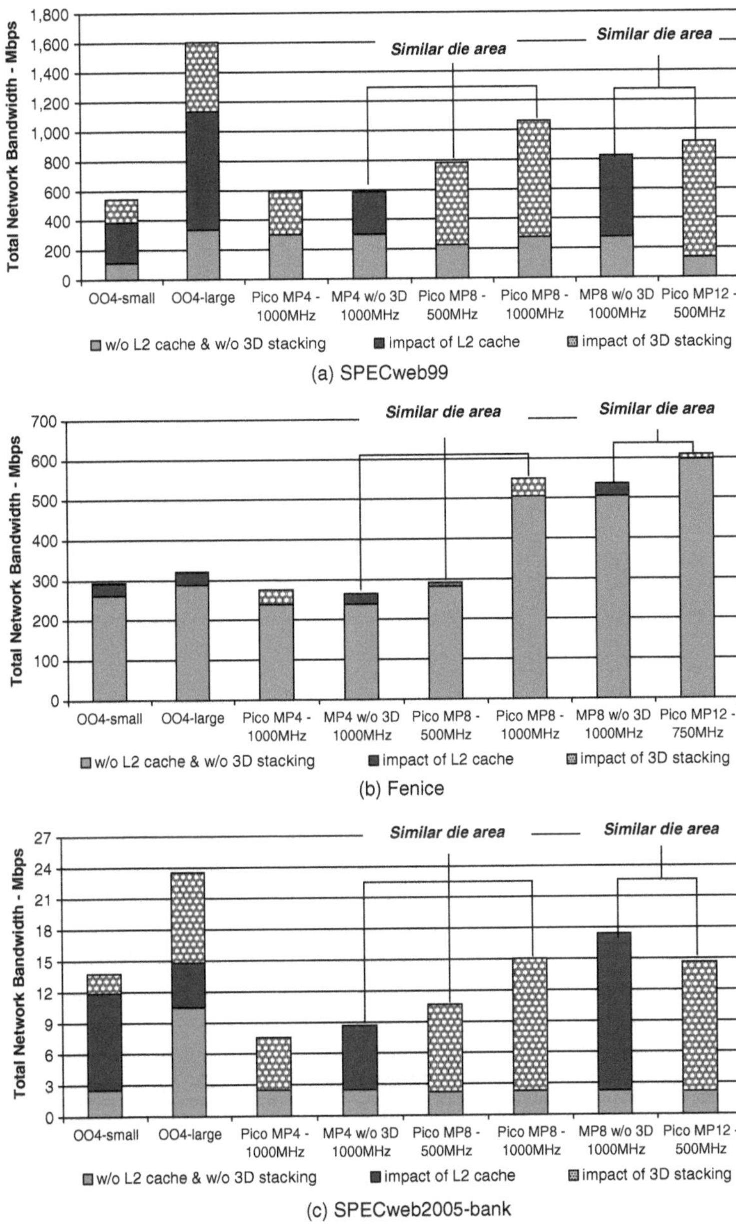

Fig. 9.17 Throughput measured for varying processor frequency and processor type. For PicoServer CMPs, we fixed the on-chip data bus width to 1024 bits and bus frequency to 250 MHz. For a Pentium 4-like configuration, we placed the NIC on the PCI bus and assumed the memory bus frequency to be 400 MHz. For a MP4, MP8 without 3D stacking configuration, to be fair we assumed no support for multithreading and an L2 cache size of 2 MB. The external memory bus frequency was assumed to be 250 MHz (SPECweb99, Fenice, SPECweb2005-bank)

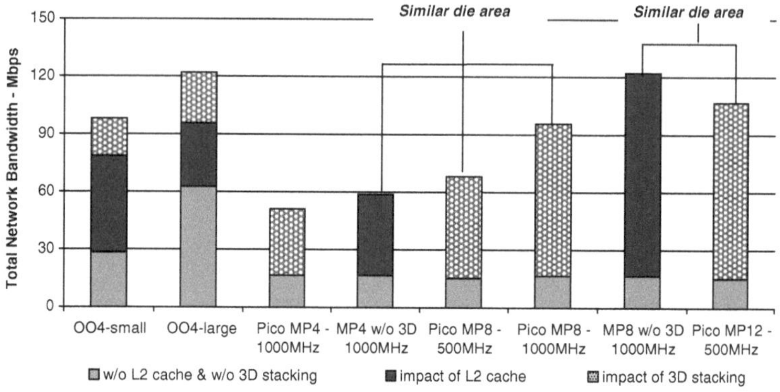

(a) SPECweb2005-ecommerce

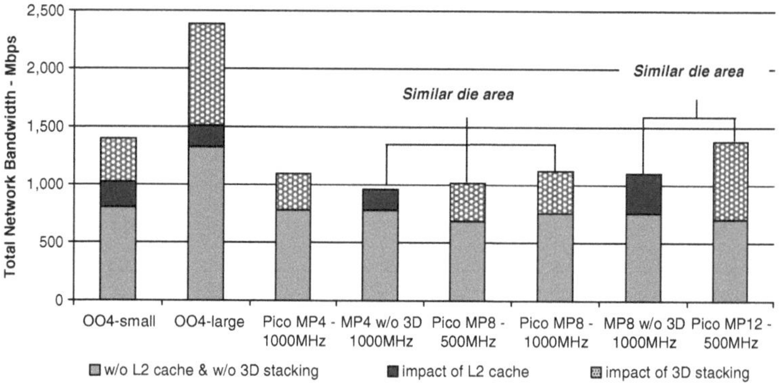

(b) dbench

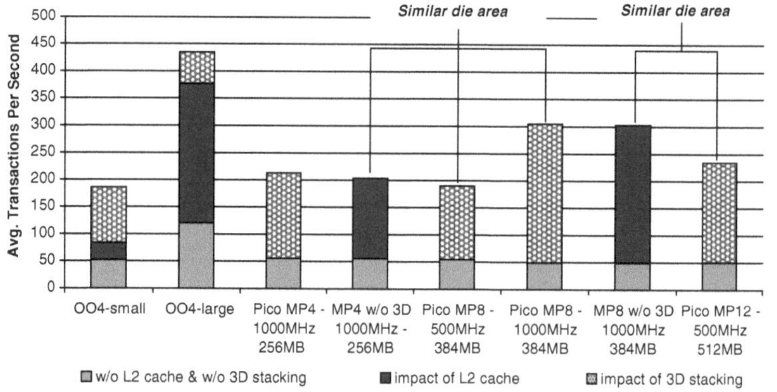

(c) TPC-C

Fig. 9.18 Throughput measured for varying processor frequency and processor type (SPECweb2005-ecommerce, dbench, TPC-C). We applied the same assumptions used in Fig. 9.17

throughput for similar die area can be conducted on the following systems: (1) a Pico MP8-500 MHz vs. a conventional MP4 without 3D-1000 MHz and (2) a Pico MP12-500 MHz vs. a conventional MP8 without 3D-1000 MHz (for Fenice, compare with a Pico MP12-750 MHz). Our results suggest that on average, additional processing elements and reducing core clock frequency by half improve throughput and significantly saves on power – shown in Section 9.5.2. For compute-bound workloads like Fenice, SPECWeb2005-bank and SPECWeb2005-ecommerce, however Pico MP12-500 MHz did not do better than a conventional MP8 without 3D-1000 MHz. For SPECWeb2005-bank and ecommerce, introducing a 2 MB L2 cache dramatically cuts the number of cache misses reducing the benefit of adding more cores while lowering core clock frequency. Pico MP12-500 MHz also did not perform well for TPC-C because of the I/O scheduler. However, we expect Pico MP12-500 MHz to perform better for OS kernels with TPC-C optimized I/O scheduling algorithms. Our estimated area for adding extra cores is quite conservative, suggesting more cores could be added to result in even more improvement in throughput.

9.5.2 Overall Power

Processor power still dominates overall power in PicoServer architectures. Figure 9.19 shows the average power consumption based on our power estimation techniques for server application runs. We find that PicoServer with a core clock frequency of 500 MHz is estimated to consume between 2 and 3 Watts for 90 nm process technology. Much of the total power is consumed by the simple in-order

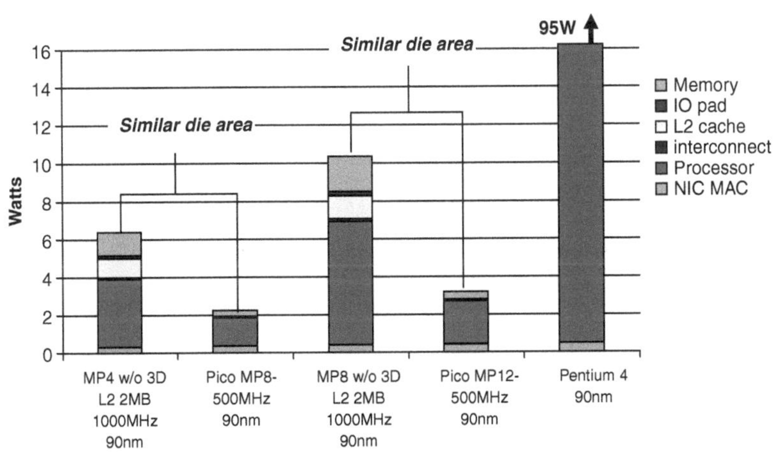

Fig. 9.19 Breakdown of average power for 4, 8, and 12 processor PicoServer architectures using 3D stacking technology for 90 nm process technology. Estimated power per workload does not vary by a lot because the cores contribute to a significant portion of power. We expect 2–3 W to be consumed at 90 nm. An MP8 without 3D stacking operating at 1 GHz is estimated to consume 8 W at 90 nm

cores. NIC power also consumes a significant amount due to the increase in the number of NICs when increasing the number of processors. However, as described in Section 9.4.4, an intelligent NIC designed for this architecture could be made more power efficient as a more advanced design would only need one. An appreciable amount of DRAM power reduction is also observed due to 3D stacking. The simplified on-chip DRAM interface requires that fewer DRAM subbanks need to be simultaneously accessed per request. Other components, such as the interconnect make marginal contributions to overall system power due to the modest access rates and toggle rates of these components.

Comparing our PicoServer architecture with other architectures, we see that for a similar die area comparison, we use less than half the power when we compare Pico MP8/12-500 MHz with a conventional MP4/8 without 3D stacking but with an L2 cache at 1000 MHz. We also recall in Section 9.5.1 that performance-wise for a similar die area, the PicoServer architectures perform on average 10–20% better than conventional CMP configurations. Furthermore, we use less than 10% of the power of a Pentium 4 processor and, as we showed in the previous section, perform comparably. At 90-nm technology, it can be projected that the power budget for a typical PicoServer platform satisfies mobile/handheld power constraints noted in ITRS projections. This suggests the potential for implementing server-type applications in ultra-small form factor platforms

9.5.3 Energy Efficiency Pareto Chart

In Figs. 9.20 and 9.21, we present a pareto chart for PicoServer depicting the energy efficiency (in Mbs per Joule) and throughput (we only list the major workloads). The points on this plot show the large out-of-order cores, the conventional CMP MP4/8 processors without 3D stacking, and the PicoServer with 4, 8, and 12 cores. On the y-axes we present Mbps and transactions per second, and on the x-axes we show Mb/J and transactions per Joule. From Figs. 9.20 and 9.21, it is possible to find the optimal configuration of processor number and frequency for a given energy efficiency/throughput constraint.

Additionally from Figs. 9.20 and 9.21, we find the PicoServer architectures clocked at modest core frequency – 500 MHz are 2–4× more energy efficient than conventional chip-multiprocessor architectures without 3D stacking technology. The primary power savings can be attributed to 3D stacking technology that enables a reduction in core clock frequency while providing high throughput. A sweetspot in system-level energy efficiency for our plotted datapoints can also be identified among the PicoServer architectures when comparing Pico MP4-500 MHz, MP8-500 MHz, and MP12-500 MHz. These sweetspots in energy efficiency occur just before diminishing returns in throughput are reached as parallel processing is increased by adding more processors. The increase in parallel processing raises many issues related to inefficient interrupt balancing, kernel process/thread scheduling, and resource allocation that result in diminishing return. Independent studies

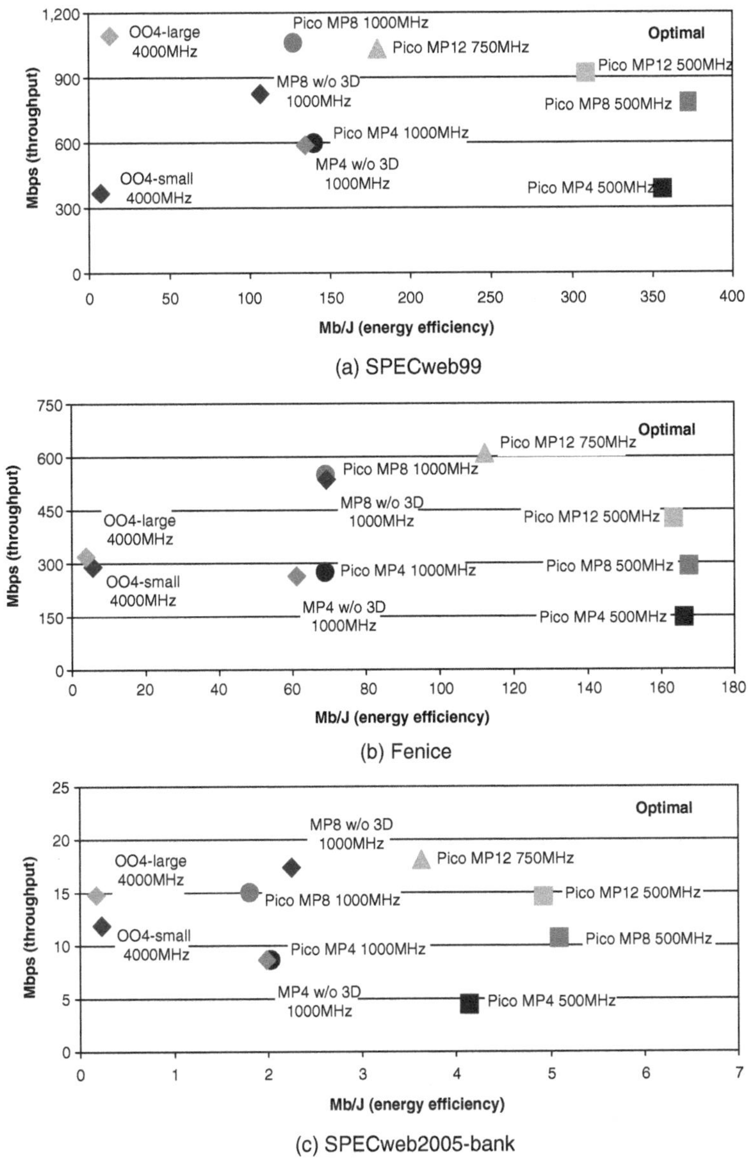

Fig. 9.20 Energy efficiency, performance pareto chart generated for 90-nm process technology. Three-dimensional stacking technology enables new CMP architectures that are significantly energy efficient. (SPECWeb99, Fenice, SPECweb2005-bank)

have shown the OS can be tuned to scale with many cores. The works [18] and [54] are examples of such an implementation. However, we feel further investigation is necessary and leave such work for future research.

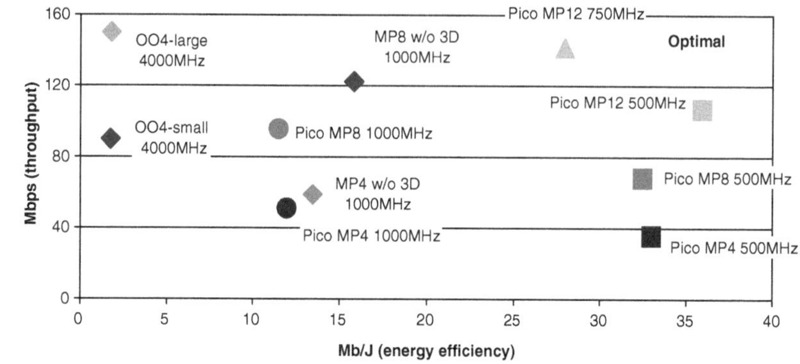

(a) SPECweb2005-ecommerce

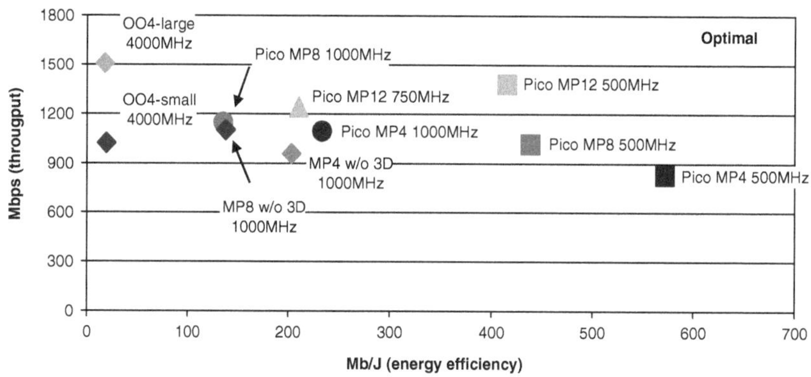

(b) dbench

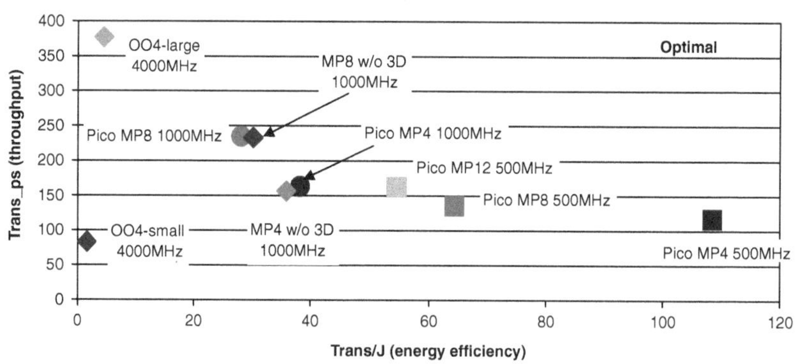

(c) TPC-C

Fig. 9.21 Energy efficiency, performance pareto chart generated for 90-nm process technology. Three-dimensional stacking technology enables new CMP architectures that are significantly energy efficient. (SPECweb2005-ecommerce, dbench, TPC-C)

9.6 Conclusions

Throughout this chapter, we showed that 3D stacking technology can be leveraged to build energy efficient servers. For a wide range of server workloads, the resulting systems have significant energy efficiency in a compact form factor. A 12-way PicoServer running at 500 MHz can deliver 1 Gbps of network bandwidth within a 3 W power budget using a 90-nm process technology. These power results are two to three times better than a multicore architecture without 3D stacking technology and an order of magnitude better than what can be achieved using a general purpose processor. Compared to a conventional 8-way 1 GHz chip multiprocessor with a 2 MB L2 cache, an area-equivalent 12-way PicoServer running at 500 MHz yields an improvement in energy efficiency of more than $2\times$. The absolute power values are also expected to scale with process technology. We expect to see additional cores and even lower power for 65-nm and 45-nm process technology-based PicoServer platforms.

The ability to tightly couple large amounts of memory to the cores through wide and low-latency interconnect pays dividends by reducing system complexity and creates opportunities to implement system memory with nonuniform access latency. Three dimensional technology enables core to DRAM interfaces that are high throughput while consuming low power. With the access latency of on-chip DRAM being comparable to the L2 cache, the L2 cache die area can be replaced with additional cores resulting in core clock frequency reduction while achieving higher throughput.

Acknowledgments This work is supported in part by the National Science Foundation, Intel, and ARM Ltd.

References

1. ARM 11 MPcore. http://www.arm.com/products/CPUs/ARM11MPCoreMultiprocessor.html.
2. FaStack 3D RISC super-8051 microcontroller. http://www.tachyonsemi.com/OtherICs/datasheets/TSCR8051Lx_1_5Web.pdf.
3. The Micron system-power calculator. http://www.micron.com/products/dram/syscalc.html.
4. National semiconductor DP83820 10 / 100 / 1000 Mb/s PCI ethernet network interface controller.
5. OSDL DataBase Test Suite. http://www.osdl.net/lab_activities/kernel_testing/osdl_database_test_suite/.
6. $(LS)^3$-libre streaming, libre software, libre standards an open multimedia streaming project. http://streaming.polito.it/.
7. RLDRAM memory. http://www.micron.com/products/dram/rldram/.
8. Seagate Barracuda. http://www.seagate.com/products/personal/index.html.
9. SPECweb2005 benchmark. http://www.spec.org/web2005/.
10. SPECweb99 benchmark. http://www.spec.org/osg/web99/.
11. Sun Fire T2000 Server Power Calculator. http://www.sun.com/servers/coolthreads/t2000/calc/index.jsp.
12. ITRS roadmap. Technical report, 2005.

13. K. Banerjee, S. J. Souri, P. Kapur, and K. C. Saraswat. 3-D ICs: A novel chip design for improving deep-submicrometer interconnect performance and systems-on-chip integration. *Proceedings of the IEEE*, 89(5):602–633, May 2001.
14. P. Barford and M. Crovella. Generating representative web workloads for network and server performance evaluation. In *Measurement and Modeling of Computer Systems*, pp. 151–160, 1998.
15. N. L. Binkert, R. G. Dreslinski, L. R. Hsu, K. T. Lim, A. G. Saidi, and S. K. Reinhardt. The M5 simulator: Modeling networked systems. *IEEE Micro*, 26(4):52–60, Jul/Aug 2006.
16. B. Black, M. Annavaram, N. Brekelbaum, J. DeVale, L. Jiang, G. H. Loh, D. McCaule, P. Morrow, D. W. Nelson, D. Pantuso, P. Reed, J. Rupley, S. Shankar, J. P. Shen, and C. Webb. Die stacking (3D) microarchitecture. In *International Symposium on Microarchitecture*, December 2006.
17. B. Black, D. Nelson, C. Webb, and N. Samra. 3D processing technology and its impact on iA32 microprocessors. In *Proceedings of International Conference on Computer Design*, pp. 316–318, 2004.
18. R. Bryant, J. Hawkes, J. Steiner, J. Barnes, and J. Higdon. Scaling Linux to the Extreme From 64 to 512 Processors. In *Linux Symposium*, July 2004.
19. T.-Y. Chiang, S. J. Souri, C. O. Chui, and K. C. Saraswat. Thermal analysis of heterogeneous 3-D ICs with various integration scenario. In *IEDM Technical Digest*, pp. 681–684, December 2001.
20. L. T. Clark, E. J. Hoffman, J. Miller, M. Biyani, Y. Liao, S. Strazdus, M. Morrow, K. E. Verlarde, and M. A. Yarch. An embedded 32-b microprocessor core for low-power and high-performance applications. *IEEE Journal of Solid State Circuits*, 36(11):1599–1608, November 2001.
21. E. L. Congduc. Packet classification in the NIC for improved SMP-based internet servers. In *Proceedings of International Conference on Networking*, February 2004.
22. W. R. Davis, J. Wilson, S. Mick, J. Xu, H. Hua, C. Mineo, A. M. Sule, M. Steer, and P. D. Franzon. Demystifying 3D ICs: The pros and cons of going vertical. *IEEE Design & Test of Computers*, 22(6):498–510, 2005.
23. M. J. Flynn and P. Hung. Computer architecture and technology: Some thoughts on the road ahead. In *Proceedings on International Conference on Engineering of Reconfigurable Systems and Algorithms*, pp. 3–16, 2004.
24. M. Ghosh and H.-H. S. Lee. Smart Refresh: An Enhanced Memory Controller Design for Reducing Energy in Converntional and 3D Die-Stacked DRAMs. In *International Symposuim on Microarchitecture*, December 2007.
25. B. Goplen and S. S. Sapatnekar. Thermal via placement in 3D ICs. In *Proceedings of International Symposium on Physical Design*, pp. 167–174, April 2005.
26. S. Gupta, M. Hilbert, S. Hong, and R. Patti. Techniques for producing 3D ICs with high-density interconnect. www.tezzaron.com/about/papers/ieee_vmic_2004_finalsecure.pdf.
27. R. Ho and M. Horowitz. The future of wires. *Proceedings of the IEEE*, 89(4), April 2001.
28. W. Huang, M. R. Stan, K. Skadron, K. Sankaranarayanan, S. Ghosh, and S. Velusam. Compact thermal modeling for temperature-aware design. In *Proceedings Design Automation Conference*, June 2004.
29. T. Kgil. *Architecting Energy Efficient Servers*. PhD thesis, University of Michigan, 2007.
30. T. Kgil and T. Mudge. FlashCache: a NAND flash memory file cache for low power web servers. In *Proceedings of International Conference on Compilers, Architecture and Synthesis for Embedded Systems*, October 2006.
31. T. Kgil, D. Roberts, and T. Mudge. Improving NAND flash based disk caches. In *Proceedings of International Symposium on Computer Architecture*, June 2008.
32. T. Kgil, A. Saidi, N. Binkert, S. Reinhardt, K. Flautner, and T. Mudge. PicoServer: Using 3D stacking technology to build energy efficient servers. *ACM Journal on Emerging Technologies in Computing Systems*, 2009.

33. M. G. Khatib, B. J. van der Zwaag, P. Hartel, and G. J. M. Smit. Interposing flash between disk and DRAM to save energy for streaming workloads. In *ESTIMedia*, 2007.
34. K. Kim and J. Choi. Future outlook of NAND flash technology for 40 nm node and beyond. In *Workshop on Non-Volatile Semiconductor Memory*, pp. 9–11, February 2006.
35. I. Koltsidas and S. D. Viglas. Flashing up the storage layer. In *VLDB*, August 2008.
36. P. Kongetira, K. Aingaran, and K. Olukotun. Niagara: A 32-way multithreaded Sparc processor. *IEEE Micro*, 25(2):21–29, March 2005.
37. M. Koyanagi. Different approaches to 3D chips. http://asia.stanford.edu/events/Spring05/slides/051205-Koyanagi.pdf.
38. S. R. Kunkel, R. J. Eickemeyer, M. H. Lipasti, T. J. Mullins, B. O'Krafka, H. Rosenberg, S. P. VanderWiel, P. L. Vitale, and L. D. Whitley. A performance methodology for commercial servers. *IBM Journal of Research and Development*, 44(6):851–872, 2000.
39. J. Laudon. Performance/watt: the new server focus. *SIGARCH Computer Architecture News*, 33(4):5–13, 2005.
40. K. Lee, T. Nakamura, T. Ono, Y. Yamada, T. Mizukusa, H. Hashimoto, K. Park, H. Kurino, and M. Koyanagi. Three-dimensional shared memory fabricated using wafer stacking technology. In *IEDM Technical Digest*, pp. 165–168, December 2000.
41. K. Lim, P. Ranganathan, J. Chang, C. Patel, T. Mudge, and S. Reinhardt. Understanding and designing new server architectures for emerging warehouse-computing environments. In *Proceedings of International Symposium on Computer Architecture*, June 2008.
42. J.-H. Lin, Y.-H. Chang, J.-W. Hsieh, T.-W. Kuo, and C.-C. Yang. A NOR emulation strategy over NAND flash memory. In *13th IEEE International Conference on Embedded and Real-Time Computing Systems and Applications*, August 2007.
43. G. L. Loi, B. Agrawal, N. Srivastava, S.-C. Lin, T. Sherwood, and K. Banerjee. A thermally-aware performance analysis of vertically integrated (3-D) processor-memory hierarchy. In *Proceedings Design Automation Conference*, June 2006.
44. J. Lu. Wafer-level 3D hyper-integration technology platform. www.rpi.edu/ luj/RPI_3D_Research_0504.pdf.
45. G. MacGillivray. Process vs. density in DRAMs. http://www.eetasia.com/ARTICLES/2005SEP/B/2005SEP01_STOR_TA.pdf.
46. D. A. Maltz and P. Bhagwat. TCP splicing for application layer proxy performance. Research Report RC 21139, IBM, March 1998.
47. R. E. Matick and S. E. Schuster. Logic-based eDRAM: origins and rationale for use. *IBM Journal of Research and Development*, 49(1):145–165, January 2005.
48. T. Ohsawa, K. Fujita, K. Hatsuda, T. Higashi, T. Shino, Y. Minami, H. Nakajima, M. Morikado, K. Inoh, T. Hamamoto, S. Watanabe, S. Fujii, and T. Furuyama. Design of a 128-Mb SOI DRAM using the floating body cell (FBC). *IEEE Journal of Solid State Circuits*, 41(1), January 2006.
49. C. Park, J.-U. Kang, S.-Y. Park, and J.-S. Kim. Energy-aware demand paging on NAND flash-based embedded storages. In *ISLPED*, pp. 338–343, 2004.
50. A. Rahman and R. Reif. System-level performance evaluation of three-dimensional integrated circuits. *IEEE Transactions on VLSI*, 8(6):671–678, December 2000.
51. F. Ricci, L. T. Clark, T. Beatty, W. Yu, A. Bashmakov, S. Demmons, E. Fox, J. Miller, M. Biyani, and J. Haigh. A 1.5 GHz 90 nm embedded microprocessor core. In *Proceedings of the IEEE Symposium on VLSI Circuits*, pp. 12–15, June 2005.
52. J. Scaramella. Enabling technologies for power and cooling. http://h71028.www7.hp.com/enterprise/downloads/Thermal_Logic.pdf.
53. J. Schutz and C. Webb. A scalable X86 CPU design for 90 nm process. In *Proceedings of IEEE International Solid-State Circuits Conference*, February 2004.
54. M. Shah, J. Barreh, J. Brooks, R. Golla, G. Grohoski, N. Gura, R. Hetherington, P. Jordan, M. Luttrell, C. Olson, B. Saha, D. Sheahan, L. Spracklen, and A. Wynn. UltraSPARC T2: A highly-threaded, power-efficient, SPARC SOC. In *Asian Solid-State Circuirts Conference*, November 2007.

55. J. Truong. Evolution of network memory. http://www.jedex.org/images/pdf/jack_troung_samsung.pdf.
56. D. Wendell, J. Lin, P. Kaushik, S. Seshadri, A. Wang, V. Sundararaman, P. Wang, H. McIntyre, S. Kim, W. Hsu, H. Park, G. Levinsky, J. Lu, M. Chirania, R. Heald, and P. Lazar. A 4 MB on-chip l2 cache for a 90 nm 1.6 GHz 64b SPARC microprocessor. In *Proceedings of IEEE International Solid-State Circuits Conference*, February 2004.
57. L. Xue, C. C. Liu, H.-S. Kim, S. Kim, and S. Tiwari. Three-dimensional integration: Technology, use, and issues for mixed-signal applications. *IEEE Transactions on Electron Devices*, 50:601–609, May 2003.

Chapter 10
System-Level 3D IC Cost Analysis and Design Exploration

Xiangyu Dong and Yuan Xie

Abstract The majority of the existing 3D IC research has focused on how to take advantage of the performance, power, smaller form-factor, and heterogeneous integration benefits offered by 3D integration. However, all such advantages will ultimately have to be translatee into cost savings when a design strategy has to be decided. Consequently, system-level cost analysis at the early design stage is imperative to help the decision making on whether 3D integration should be adopted. In this chapter, we discuss the design estimation method for 3D ICs at the early design stage. We also describe a cost analysis model to study the cost implication for 3D ICs and address cost-related problems for 3D IC design.

10.1 Introduction

The majority of the 3D IC research so far has focused on how to take the advantage of the performance, power, smaller form-factor, and heterogeneous integration benefits offered by 3D integration. For example, Section 7.9 have demonstrated such benefits of 3D designs. However, when it comes to discussions on the adoption of such emerging technology as a mainstream design approach, it all comes down to the question of the cost for 3D integration. All the advantages of 3D ICs ultimately have to be translated into cost savings when a design strategy has to be decided [18]. For example, designers may ask themselves questions like:

X. Dong (✉)
Pennsylvania State University, University Park, PA 16802, USA
e-mail: xydong@cse.psu.edu

This chapter includes portions reprinted with permission from the following publication: (a) Xiangyu Dong and Yuan Xie. System-level cost analysis and design exploration for 3D ICs. *Proceedings of Asia- South Pacific Design Automation Conference (ASP-DAC)* (2009). Copyright 2009 IEEE.

Y. Xie et al. (eds.), *Three-Dimensional Integrated Circuit Design*,
Integrated Circuits and Systems, DOI 10.1007/978-1-4419-0784-4_10,

- *Do all the benefits of 3D IC design come with a much higher cost?* For example, 3D bonding incurs extra process cost, and the through silicon vias (TSVs) may increase the total die area, which has a negative impact on the cost; However, smaller die sizes in 3D ICs may result in higher yield than that of a larger 2D die and reduce the cost.
- *How to do 3D integration in a cost-effective way?* For example, redesigning a small chip may not gain the cost benefits of improved yield resulting from 3D integration. In addition, if a chip is to be implemented in 3D, how many layers of 3D integration would be cost-effective? And should one use wafer-to-wafer or die-to-wafer stacking [20]?
- *Are there any design options to compensate for the extra 3D bonding cost?* For example, in a 3D IC, since some global interconnects are now implemented by TSVs, it may be feasible to use a fewer number of metal layers for each 2D die. In addition, heterogeneous integration via 3D could also help cost reduction.

Cost analysis for 3D ICs at the early design stage is critical to answer these questions and helps the decision making on whether 3D integration should be used and what design options should be adopted (such as the number of layers and the bonding approaches). Cost-efficient design is the key for the future wide adoption of the emerging 3D IC design, and 3D IC cost analysis needs close coupling between 3D IC design and 3D IC process.

In this chapter, we first describe the design estimation method for 3D ICs at the early design stage (Section 10.2) and propose a cost analysis model to study the cost implication for 3D ICs (Section 10.3). Using the design estimation method and the 3D cost analysis model, we compare the estimated cost between 2D and 3D designs and investigate the impact of various factors on the cost, as well as possible ways that 3D integration offers to reduce the cost; a cost-driven 3D IC design flow is also proposed to guide the design space exploration for 3D ICs toward a cost-effective direction (Section 10.4).

10.2 Early Design Estimation for 3D ICs

To facilitate the design decision of whether 3D integration should be used, from a cost perspective, it is necessary to perform cost analysis at the early design stage, when there is not much detailed design information available. The cost of an IC chip is closely related to the die area. In 3D ICs, TSVs may incur extra area overhead. However, it is possible to use a fewer number of metal layers for routing in 3D ICs, which can help cost reduction.

In this section, we describe how to estimate the die area, the number of metal layers for feasible routing, and the impact of TSVs on die area, at the very early design stage, when only limited information about the design (such as the estimation of the gate counts in the design) is available. Such an early estimation will facilitate the 3D IC cost analysis discussed in Section 10.3.

10.2.1 The Preliminary on Rent's Rule

Our early design estimation is based on the well-known *Rent's Rule* [14]. Rent's Rule reveals the trend between the number of signal terminals and the number of internal gates. It is an empirical result based on the observations of existing designs and can be expressed as

$$T = kN_{\mathrm{g}}^{p} \tag{10.1}$$

where the parameters k and p are Rent's coefficient and exponent, N_{g} is the gate counts, and T is the number of signal terminals.

Using Rent's Rule, it becomes possible to further estimate the average wire length [11] and the wire length distribution [9]. The average wire length can be given by

$$\bar{R_m} = \frac{2}{9}\frac{1-4^{p-1}}{1-N_{\mathrm{g}}^{p-1}}\left(7\frac{N_{\mathrm{g}}^{p-0.5}-1}{4^{p-0.5}-1} - \frac{1-N_{\mathrm{g}}^{p-1.5}}{1-4^{p-1.5}}\right) \tag{10.2}$$

when p=0.5, the expression can be calculated using *L'Hospital Law* [11].

Derived from Rent's rule, the wire length distribution function $i(l)$ has the following forms:

Region I: $1 \leq l \leq \sqrt{N_{\mathrm{g}}}$

$$i(l) = \frac{\alpha k}{2}\Gamma\left(\frac{l^3}{3} - 2\sqrt{N_{\mathrm{g}}}l^2 + 2N_g l\right) l^{2p-4} \tag{10.3}$$

Region II: $\sqrt{N_g} \leq l < 2\sqrt{N_{\mathrm{g}}}$

$$i(l) = \frac{\alpha k}{6}\Gamma\left(2\sqrt{N_{\mathrm{g}}} - l\right)^3 l^{2p-4} \tag{10.4}$$

where l is the interconnect length in units of gate pitches, α is the fraction of the on-chip terminals that are sink terminals and is related to the average fanout of a gate (*f.o.*) as follows:

$$\alpha = \frac{f.o.}{f.o.+1} \tag{10.5}$$

and Γ is given by

$$\Gamma = \frac{2N_{\mathrm{g}}\left(1-N_{\mathrm{g}}^{p-1}\right)}{\left(-N_{\mathrm{g}}^{p}\frac{1+2p-2^{2p-1}}{p(2p-1)(p-1)(2p-3)} - \frac{1}{6p} + \frac{2\sqrt{N_{\mathrm{g}}}}{2p-1} - \frac{N_{\mathrm{g}}}{p-1}\right)} \tag{10.6}$$

10.2.2 Die Area and Metal Layer Estimator

At the early design stage, the die area can be estimated as a function of the gate counts:

$$A_{\text{die}} = N_g A_g \tag{10.7}$$

where N_g is the number of gates, and A_g is an empirical parameter that shows the proportional relationship between area and gate counts. Based on empirical data from our industrial designs, in this work we assume that A_g=3125λ^2 in which λ is the half of the feature size for a specific technology node.

The number of required metal layers for routing depends on the complexity of the interconnections. A simple metal layer estimation can be derived from the average wire length [19]:

$$n_w = \frac{f.o.\bar{R_m}p_w}{e_w}\sqrt{\frac{N_g}{A_{\text{die}}}} \tag{10.8}$$

where *f.o.* refers to the average gate fanout, p_w to wire pitch, e_w to the utilization efficiency of metal layers, $\bar{R_m}$ to the average wire length, which is formulated by Eq.(10.2), and n_w to the number of metal layers. Such a simplified model is based on the assumptions that each metal layer has the same utilization efficiency and the same wire width [19]. However, such assumptions may not be valid in real design [13]. Moreover, the model in [19] does not include the impact of TSV area overhead, which will be a considerable penalty when the complexity of 3D ICs rises.

To improve the estimation of the number of metal layers needed for feasible routing, we propose a new 3D routability model, which is based on the wire length distribution rather than a simple estimation of average wire length. The basic idea of this model is explained as follows:

- *Estimate the available routing area of each metal layer with the expression*

$$K_i = \frac{A_{\text{die}}\eta_i - 2A_v\left(N_g f.o. - I(l_i)\right)}{w_i} \tag{10.9}$$

  where η_i is the utilization efficiency, w_i is the wire pitch, A_v is the blockage area of each via, and function *I(l)* is the cumulative integral of the wire length distribution function *i(l)*, which is expressed in Eq.(10.4).
- *Assume that shorter interconnects are routed on lower metal layers. Starting from Metal 1, we route as many interconnects as possible on the current metal layer until the available routing area is used up.* The interconnects routed on each metal layer can be expressed as

$$\chi L(l_i) - \chi L(l_{i-1}) \leq K_i \tag{10.10}$$

where $\chi = 4/(f.o. + 3)$ is a factor accounting for the sharing of wires between interconnects on the same net [9, 8]. The function $L(l)$ is the first-order moment of $i(l)$.

- *Repeat the same calculations for each metal layer in a bottom-up manner until all the interconnects are routed properly.*

By applying the estimation methodology introduced above, we can predict the die area and the number of metal layers at the early design stage where we only have the number of gates as the input. Figure 10.1 shows an example which estimates the area and the number of metal layers of 65 nm designs with different scale of gates.

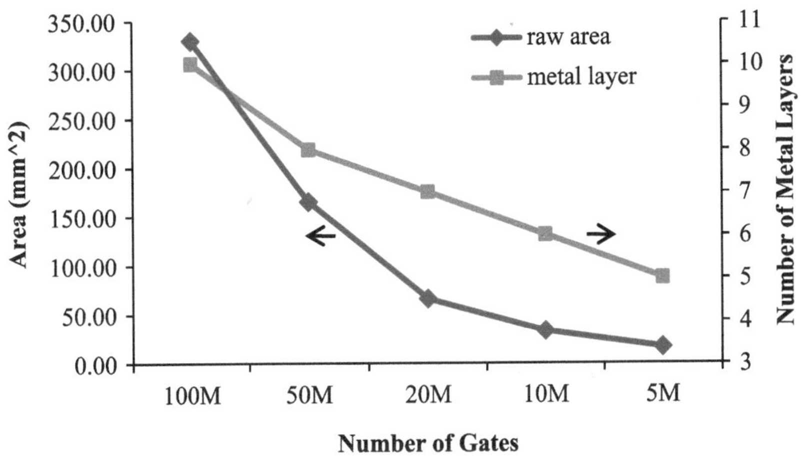

Fig. 10.1 Early design estimation of die area and metal layer (65-nm process). (The estimation is well correlated with the state-of-the-art microprocessor designs. For example, Sun SPARC T2 [16] contains about 500 M transistors (equivalent to 125 M gates), with an area of 342 mm^2 and 11 metal layers)

Figure 10.1 shows an important implication for 3D IC cost reduction: when a large 2D chip is partitioned into multiple smaller dies with 3D stacking, each smaller die requires a fewer number of metal layers to satisfy the interconnect routability requirements. Such metal layer reduction could offset the extra cost resulting from 3D stacking.

10.2.3 The Impact of TSVs

The impact of through silicon vias (TSVs) used in 3D stacking on the cost analysis is twofold:

- In 3D ICs, some global interconnects are now implemented by TSVs, going between stacked dies. This could lead to the reduction of the total wire length, and provide opportunities for metal layer reduction for each smaller die.
- On the other hand, 3D stacking with through silicon vias (TSVs) may increase the total die area, since the silicon area where TSVs punch through may not be utilized for building devices or 2D metal layer connections. (Based on current TSVs technologies, the diameter of TSVs ranges from 0.2 to 10μ m [15].)

Consequently, it is important to estimate the number of TSVs and the impact on the die area increase.

To predict the number of required TSVs for a certain partition pattern, the relationship between the interconnections (X) and gates (N_g) (a derivation of Rent's rule) can be used [11]:

$$X = \alpha k N_g \left(1 - N_g^{p-1}\right) \tag{10.11}$$

As illustrated in Fig. 10.2, the number of TSVs can be estimated by

$$\begin{aligned} X_{TSV} = {} & \alpha k_{1,2}(N_1 + N_2)(1 - (N_1 + N_2)^{p_{1,2}-1}) \\ & - \alpha k_1 N_1 \left(1 - N_1^{p_1-1}\right) \alpha k_2 N_2 \left(1 - N_2^{p_2-1}\right) \end{aligned} \tag{10.12}$$

where $k_{1,2}$ and $p_{1,2}$ are the equivalent Rent's coefficient and exponent. The area overhead caused by TSVs can be modeled as follows:

$$A_{3D} = A_{die} + N_{TSV/die} A_{TSV} \tag{10.13}$$

where A_{die} is calculated by die area estimator, $N_{TSV/die}$ is the equivalent number of TSVs on each die, A_{TSV} is the size of TSVs, and A_{3D} is the final 3D component die area.

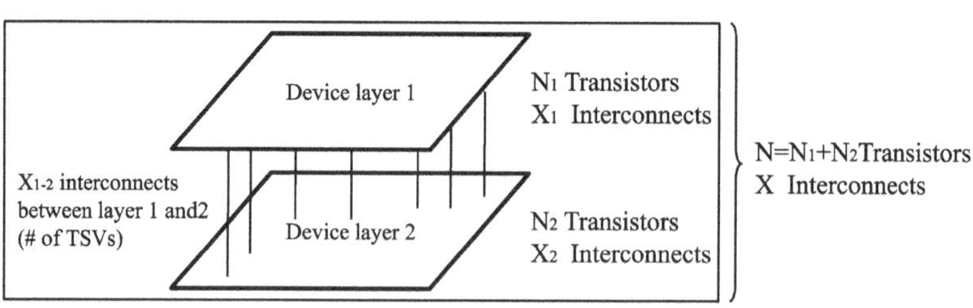

Fig. 10.2 The basic idea of how to estimate the number of TSVs

10.3 Three-Dimensional Cost Model

Three-dimensional integration involves stacking multiple dies through traditional fabrication processes. There are several different ways to stack separate dies together [10], and the TSV-based approach is the most promising. In addition to conventional 2D processes, 3D integration needs extra fabrication steps such as forming TSVs via laser drilling or etching, wafer thinning, and wafer bonding.

We model the cost brought by each step during the 3D fabrication; and the cost analysis can be divided into the die cost model and the 3D bonding cost model, as shown in Fig. 10.3.

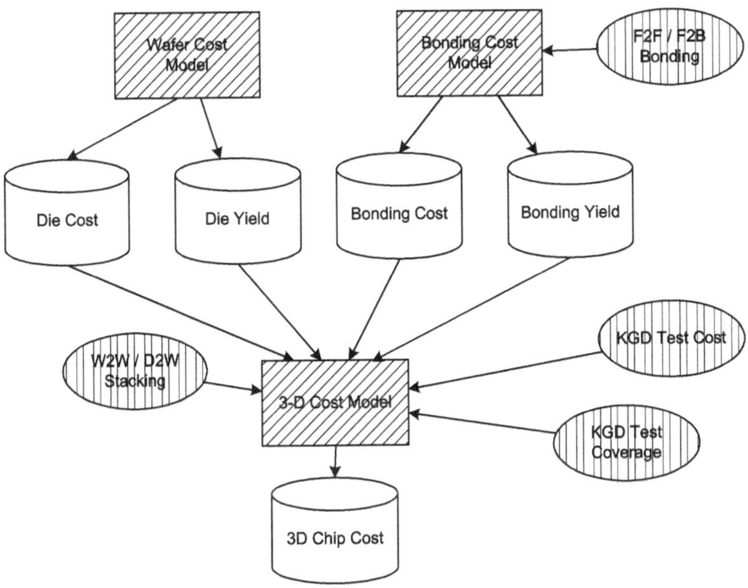

Fig. 10.3 The overview of the proposed 3D cost model

- *Wafer Cost Model.* The key factor of the die cost model is the die area. If we assume that the wafer cost, the wafer yield, and the defect density are constant for a specific foundry using a specific technology node, the impact of die areas can be formulated by two expressions [17] as the follows:

$$N_{\text{die}} = \frac{\pi \times (\phi_{\text{wafer}}/2)^2}{A_{\text{die}}} - \frac{\pi \times \phi_{\text{wafer}}}{\sqrt{2 \times A_{\text{die}}}} \tag{10.14}$$

$$Y_{\text{die}} = Y_{\text{wafer}} \times \frac{\left(1 - e^{-2A_{\text{die}}D_0}\right)}{2A_{\text{die}}D_0} \tag{10.15}$$

where N_{die} is the number of dies per wafer, ϕ_{wafer} is the diameter of the wafer, Y_{die} and Y_{wafer} are the yields of dies and wafers, respectively, and D_0 is the defect density of the wafer.

Our wafer cost model obtained from different foundries includes material cost, labor cost, foundry margin, number of reticles, cost per reticle, and other miscellaneous cost [3]. Figure 10.4 shows the predicted wafer cost of 90 nm, 65 nm, and 45 nm processes, with 9 or 10 layers of metal, for three different foundries.

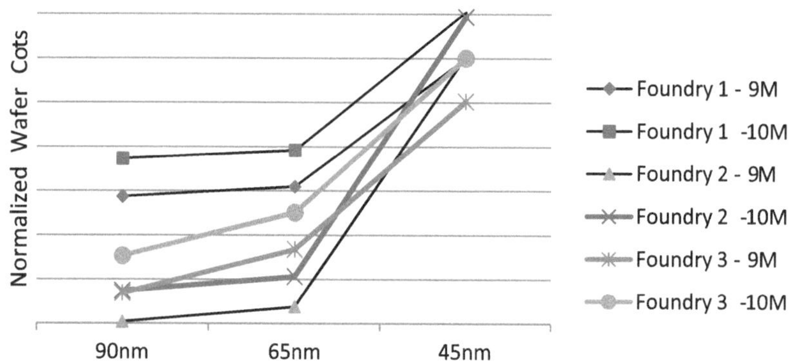

Fig. 10.4 A batch of data calculated by the wafer cost model. The wafer cost varies for different processes, different number of metal layers, different foundries, and some other factors

- *Three-Dimensional Bonding Cost.* Chapter 2 described various 3D bonding methods. The extra steps required by 3D integration consists of TSV forming, thinning, and bonding. In this work, we model two approaches to build 3D TSVs: *laser drilling or etching*. Laser drilling is only suitable for a small number of TSVs (hundreds to thousands) while etching is suitable for a large number of TSVs. The TSV etching process is similar to building conventional vias between metal layers, but as its name implies, TSV is "through silicon." There are two approaches for TSV etching: (1) *TSV-first approach.* TSVs can be formed during the 2D die fabrication process, before the back-end-of-line (BEOL) processes. Such an approach is called *TSV-first* approach and is shown in Fig. 10.5a; (2) *TSV-later approach.* TSVs can also be formed after the completion of 2D fabrications, after the BEOL processes. Such an approach is called *TSV-later* approach and is shown in Fig. 10.5b. Our 3D bonding cost model is based on the 3D process from our industry partners, with the assumption that the yield of each 3D process step is 99%.
- *Overall 3D Cost Model.* In addition to the wafer cost model and the bonding cost model, the entire 3D cost model also depends on some design options, such as die-to-wafer/wafer-to-wafer bonding, face-to-face/face-to-back bonding, and known-good-die cost [18].

  For D2W bonding, the bare chip cost before packaging is calculated by

$$C_{\mathrm{D2W}} = \frac{\sum_{i=1}^{N}\left(C_{die_i}+C_{\mathrm{KGDtest}}\right)/Y_{\mathrm{die_i}} + (N-1)C_{\mathrm{bonding}}}{Y_{\mathrm{bonding}}^{N-1}} \tag{10.16}$$

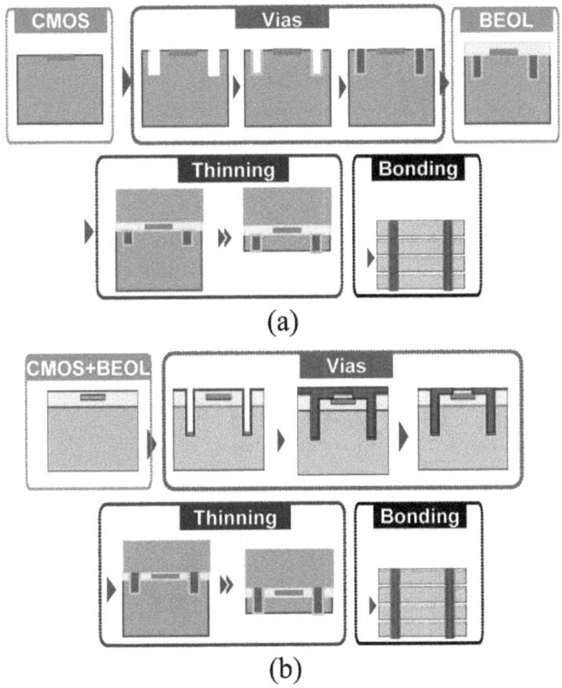

Fig. 10.5 Fabrication steps for 3D ICs: (**a**) TSVs are formed before BEOL process, thus TSVs only punch through the silicon substrate but not the metal layers; (**b**) TSVs are formed after BEOL process, thus TSVs punch through not only the silicon substrate but the metal layers as well

For W2W bonding, the calculation becomes

$$C_{\mathrm{W2W}} = \frac{\sum_{i=1}^{N} C_{\mathrm{die}_i} + (N-1)C_{\mathrm{bonding}}}{\left(\Pi_{i=1}^{N} Y_{\mathrm{die}_i}\right) Y_{\mathrm{bonding}}^{N-1}} \tag{10.17}$$

In order to support multiple-layer bonding, the default bonding mode is face-to-back. If the face-to-face mode is used, there is one more component die that does not need the thinning process, and the thinning cost of this die is subtracted from the total cost.

10.4 System-Level 3D IC Design Exploration

Based on the early design estimation methods and the 3D cost analysis model described in the previous sections, we use the IBM common platform foundry cost model as an example to perform a series of design analysis at the system level, investigating the impact of different design options on the 3D IC cost and inducing a few rules of thumb as 3D IC design guidelines from cost perspectives.

10.4.1 Evaluation of the TSV's Impact on Die Area

As mentioned in Section 10.2, building TSVs in 3D ICs not only incurs additional process cost but also causes area overhead. The area overhead affects the die yield and the wafer utilization. Based on the TSV estimation Eq.(10.12), we set the exponent parameter $p=0.63$ and the coefficient parameter $k=1.4$ according to *Bakoglu's* research [5]. We further assume that the number of 3D layers is N, and all the gates are uniformly partitioned into N layers. We choose the pitch size of TSVs to be 8μ m. Using the early design estimation, the predicted TSV impacts under the 65 nm process are shown in Fig. 10.6.

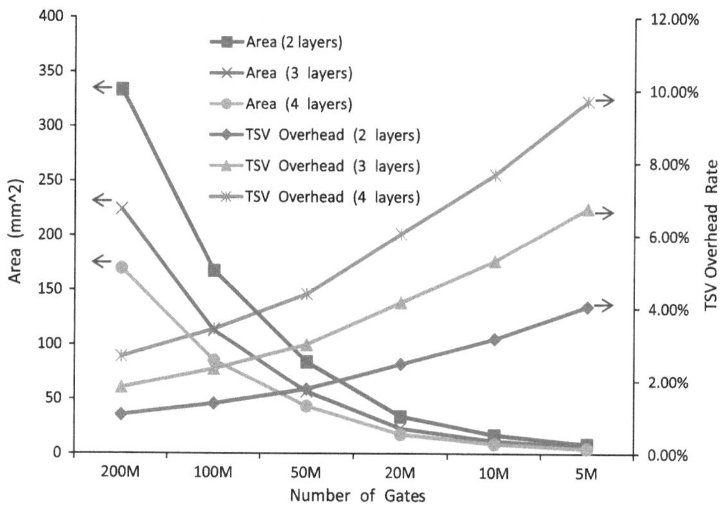

Fig. 10.6 The total area and the percentage of area occupied by TSVs: for small designs, the TSV area overhead is near to 10%, but for large designs, the TSV area overhead is less than 4%

Consistent with what Eq.(10.13), the TSV area overhead increases with the increase of 3D layers. The overhead can reach as high as 10% for a small design (5 M gates) using four layers. However, when the design is sufficiently large (200 M gates) and 3D integration only stacks two dies together, the TSV area overhead is usually below 2%. To summarize, *for large design, the area overhead due to TSVs is acceptable.*

10.4.2 The Potential of Metal Layer Reduction in 3D ICs

Theoretically, when the number of gates is distributed evenly to multiple dies in 3D stacking, the total wire length on each smaller die equals the total wire length for a large 2D chip divided by the number of dies. In addition, as discussed in

Section 10.2, the total wire length decreases as the number of 3D layers increases, due to the existence of TSVs. Comparing these two factors together, the routing complexity on each 3D component die is much less than that of the 2D baseline design. As a result, it becomes possible to remove one or two metal layers on each smaller die in 3D stacking.

Estimated by the 3D routability model discussed in Section 10.2, we can obtain the prediction of the metal layer reduction effect; the result is listed in Table 10.1.

Table 10.1 The number of required metal layers per die (65-nm microprocessor)

Gate counts	One-layer 2D	Two-layer 3D	Three-layer 3D	Four-layer 3D
5 M	5	5	5	4
10 M	6	5	5	5
20 M	7	6	5	5
50 M	8	7	7	6
100 M	10	8	7	7
200 M	12	10	9	8

Although the result shows that there is little opportunity to reduce metal layers in a relatively small design (such as the 5 M-gate design), the metal layer reduction becomes more and more obvious with the growth of design complexities. For instance, the number of required metal layers can be reduced by 2, 3, and 4 when a large 2D design (i.e., 200 M gates) is uniformly partitioned into 2, 3, and 4 separate dies, respectively.

To summarize, *in 3D IC design, it is possible to use a fewer number of metal layers for each smaller die, compared to the baseline 2D design. Such metal layer reduction could offset the extra bonding cost in 3D integration.*

10.4.3 Bonding Techniques: D2W or W2W

Die-to-wafer (D2W) and wafer-to-wafer (W2W) are two different ways to bond multiple dies together in 3D integration [10]. Section 10.3 discusses the modeling for these two methods. D2W bonding can achieve a higher yield by introducing the known-good-die (KGD) test, while W2W bonding does not need any test before bonding and it is easy for die alignments with higher throughput, at the expense of yield loss [10]. Since both D2W and W2W have their pros and cons, we use our 3D cost model to find out which one is more suitable for 3D integration technologies.

Figure 10.7 shows the cost comparison among the conventional 2D process, the two-layer D2W bonding and the two-layer W2W bonding. It can be observed that, although the cost of W2W is lower than that of D2W in the cases of small design, the cost of W2W is always higher than that of 2D processes. This phenomena can be explained by the relationship between areas and yields, which is expressed by Eq.(10.15). For the W2W bonding, the yield of each component die does increase

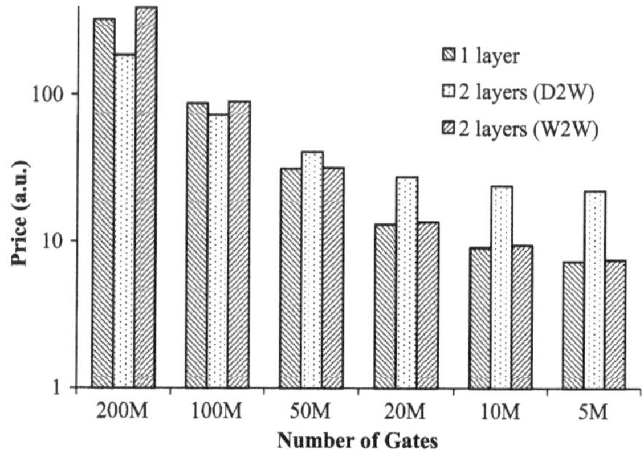

Fig. 10.7 The cost comparison among 2D, two-layer D2W, and two-layer W2W under 65-nm process (with the consideration of TSV area overheads and metal-layer reductions)

due to the area reduction, but when all the dies are stacked together without a pre-stacking test, the chip yield equals the product of the yield of each component die. Thus the overall yield of W2W-bonded 3D chips becomes as low as the one of 2D chips. After the extra bonding cost is included, it becomes reasonable to understand why W2W is always more expensive than conventional 2D.

To summarize, *from a yield perspective, die-to-wafer (D2W) stacking has a cost advantage over wafer-to-wafer (W2W) stacking* based on our wafer cost model and 3D-bonding cost model.

10.4.4 Cost vs. Number of 3D Layers

Based on the early design estimation methods that predict the 3D IC die area, the 3D TSV impact, and the metal layer reduction effect, we can further use these design-related parameters as the inputs to the 3D IC cost model proposed in Section 10.3, and estimate the cost of each 3D design option.

First of all, we select the IBM common platform 65-nm model and compare the cost of the 2D baseline design with its 3D counterparts, which have 2, 3, and 4 dies stacked together. Figure 10.8 shows the cost estimations based on the assumption that the 2D baseline design is partitioned uniformly.

It can be observed from Fig. 10.8 that the cost increases with the growth of the chip size dramatically due to the exponential relationship between die sizes and die yields. Because the yield becomes more sensitive to the die area when the area is large, splitting a large 2D design into multiple dies is more likely to reduce the total cost than splitting a small 2D design.

Another observation is that the optimal number of 3D layers (in terms of cost) may vary, depending on how large the design is. For example, the most cost-effective way for a 200 M-gate design is to do a 3D stacking with 4-layer 3D partitioning; but for a 100 M-gate design, the most cost-effective option is to use 2-layer 3D integration; finally, when the original 2D design is relatively small (< 50 M gates),

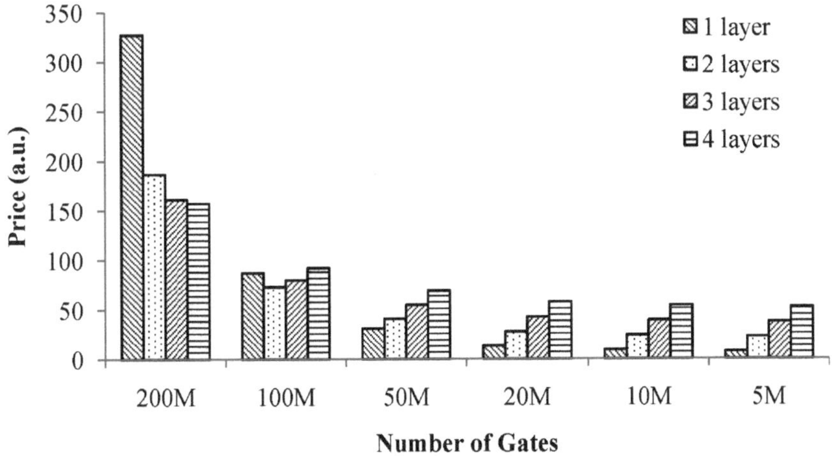

Fig. 10.8 The cost of 3D ICs with different number of layers under 65-nm process (with the consideration of TSV area overheads and metal-layer reductions)

the conventional 2D fabrication is always the cheapest one, because the 3D bonding cost starts to dominate and the yield improvement due to 3D stacking is small.

Repeating the experiment using different technology nodes with IBM common platform technology models, we estimate a set of boundaries that indicate where the two-layer 3D stacking starts to be more cost-effective than the traditional 2D process. The data are listed in Table 10.2. If we convert the number of gates into chip size, the enabling point of the two-layers 3D process is about 250 mm^2. The enabling point of more than two layers of stacking can be even larger.

Table 10.2 The enabling point of 3D fabrications

Process (IBM common platform)	45 nm	65 nm	90 nm	130 nm
Enabling point (number of gates)	143 M	76 M	40 M	21 M

To summarize, *3D integration is cost-effective for large design, but not for small design; The optimal number of 3D layers (from a cost perspective) increases as the gate count increases.*

10.4.5 Heterogenous Stacking

All the discussions above are focused on homogeneous stacking. However, one of the biggest advantages of 3D integration is that it supports heterogeneous stacking because different types of components can be fabricated separately.

Using today's high-performance microprocessors as an example, a large portion of the silicon area is occupied by on-chip SRAM or DRAM, and nonvolatile memory can also be integrated as on-chip memory [12]. However, the fabrication processes for these different modules are different. For instance, while the underlying conventional CMOS logic circuits require 1-poly-9-copper-1-aluminum interconnect layers, the one needed by DRAM modules is 7-poly-3-copper and the one needed by Flash modules is 4-poly-1-tungsten-2-aluminum. As a result, heterogeneous integration will dramatically increase the cost. As an example, Intel shows that heterogenous integration for large 2D SoC could boost the chip cost by three times [7].

Separating the fabrication of heterogenous technology and stacking components with 3D integration could be a cost-effective way for such systems. Here, we take the OpenSPARC T2 [16] as a case study. The original 2D OpenSPARC T2 chip has an area of 342 mm^2 and is fabricated with a TI 65-nm process with 11 metal layers. About half of the die area is attributed to on-chip SRAM cache. One way of using 3D integration for such a microprocessor is to partition all SRAM modules on one die and all the rest of the modules on the other die similar to the recent Intel 80-core Tera-scale chip [1]. Applying the early design estimation method in Section 10.2 and choosing the Rent's parameters for SRAM as p=0.12, k=6, we estimate that the number of metal layers for the SRAM modules can be reduced to 5. And we further estimate the total chip cost by using our 3D IC cost model. The comparison is shown in Fig. 10.9.

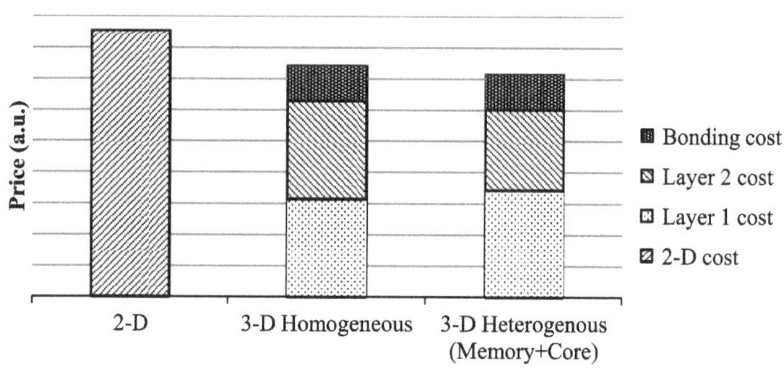

Fig. 10.9 The estimated cost of OpenSPARC T2 by using conventional 2D, homogeneous 3D partitioning, and heterogeneous 3D partitioning: fabricating the memory and the core part separately can further reduce the cost

To summarize, *the ability to enable heterogenous integration offers extra opportunities to reduce the total cost in 3D IC designs.*

10.5 Cost-Driven 3D Design Flow

The 3D IC cost analyses discussed above are all conducted before the real design, and all the inputs of the cost model are predicted from early design estimation. However, if the same cost analysis methodology is applied during the design time, and uses the real design data, such as die area, TSV interconnects, and metal interconnects as the inputs to the cost model, then a cost-driven 3D IC design flow becomes possible. Figure 10.10. shows a proposed cost-driven 3D IC design flow. The integration of 3D IC cost models into design flows guides the designer to optimize their 3D IC design and eventually manufacture a low-cost product.

Such a cost analysis/reduction EDA flow consists of three groups of operations: design-related operations, cost-modeling-related operations, and cost reduction operations.

- **Design-related operations** include *3D Partitioning*, *Timing Analysis*, and *Placement & Routing*, all of which are part of a typical 3D chip design flow. Such operations can affect the cost estimation. For example, different partitioning strategies may result in different components on a die with a different number of I/O interfaces. Placement & routing determines the interconnect topology of each layer, and results in a different number of through silicon vias needed for 3D integration, which impacts the bonding overhead.
- **Cost-modeling operations** include *die area estimation* (which evaluates the area of the die in each layer), *wafer cost modeling* (which evaluates the cost of each stacking layer), *3D bonding cost model* (which evaluates the cost of stacking multiple layers of dies together and the cost of fabricating 3D vias), as well as *stacking cost model*, which evaluates the cost related to different stacking options. For example, die-to-wafer (D2W) stacking requires a known-good-die test before stacking a die on top of other dies and incurs additional cost for die test; but it can improve the yield of stacked chips. These models are described in previous sections.
- **Cost reduction operations** include possible ways to reduce the cost. For example, one approach is called *Heterogenous Process Technology Stacking*: components that are not critical can be partitioned onto a die with a slower (but cheaper) technology fabrication (such as 0.18 μm CMOS), while critical components are partitioned onto a die with a more advanced (faster but expensive) technology fabrication (such as 65-nm CMOS). The second approach is called *Metal-Layer Reduction*: when moving from 2D to 3D design, each die itself may be able to use a fewer number of the metal layers to do routing, which can save the back-end process cost.

Such a unique and close integration of cost analysis with 3D EDA design flow has two advantages. First, as we discussed earlier, many design decisions (such as partitioning and placement & routing) can affect cost analysis. Closely coupling cost analysis with 3D EDA flow can result in more accurate cost estimation. Second, cost analysis results can drive 3D EDA tools to carry out a more cost-effective optimization in addition to considering other design goals (such as performance and power).

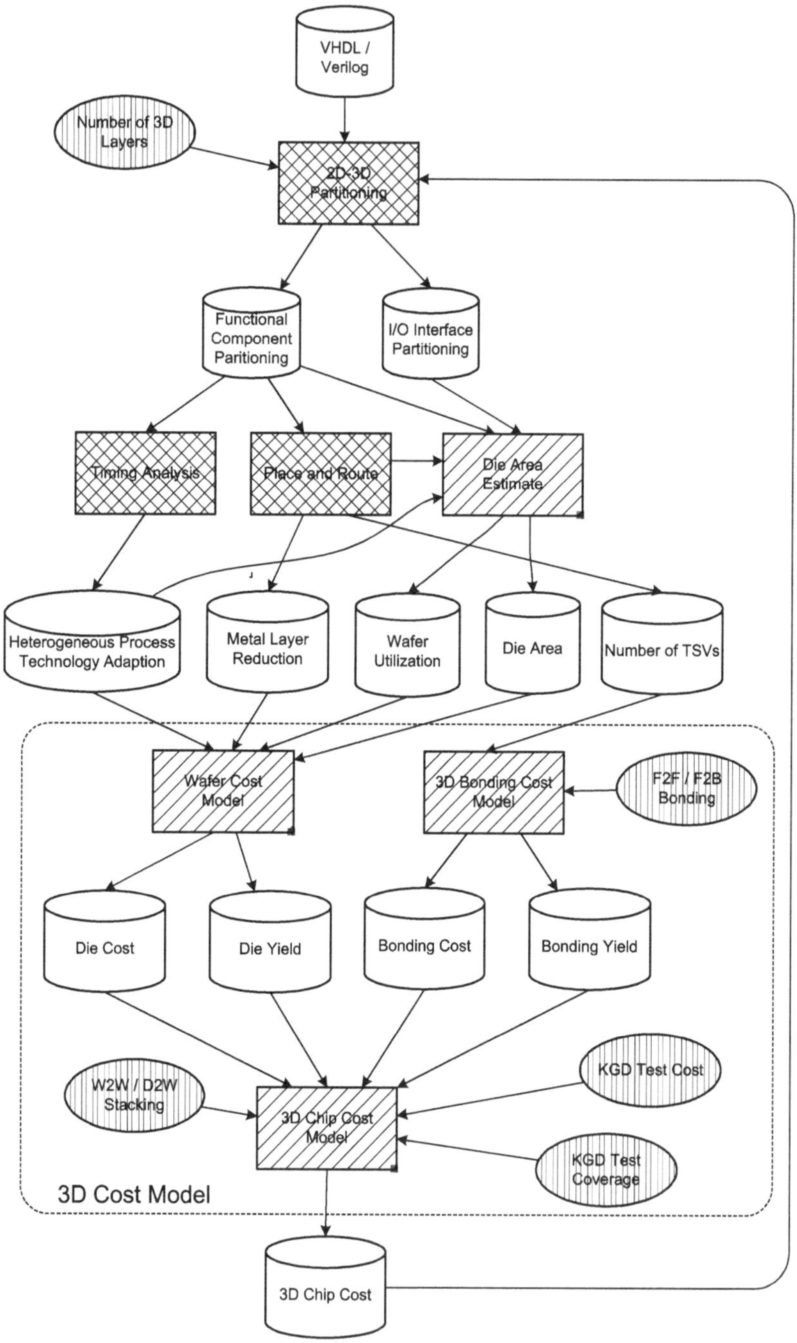

Fig. 10.10 The scheme of a cost-driven 3D IC design flow

10.5.1 Case Study: Two-Layer OpenSPARC T1 3D Processor

We use the Sun OpenSPARC T1 processor [2][1] as a case study to demonstrate how the extra manufacturing cost related to 3D technology could be offset by the cost reduction methods mentioned in the last section. As a result of cost reduction, the total cost of a 3D chip could be lower than that of its 2D counterpart.

As we mentioned in the previous section, the two major 3D cost reduction methods are (1) *Metal-Layer Reduction*, which reduces the number of metal layers during fabrication by taking advantage of the third routing dimension introduced by 3D technology; (2) *Heterogenous Process Technology Stacking*, which partitions the noncritical components into a specific layer manufactured using an older and cheaper process node.

There are two partitioning approaches that could help reduce the cost using 3D stacking: (1) coarse-granularity partitioning. The OpenSPARC T1 processor can be split into processor cores and cache banks; in Section 10.4.5, we saw that such partitioning can help the cost reduction because separating memory from logic layer can reduce the cost. (2) Fine-granularity partitioning. In this approach, we partition the components at the unit level [6] using our cost-driven design flow proposed in Fig. 10.10. Following such flow, we divide the entire 8-core OpenSPARC T1 processor into two-layer fine-grained partitioning and ensure that the timing requirements are not changed. With such fine-granularity partitioning, there are two possible ways to save cost:

- *90 nm-90-nm stacking*. In this approach, both layers are implemented using 90-nm technology. Carefully partitioning these units into two layers, we can make the area of the resulting layers equal to each other and the critical path remains unchanged. From the synthesis results using a 90-nm standard cell library, we observe that the total area of a single core in the 8-core SPARC T1 is about 10.63 mm^2. With two-layer 3D partitioning, the area of a single core is reduced to 7.18 and 7.03 mm^2. Based on our cost model, the 3D implementation cost is \$125, compared to the original 2D cost of \$146.
- *90 nm-130-nm heterogenous process technology stacking*. In this approach, the timing analysis results are used to find out which sets of components are not on the critical paths and can be moved to a slower layer that is synthesized with 130-nm standard cell library. Based on the synthesis results and the cost analysis, the cost is further reduced to \$121.

10.6 Reciprocal Design Symmetry for Mask Reuse in 3D ICs

The cost model discussed in the previous sections only models the fabrication cost and does not include the mask cost. Usually, each die in the stacked chip requires a

---

[1]The design has open-source design verilog code and synthesis scripts on http://www.opensparc.net. The original T1 chip was fabricated at 90-nm technology, with 300 M transistors and an area of 340 mm^2.

unique mask set, and therefore the mask cost for a 3D chip could be increased compared to its 2D counterpart. The mask cost has increased dramatically as technology scales; consequently, the cost of mask sets for 3D ICs would have a significant impact on the final cost of a 3D IC chip. However, because the lack of mask cost models, and the mask cost is usually application-specific, our 3D cost model described in this chapter does not include the cost of mask sets.

Even though, in general, a 3D IC chip needs multiple mask sets, there are special design cases, such as memory stacking, which allow the re-use of mask sets for all memory layers. In addition, in Section 10.4.5, we have shown that if the SRAM-based L2 cache is stacked on top of the processor core, the number of metal layers for the cache module is reduced to five layers, instead of 11 layers for the processor core, which results in a substantial cost reduction, as shown in Fig. 10.9. Furthermore, in Section 10.4.2, the potential of metal layer reduction for 3D ICs was demonstrated. Such metal layer reduction not only reduces the manufacturing cost, but also helps offset the extra mask cost due to 3D stacking.

Alam et al. [4] recently proposed a novel design technique using reciprocal design symmetry (RDS) that allows a mask set to be re-used for other layers in the 3D-stacked dies. The idea of RDS can be described using the dual-core memory stacking example shown in Fig. 10.11.

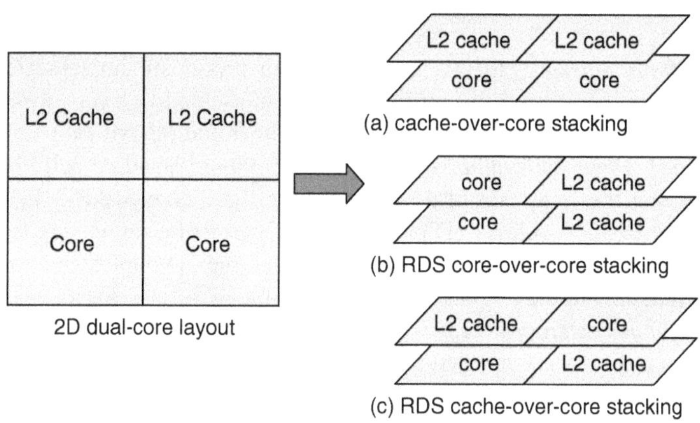

Fig. 10.11 Three ways to implement a 3D dual-core microprocessor

As shown in Fig. 10.11, a 3D dual-core microprocessor can be implemented in three different approaches: (a) the first is to move all the L2 caches for both cores and stack them on top of cores. This is the same approach described in Section 7.2 (Fig. 7.2a). Such a approach needs two mask sets: one for the core layer and the other for the cache layer. (b) The second approach is to have a core-over-core (cache-over-cache) stacking. This approach will result in a single mask set, which can be used for both layers. However, such a stacking approach could result in much higher temperature increases because the power density of the core is usually much higher than the cache; (c) the third approach is to rotate one layer in (b) to achieve a cache-over-core (core-over-cache) stacking. This approach can result in mask reuse with reciprocal design symmetry and at the same time minimize the impact of thermal increase.

10.7 Conclusion

To overcome the barriers in technology scaling, 3D integrated circuit (3D IC) is emerging as an attractive option for future IC design. However, fabrication cost is one of the important considerations for wide adoption of the 3D integration. System-level cost analysis at the early design stage to help the decision making on whether 3D integration should be used for the application is very critical.

To facilitate the system-level cost analysis, we study the design estimation method for 3D ICs at the early design stage and propose a cost analysis model to study the cost implications. Based on the cost analysis, we identify the design opportunities for cost reduction in 3D ICs and provide a few design guidelines on cost-effective designs. Our research is complementary to the existing research on the 3D IC benefits analysis on other design goals (such as performance and power analysis).

Acknowledgments The authors would like to thank Dr. Larry Smith from SEMATECH, Dr. Mike Ignatowski from IBM, Dr. Sam Gu from Qualcomm, and Dr. Pol Marchal from IMEC for the valuable discussions and guidance on this research. This work was supported in part by NSF grants of CAREER 0643902 and CCF 0702617, a grant from Qualcomm, and an IBM Faculty Award.

References

1. http://techresearch.intel.com/articles/Tera-Scale/1421.htm. 2007.
2. http://www.opensparc.net/. 2008.
3. IC Cost Model, 2008 revision 0808a. In *IC Knowledge LLC*, 2008.
4. S. Alam, R. Jones, S. Pozder, and A. Jain. Die/wafer stacking with reciprocal design symmetry (rds) for mask reuse in three-dimensional (3D) integration technology. In *International Symposium on Quality Electronic Devices*, 2009.
5. H. B. Bakoglu. *Circuits, Interconnections, and Packaging for VLSI*. Addison-Wesley, Reading, MA, 1990.
6. K. Bernstein. New dimension in performance. *EDA Forum*, 3(2), 2006.
7. S. Borkar. 3D-Technology: a system perspective. In *International 3D-System Integration Conference*, 2008.
8. P. Chong and R. K. Brayton. Estimating and optimizing routing utilization in DSM design. In *Workshop System-Level Interconnect Prediction*, 1999.
9. J. A. Davis, V. K. De, and J. D. Meindl. A stochastic wire-length distribution for gigascale integration derivation and validaton. derivation and validation. *IEEE Transaction on Electron Devices*, 45(3):580–589, 1998.
10. W. R. Davis, J. Wilson, S. Mick, J. Xu, H. Hua, C. Mineo, A. M. Sule, M. Steer, and P. D. Franzon. Demystifying 3d ics: the pros and cons of going vertical. *IEEE Design and Test of Computers*, 22(6):498–510, 2005.

11. W. E. Donath. Placement and average interconnection lengths of computer logic. *IEEE Trans. on Circuits and Systems*, 26(4):272–277, 1979.
12. X. Dong, X. Wu, G. Sun, Y. Xie, H. Li, and Y. Chen. Circuit and microarchitecture evaluation of 3D stacking magnetic RAM (MRAM) as a universal memory replacement. In *Design Automation Conference*, 2008.
13. A. B. Kahng, S. Mantik, and D. Stroobandt. Toward accurate models of achievable routing. *IEEE Transaction on Computer-Aided Design of Integrated Circuits and Systems*, 20(5):648–659, 2001.
14. B. S. Landman and R. L. Russo. On a pin versus block relationship for partitions of logic graphs. *IEEE Trans. on Computers*, C-20(12):1469–1479, 1971.
15. G. H. Loh, Y. Xie, and B. Black. Processor design in 3D die-stacking technologies. MICRO, 27(3):31–48, 2007.
16. S. C. Marc Tremblay. A third-generation 65 nm 16-core 32-thread plus 32-scout-thread CMT SPARC(R) Processor. In *International Solid State Circuit Conference*, pp. 82–83, 2008.
17. J. Rabaey, A. Chandrakasan, and B. Nikolic. *Digital Integrated Circuits*. Prentice-Hall, Englewood Cliffs, NJ, 2003.
18. L. Smith, G. Smith, S. Hosali, and S. Arkalgud. 3D: it all comes down to cost. *Proceedings of RTI Conference of 3D Architecture for Semiconductors and Packaging*, 2007.
19. R. Weerasekera, L.-R. Zheng, D. Pamunuwa, and H. Tenhunen. Extending systems-on-chip to the third dimension: performance, cost and technological tradeoffs. In *ICCAD*, pp. 212–219, 2007.
20. Y. Xie, G. H. Loh, B. Black, and K. Bernstein. Design space exploration for 3D architectures. *J. Emerg. Technol. Comput. Syst.*, 2(2):65–103, 2006.

Index

3D IC, 12, 16–18, 34–58, 64, 66, 104, 122–123, 128, 137, 139–141, 145–159, 226, 261–279
3D integration, 4, 7–9, 12, 15–18, 20, 22, 24–25, 27–28, 30, 163, 169–170, 172–173, 176, 181, 191, 203–204, 261–262, 267, 270–271, 274
3D placement, 12, 89, 91, 103–141
3D slicing tree, 78, 82, 83
3D stacking, 12, 162–163, 166, 173, 175, 219–257, 265–266, 270–271, 273, 277, 278

A
Access latency, 6, 163–165, 176, 221, 234–236, 238–240, 244, 248, 250
Alignment, 1, 18–21, 24, 28, 30, 66, 77, 148, 198, 204, 271
Analytical floorplanning, 88, 93
Analytical placement, 115–121, 131, 133, 136
Arbiter, 195, 198, 200, 206–207

B
B*-tree, 68, 72–75, 77
Bandwidth, 4, 6–11, 17–18, 192–194, 196, 198, 204–206, 220–221, 228, 234, 236–237, 245, 248, 250
Bell-shaped function, 117
Bitline, 98, 177–179, 227
Block folding, 96, 98–99
Bonding, 1, 15–25, 44, 46–47, 95, 146, 163, 166, 170, 203–204, 226, 262, 267–268, 271–272, 275
Bus
  bandwidth, 10
  frequency, 10, 233, 251
  occupancy, 9, 11
  transaction, 11

C
Cache
  array, 5–6
  line, 10, 165–166, 205, 208
    migration, 209, 211, 212
  size, 6, 179, 214, 236, 248, 251
Clique model, 112
Clock rate, 10
Clustering, 121, 133
Coarse-grained, 53, 176
Coarse legalization, 105, 125–127, 130
Column module, 201–202
Commercial workload, 7, 224
Compact packing, 67–68, 72–73, 75, 77
Computer throughput, 2
Concentrated mesh (Cmesh), 192–193, 203
Connection box, 199, 200
Corner block list (CBL), 69–70, 76
Crossbar, 193–202, 205–206, 208
Cut size, 57, 108
Cycle time constraint, 2

D
Datapath, 166, 184
DC-DC conversion, 49–51
Decaps, *see* Decoupling capacitors
Decomposable router, 201
Decoupling capacitors, 44, 54
Density penalty, 107, 116–121, 141
Detailed legalization, 105, 125, 127–128, 130–131
Detailed placement, 106–107, 111, 115–116, 122, 125–131, 133, 141
Device layer, 34, 54–55, 64–65, 77, 91, 96, 104–107, 117–118, 120, 122–125, 128, 139–141, 203–204, 207
Die size, 2, 167, 222, 233, 238, 262, 272

Y. Xie et al. (eds.), *Three-Dimensional Integrated Circuit Design*, Integrated Circuits and Systems, DOI 10.1007/978-1-4419-0784-4,

Die-to-wafer, 203–204, 262, 268, 271, 272, 275
Directory, 9, 11, 211
Discrete cosine transformation (DCT), 112–114

E
Electrical connectivity, 1
Etching, 17–18, 25–27, 30, 267–268

F
Fabrication, 12, 20, 27, 30, 161, 163, 185, 198, 203–204, 267–268, 273–275, 277, 279
Face-to-back, 18–19, 24, 28, 46, 181, 203–204, 207, 226, 227, 268
Face-to-face, 18–19, 47, 203–204, 226–227, 268–269
Fan out of 4 (FO4), 2, 3, 232
Fiduccia–Mattheyses (FM) heuristic, 108
Fine-grained, 53, 127, 162, 277
Finite cache, 5–6, 8–9
Finite difference method (FDM), 35–39, 109
Finite element analysis (FEA), 36, 149
Finite element method (FEM), 35–36, 39–42
Flattened butterfly topology, 193, 203
Flit, 193–194, 196, 198–199, 201–202, 208, 212
Floorplanning, 12, 43, 56, 63–77, 82–89, 91–96, 138, 146, 169, 175, 186, 201
Folding transformation scheme, 123–124
Force-directed placement, 89–93, 96, 107

G
Global placement, 88–90, 106–107, 110–111, 115–117, 125, 127–128, 130–131, 141

H
Half-perimeter wirelength, 104–105, 116, 137–138
Heat equation, 35, 39
Helmholtz equation, 120
Hop count, 193, 202–203

I
Infinite cache, 5–6

K
Known-Good-Die (KGD), 268, 271, 275

L
Laser drilling, 267–268
Layer-to-layer, 1
Legalization, 88–90, 105–107, 111, 122–131, 137, 141
Linear system, 107, 110–111
Line length, 9, 11, 171, 179, 186, 200
Logic depth, 3, 232
Low temperature bonding, 1, 24

M
Mask reuse, 277–279
Mesh, 26, 39, 41, 56, 127, 148, 192–194, 203, 205, 208
Micropad, 204
MIM (metal–insulator–metal), 55–56
Min-cut partitioning, 107–108
Miss rate, 6–9, 165, 214, 224, 235, 237, 241, 248, 250
Miss time penalty, 6
Mosaic packing, 76–78, 81, 82–83
MPU frequency, 10
Multicore system, 3–4
Multilevel placement, 106, 115–121, 141
Multistory power delivery, 51–54
Multithreaded, 7, 229–230, 248

N
Network
-on-chip, 189–216
topology, 191–193, 205
Non-slicing, 67, 70, 78
NonUniform Cache Architecture (NUCA), 205

P
Packaging, 12, 17, 26, 33, 146, 147, 161, 268
Partitioning-based placement, 107, 134
quadratic placement, 107
Partitioning-based placement, 107, 134
quadratic placement, 107
Permittivity dielectric, 1
Physical channel (PC), 193–194, 199, 208–209
Physical design flow, 103
Pipeline, 2–3, 162–163, 166, 168, 171–173, 175–176, 186, 194, 200, 221, 235
Placement
3D, 12, 89, 91, 103–141
analytical, 115–120, 121, 131, 133, 136
detailed, 106, 107, 111, 115, 116, 122, 125–130, 131, 133, 141
force-directed, 89–93, 96, 107
global, 88–89, 90, 106, 107, 110–111, 115–117, 125, 127–128, 130–131, 141
multilevel, 106, 115–121, 141

partitioning-based, 107, 134
quadratic, 107
quadratic, 110, 111, 115, 141
Place-and-route, 162
Port partitioning (PP), 96–100
Power
delivery, 12, 33–58, 145
network modeling, 44, 52, 54
supply noise, 43, 55, 168
spectrum, 43, 55, 168
Prefetch, 11, 168

Q
Quadratic placement, 107, 110–111, 115, 141

R
Radix, 192–193, 203
RCN graph, 125, 129–131, 134, 141
Reciprocal design symmetry, 277–279
Rectilinear Steiner minimal tree, 138–139
Redundancy, 68, 75–78, 82–83, 168, 200
Relaxation, 115, 116, 121, 130, 155, 158
Rent's rule, 263, 266
Repeater, 137–141, 167, 177, 233
Resonant frequency, 44–45, 47, 49
Retirement rate, 11
Router, 157, 190–209, 212
Row
decoder, 177–181
module, 201–202

S
Scientific workload, 8
Sequence pair (SP), 68, 70–72, 74–77, 80, 82
Sequence triple, 78, 80–82, 96
Silicon-on-insulator (SOI), 20, 21, 24–25, 27–28, 46, 53, 57, 105, 146, 204
Simulated annealing, 64, 68, 83–86, 88, 90–91, 96
Slicing, 67–69, 70, 73–76, 78, 82–83, 96
tree, 68–69, 75, 78, 82–83, 96
Smoothed density, 120
Stacking transformation scheme, 122–123
Star model, 112
Steady-state analysis, 35
Switch allocation unit (SA), 85–88, 93, 178, 194, 208

T
Terminal propagation, 108
Tetris-style legalization, 123, 125, 127–128, 130–131, 137
Thermal analysis, 12, 34, 35–39, 91, 104, 114, 148, 149, 241
Thermal awareness, 91, 106, 108, 109, 131, 136, 137
Thermal conductivity, 34–35, 41, 43, 104, 146, 148–151, 242
Thermal gradient, 40, 89, 149–151
Thermal management, 34, 145
Thermal optimization, 34, 42, 131, 136, 141, 148
Thermal PDE, 34
Thermal resistance, 38, 42, 95, 104, 109–110, 147
Thermal via
density, 150–151
regions, 148–151
Thinning, 15, 17–19, 21, 24–27, 29–30, 267–269
Thread, 6–7, 195, 223,–225, 229, 235–236, 248, 251, 254
Three-dimensional, 4, 16, 19, 45, 50, 63–64, 66, 78, 89, 93, 97, 103, 132–134, 140
Through-silicon-via (TSV), 34, 148, 162, 226, 262, 265, 275
Through-silicon via (TS via), 104
Timing margin, 3
Tradeoffs, 15, 131, 175, 222
Transaction, 5, 7, 10, 11, 206, 228, 231, 237, 250, 252, 254
Transfer rate, 10
Two-way partitioning (bisection), 107

U
Uniprocessor, 2, 4, 11, 250

V
Via
density, 123–124, 150, 152, 198, 207–208, 226
impedance, 12
pitch, 19, 26, 28, 99, 204, 207
Virtual channel allocation unit (VA), 193–194, 208
Virtual channels (VC), 193–194
Voltage regulators, 44, 56, 57

W
Wafer-to-wafer (W2W), 20, 203–204, 262, 268, 271, 272
Weighted wirelength, 104, 108, 110
Wire delay, 1, 2, 66, 171–172, 186–187, 190
Wirelength, 64–66, 72–73, 77, 84, 86–87, 90–91, 96–97, 99, 104–106, 108–112, 116–117, 121–125, 131–141, 155

Y

Yield, 30, 37, 41, 53, 150, 165–166, 168, 193, 226–227, 257, 262, 267–268, 270–273, 275

Z

Z-axis power delivery, 50

Z-plane, 10

FSC
www.fsc.org
MIX
Papier | Fördert
gute Waldnutzung
FSC® C083411

Kolibri 360 GmbH
Ferdinand-Jühlke-Straße 7
99095 Erfurt, Deutschland
produktsicherheit@kolibri360.de